This book presents a compendium of views on the negotiations leading to the Framework Convention on Climate Change. It brings together the recollections of key participants in those negotiations, each representing a different constituency. To situate these recollections in the evolving policy context of the international debate on the climate problem, it is important to note that these views were first compiled during late 1992 and 1993. Half of these participants bring perspectives from industrialized countries and half from developing countries. Each eyewitness was asked to respond in their personal capacity to two questions:

(1)   From your perspective, what were the key events, decisions, or initiatives that were most important in shaping the final structure of the Climate Convention?

(2)   What lessons have been learned by the international community from the process of the climate negotiations that might suggest a structure for the next phase of the process, so as to maximize the probability of reaching additional agreements that are both practical and equitable?

These ten "insider views" are supplemented by a legal history of the events leading up to the opening of negotiations, by an analysis of the structure and dynamics of the process, and by the recommendations of the editors concerning lessons from the climate debate which may be of value to those involved in other international environmental negotiations.

Negotiating Climate Change:

The Inside Story of the Rio Convention

CAMBRIDGE STUDIES IN ENERGY AND ENVIRONMENT

Editors
Chris Hope, Judge Institute of Management, Cambridge University
Jim Skea, Science Policy Research Unit, University of Sussex

We live at a time when people are more able than ever to affect their environment, and when the pace of technological change and scientific discovery continues to increase. Vital questions must continually be asked about the allocation of resources under these conditions. This series is intended to provide readers interested in public policies on energy and the environment with the latest scholarship in the field. The books will address the scientific, economic, and political issues which are central to our understanding of energy use and its environmental impact.

Negotiating Climate Change:
The Inside Story of the Rio Convention

Edited by
Irving M. Mintzer and J. Amber Leonard
Stockholm Environment Institute

Foreword by
Michael J. Chadwick
Director
Stockholm Environment Institute

Stockholm Environment Institute

CAMBRIDGE
UNIVERSITY PRESS

Published by the Press Syndicate of the University of Cambridge
The Pitt Building, Trumpington Street, Cambridge, CB2 1RP
40 West 20th Street, New York, NY 10011–4211, USA
10 Stamford Road, Oakleigh, Melbourne 3166, Australia

© Cambridge University Press and
Stockholm Environment Institute 1994

First Published 1994

Printed in Great Britain at the University Press, Cambridge

A catalogue record for this book is available from the British Library

Library of Congress cataloguing in publication data
Negotiating climate change: the inside story of the Rio Convention /
edited by Irving M. Mintzer and J. Amber Leonard; foreword by
Michael Chadwick.
p.   cm.
ISBN 0 521 47355 1. – ISBN 0 521 47914 2 (pbk.)
1. Environmental policy – International cooperation.   2. Climatic
change.   I. Mintzer, Irving M.   II. Leonard, J. Amber.
HC79.E5N44   1994
363.7 – dc20
94-13041
CIP

Includes Index

ISBN 0 521 47355 1 hardback
ISBN 0 521 47914 2 paperback

UP

Contents

Acknowledgements   ix
Commonly Used Acronyms   xi
Foreword Michael J. Chadwick   xiii

Part I: Background
 1 Visions of a Changing World   3
      Irving M. Mintzer and J. Amber Leonard
 2 Prologue to the Climate Change Convention   45
      Daniel Bodansky

Part II: Views from Within the Ring
 3 Exercising Common but Differentiated Responsibility   77
      Delphine Borione and Jean Ripert
 4 The Beginnings of an International Climate Law   97
      Ahmed Djoghlaf
 5 Constructive Damage to the Status Quo   113
      Elizabeth Dowdeswell and Richard J. Kinley
 6 The Climate Change Negotiations   129
      Chandrashekhar Dasgupta
 7 A Personal Assessment   149
      Bo Kjellen
 8 The Road to Rio   175
      José Goldemberg
 9 A Failure of Presidential Leadership   187
      William A. Nitze
10 Looking Back to See Forward   201
      Tariq Osman Hyder

Part III: The Outside Edges In
11 Some Comments on the INC Process   229
      Hugh Faulkner
12 A View from the Ground Up   239
      Atiq Rahman and Annie Roncerel

Part IV: Prospects for the Future
13 Towards a Winning Climate Coalition   277
      James K. Sebenius
14 Visions of the Past, Lessons for the Future   321
      Irving M. Mintzer and J. Amber Leonard

Appendix: The Framework Convention on Climate Change   335
Index   367

Acknowledgements

This book reflects the contributions of many individuals. We would like to acknowledge our appreciation for the encouragement, patience, and support of our colleagues at the Stockholm Environment Institute, especially Professor Michael Chadwick, Executive Director of SEI, and Dr. Lars Kristoferson, Vice Director of SEI. We are also grateful for the help of Dr. Arno Rosemarin and Heli Pohjolainen of the Information Department of the Institute who helped us to overcome many of the logistical and technical problems involved in producing this book. For their astute comments and practical guidance, we want to thank Dr. Paul Raskin and Gerald Leach.

Here in Washington, we would like to thank Alan Miller for his advice and counsel, and Alma Stone, Dale Hopper, Lorraine Wells, Marilyn Powell, Noreen O'connor-Abel, Pamela Wexler, and Susanne Kamalieh for their help in the book's production. We are also grateful to Patricia Feuerstein of Integral Publishing for preparing the index that will guide the diligent reader through this story. Without their help, this work could not have reached its final audience.

We are grateful to Patrick McCartan, our careful and patient editor at the Press Syndicate whose steady hand helped to guide us in our efforts to bring these events to light. We have benefited as well from the constructive criticism provided by our reviewers, Professor Jim Skea, the series editor, and Dr. Michael Grubb, Royal Institute of International Affairs, London.

Beyond all this, we are deeply indebted to both our immediate and wider families. We have been the frequent beneficiaries of the time and support of our children and their spouses, as well as of the families of all our contributing authors, each of whom tolerated more than their share of intense, late-night calls as we sought to iron out the creases that crept inadvertently into these pages. And finally, we want to thank the men and women who have struggled for the last five years and more to build a practical, equitable, and efficient international regime to reduce the risks of rapid climate change and promote sustainable development. They are the true heroes of this continuing drama.

Irving M. Mintzer
J. Amber Leonard
STOCKHOLM ENVIRONMENT INSTITUTE
March 1994

Commonly used acronyms

AOSIS	Alliance of Small Island States
ASEAN	Association of South East Asian Nations
BCSD	Business Council for Sustainable Development
CAN	Climate Action Network
CAN-A	Climate Action Network-Africa
CAN-LA	Climate Action Network-Latin America
CAN-SA	Climate Action Network-South Asia
CAN-SEA	Climate Action Network-South East Asia
CFCs	Chlorofluorocarbons
CNE	Climate Network Europe
CO_2	carbon dioxide
COP	Conference of the Parties
EC	European Community
ECO	*ECO: The Daily News*, an NGO journal
EEA	European Economic Area
EFTA	European Free Trade Association
ENDA	Environment and Development-Africa
EPA	US/Environmental Protection Agency
FCCC	Framework Convention on Climate Change
G-7	Group of Seven Industrial Nations
G-77	Group of Seventy-seven Developing Nations
GATT	General Agreement on Tariffs and Trade
GDP	gross domestic product
GEF	Global Environment Facility
GHG	greenhouse gas
GNP	gross national product
GWPs	global warming potential factors
HCFCs	Hydrochlorofluorocarbons
ICSU	International Council of Scientific Unions
IEA	International Energy Agency
INC	Intergovernmental Negotiating Committee for a Framework Convention on Climate Change

IPCC	Intergovernmental Panel on Climate Change
LDCs	less developed countries
LOS	Law of the Sea
NATO	North Atlantic Treaty Organization
NGO	nongovernmental organization
NIEO	new international economic order
ODA	official development assistance
OECD	Organization for Economic Cooperation and Development
OPEC	Organization of Petroleum Exporting Countries
PrepCom	Preparatory Committee of the UN Conference on the Environment and Development
SBI	Subsidiary Body on Implementation
SBSTA	Subsidiary Body for Scientific and Technological Advice
SEI	Stockholm Environment Institute
STAP	Scientific and Technical Advisory Panel of the GEF
SWCC	Second World Climate Conference
UNCED	UN Conference on Environment and Development (also called the Earth Summit or Rio '92)
UNDP	UN Development Programme
UNEP	UN Environment Programme
UNITAR	UN Institute for Training and Research
WG	Working Group
WHO	World Health Organization
WMO	World Meteorological Organization

Foreword

The Framework Convention on Climate Change, signed by 154 heads of State and governments at the Earth Summit in Rio de Janeiro, Brazil, is a path-breaking document. It represents one of the first instances in which national governments have signalled their intention to control widespread and economically important activities at such a scale in order to reduce a well-documented but incompletely understood set of risks to the global environment. The treaty is likely to be an important model for future international agreements in many areas.

The negotiations that led to the signing of this Convention were remarkably brief and have been generally unmarred by the traditional, partisan attacks which have disrupted many UN negotiations in the past. This pattern of success on such a complex issue suggests that the climate negotiations may offer some significant lessons for both those who seek to make further progress in reducing the risks of rapid climate change and those who are now designing and implementing international environmental negotiations on other issues.

This book offers fascinating insights into the structure and dynamics of the climate negotiations. It reflects the perceptions of central participants in these negotiations, individuals who were participants in the key decisions, framed the most important initiatives, and represented the pivotal constituencies in the Intergovernmental Negotiating Committee (INC). The considered reflections they offer here represent a series of diverse visions of a world attempting to anticipate and deal with a potentially far reaching agent of long-term environmental change. Supplementing these visions are a set of retrospective analyses of the events which shaped the negotiations and the structural elements which influenced their outcome.

The insights presented here will be useful to those who continue the climate negotiations under the auspices of the INC for the Framework Convention on Climate Change. But equally, or perhaps more important, they offer some broad lessons on the dynamic of international environmental negotiations in the 1990s. Appropriately applying these lessons may ease the tasks of others in the future who will be concerned with the institution of environmental accords. The audience for this volume will include those now involved in the INC for Desertification, those hoping to organize an INC for Forests, academics wishing to improve their general understanding of negotiating processes, business leaders who find their investment and operating choices increasingly shaped by the results of international environmental negotiations and those members of the general public who wish to understand more fully the institutional background to global environmental protection.

This volume is just one of the products of the Climate and Sustainable Development Programme of the Stockholm Environment Institute, which seeks to contribute to a secure, equitable and lasting collaboration between peoples of all countries in working towards the development of a sustainable world.

Professor Michael J. Chadwick
Director
STOCKHOLM ENVIRONMENT INSTITUTE
Stockholm, Sweden
March 1994

PART I

Background

CHAPTER 1

Visions of a Changing World

Irving M. Mintzer
Stockholm Environment Institute
Washington, DC

and

J. Amber Leonard
Stockholm Environment Institute
Washington, DC

Introduction

Farmers, villagers, and city-dwellers worldwide have recently begun to notice what atmospheric scientists have known for decades: the Earth's climate is changing. During the last five years, extreme and often catastrophic weather events have captured the attention of scientists, public citizens, politicians, and the media. Twice in five years, the "hundred-year" windstorm blew across the United Kingdom, overturning historic structures, uprooting centuries-old trees, and reshaping the countryside. In the same period, the United States, France, and parts of Africa were racked by droughts of historical proportions. Major floods swept across India, Bangladesh, the Sudan, and China, leaving thousands homeless and destitute. Hot spells killed hundreds in Greece. Fierce storms, as harsh as the "Hurricane of the Century," twice battered the coasts of the United States, thrashed vulnerable islands in the Caribbean, and wreaked havoc on the Pacific Island of Guam.

During earlier times and in most cultures, events like these would have been viewed with cosmic significance. In the history of ancient China, for example, the confluence of natural disasters was interpreted as a divine message. A series of unusual and extreme weather events might signal the imminent end of an aging dynasty. It marked the disintegration of the old order, the emperor's loss of "the mandate of Heaven," and the opportunity to establish a new regime.

In recent years, abrupt atmospheric changes have similarly captured public attention and influenced human affairs, reshaping the global politics of the environment. Although the last string of weather anomalies may not demonstrate divine intervention in human affairs, these events helped stimulate worldwide agreement to an innovative political regime and to a process for its future evolution. The new regime was inaugurated in June 1992 when representatives of 154 nations signed the Framework Convention on Climate Change (the Climate Convention or FCCC). The signing occurred at the United Nations Conference on Environment and Development (known popularly as UNCED or the Earth Summit) in Rio de Janeiro, Brazil. Its principal goals are to protect the global atmosphere and to promote the prospects for sustainable economic development worldwide. But the impact of the Climate Convention will stretch far beyond these immediate goals, strengthening international institutions and framing new mechanisms for cooperation on environment and development.

This book explores the key events, initiatives, results, and decisions that led to the dramatic signing of the Climate Convention in Rio. Principals in that drama provide eyewitness accounts in the chapters which follow this introduction, highlighting the interactions between diplomatic negotiations and scientific research on the global environment. In the process, this volume presents the most authoritative study to date on the politics and perceptions that shaped the obscure and often ambiguous text of the Convention itself, and the recollections of these eyewitnesses bring a human dimension to that dry legal instrument. They reveal the differences in personal, national, and institutional priorities that sometimes fostered unexpected friendships and, at other times, divided old friends and former allies into contentious and fiercely competitive factions.

Studying the dynamics of these debates suggests new insights into the international process and into the politics that will shape future environmental negotiations. But the climate negotiations have implications that extend far beyond the greenhouse problem and related environmental issues. It is quite likely that the precedents set in these debates will shape the dialogue on sustainable development and the structure of North–South relations throughout the post-Cold War era.

Before turning to the actors' visions of this drama, the next section briefly reviews the scientific basis of the climate problem and the expected impacts of a greenhouse gas buildup. This background is important for understanding how the perceived risks of rapid climate change motivated the participants in the negotiations. Subsequent sections summarize the agreements contained in the Climate Convention and discuss the implications of the Convention for other international negotiations. We close this chapter with an overview of the volume and highlight the key conclusions drawn by the participants.

A Brief Introduction to the Climate Problem

The composition and behavior of Earth's atmosphere are changing. For most of the last 10,000 years the concentrations of the most important gases in the atmosphere have been fairly constant. But in the last two centuries, human activities—including industrialization and agricultural expansion—have caused the release of large quantities of gaseous pollutants into the atmosphere. These gaseous emissions have rapidly raised the atmospheric concentrations of some of the compounds which determine the atmosphere's ability to trap infra-red radiation and thus to affect the average temperature at the Earth's surface. This heat-trapping process, called the greenhouse effect, is described below, along with the impacts on global and regional climates that can be expected if current trends in greenhouse gas emissions continue.

The Greenhouse Effect Makes Earth a Habitable Home

The greenhouse effect is not new. It is a natural geophysical process, a part of Earth's history for millions of millennia. Before humans or other living species existed, this heat-trapping process transformed the character of the planet, making Earth hospitable to life, instead of a bleak, ice-covered rock.

Most of the energy available on Earth comes from the Sun, arriving as electromagnetic radiation in the short-wave part of the spectrum. More than 30 percent of the incoming energy from the Sun is reflected back to space by clouds and by water, land, and structures on the Earth's surface. Of the remainder, most is absorbed by the planet's surface. If the tale ended here, Earth's surface would long ago have heated up and melted, like an iron bar left in a furnace. But Earth's surface is kept in a perfect energy balance with its surroundings, radiating away as much energy as it absorbs. Because the Earth's surface temperature is so much lower than that of the sun, it radiates energy at a longer wavelength, principally in the infra-red (IR) part of the electromagnetic spectrum.

Beginning in primordial times, molecules of several simple gases appeared in Earth's atmosphere. Some of these molecules, including water vapor, carbon dioxide, and ozone, are transparent to the incoming short-wave solar radiation that passes through the atmosphere and is absorbed by the planet's surface. This energy is subsequently released as long-wave, infra-red radiation from the surface, which these simple gases can absorb.[1] Having absorbed the IR radiation from the surface, these molecules release energy isotropically (i.e., in all directions). Thus, for billions of years, some of the infra-red energy released from Earth has been absorbed in the atmosphere and re-emitted downward, warming the surface. This process is popularly referred to as the greenhouse effect. Gaseous molecules which are transparent to sunlight but absorb and re-emit in the IR part of the spectrum are called greenhouse gases (or GHGs).

The natural background greenhouse effect raises the planet's temperature by about 33° Centigrade (C), from -18° C to +15° C. The resulting surface warming has allowed water to exist on the planet's surface as a liquid—rather than as a solid (ice). Over the millennia, that liquid has provided a rich substrate for the biological evolution of life.

The Greenhouse Problem: Getting Too Much of a Good Thing

Water vapor is the largest contributor to the background greenhouse effect. The next largest contributor is carbon dioxide. Carbon dioxide molecules cycle among various reservoirs on Earth, including the atmosphere, oceans, and natural ecosystems or biota. The atmospheric reservoir contains approximately 750 billion tonnes (i.e., 750 gigatons or GT) of carbon in the form of carbon dioxide. The natural flux of carbon into and out of the atmospheric reservoir is approximately 100 GT annually. The atmospheric concentration of carbon dioxide has been controlled by these natural fluxes, remaining fairly stable for most of the last 10,000 years at approximately 280 parts per million by volume (ppmv).[2]

During the past century, the greenhouse effect became the green-house problem as human activities enhanced the natural greenhouse effect through the release to the atmosphere of billions of tonnes of carbon dioxide and other GHGs. These GHG emissions have caused a continuing atmospheric buildup of carbon dioxide and similar gases, amplifying the background greenhouse effect. In addition to carbon dioxide, the principal anthropogenic greenhouse gases are methane, nitrous oxide, tropospheric ozone, and the chlorofluoro-carbons. At current emission rates, the concentration of carbon dioxide in the atmospheric reservoir is increasing by approximately 0.5 percent annually (or a little more than 3 GT of carbon per year). Since the eighteenth century, human emissions have raised the atmospheric concentration of carbon dioxide by a total of approxi-mately 25 percent (to about 355 ppmv) and more than doubled the concentration of methane. Nitrous oxide has increased by about 10 percent during the same period. The most common chlorofluoro-carbons, wholly absent from the background atmosphere, have increased in concentration during the last fifty years from close to zero to about 1–2 parts per billion by volume (ppbv) in air. (See Table 1.1.)

Table 1.1

Summary of Key Greenhouse Gases Influenced by Human Activities

Parameter	CO_2	CH_4	CFC-11	CFC-12	N_2O
Pre-industrial atmospheric concentration (1750–1800)	280 ppmv[2]	0.8 ppmv	0	0	288 ppbv[2]
Current atmospheric concentration(1990)[3]	353 ppmv	1.72 ppmv	280 pptv[2]	484 pptv	310 ppbv
Current rate of annual atmospheric accumulation	1.8 ppmv (0.5%)	0.015 ppmv (0.9%)	9.5 pptv (4%)	17 pptv (4%)	0.8 ppbv (0.25%)
Atmospheric lifetime[4] (years)	(50–200)	10	65	130	150

1   Ozone has not been included in the table because of lack of precise data.
2   ppmv = parts per million by volume; ppbv = parts per billion by volume; pptv = parts per trillion by volume.
3   The current (1990) concentrations have been estimated based upon an extrapolation of measurements reported for earlier years, assuming that the recent trends remained approximately constant.
4   For each gas in the table except CO_2, the "lifetime" is defined here as the ratio of the atmospheric content to the total rate of removal. This time scale also characterizes the rate of adjustment of the atmospheric concentrations if the emission rates are changed abruptly. CO_2 is a special case since it has no real sinks but is merely circulated between various reservoirs (atmosphere, ocean, biota). The "lifetime" of CO_2 given in the table is a rough indication of the time it would take for the CO_2 concentration to adjust to changes in the emissions.

Source: IPPC 1990 Scientific Assessment, p. 7

(A) Canadian Climate Centre Model

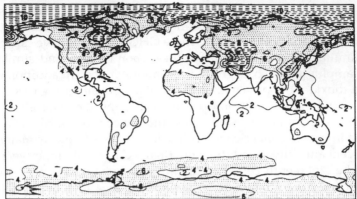

(B) Geophysical Fluid Dynamics Lab Model

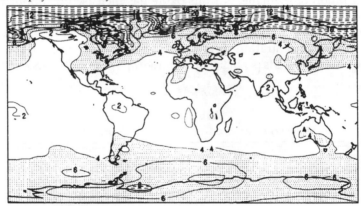

(C) UK Met Office Model

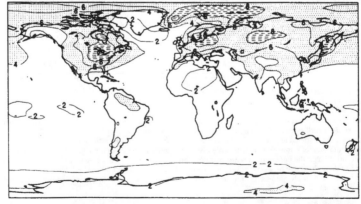

Figure 1.1: Change in surface air temperature due to doubling CO_2 for months of December-January-February.   *Source:* Houghton et al. (1990), p. 141

Global Effects of a Greenhouse Gas Buildup

Human emissions are still small compared to the natural flux of greenhouse gases into the atmosphere, but they can have significant effects. If current trends in anthropogenic emissions of carbon dioxide and other greenhouse gases continue, the combined radiative effects of this group of gases will equal, by the middle of the next century, the warming effect that could be expected from raising the concentration of carbon dioxide alone to twice its pre-industrial level.[3] The resulting change in the planet's radiation balance is widely expected to increase the average annual surface temperature of the Earth, including the sea surface temperature. The general consensus of the international scientific community has been presented in the First Assessment Report of Intergovernmental Panel on Climate Change (IPCC). This assessment supports the view that if thermal equilibrium is re-established at concentration levels equivalent to doubling the pre-industrial level of carbon dioxide, average temperatures will have increased by approximately 2–5° C.[4] This warming will cause a thermal expansion in the upper mixed layer of the oceans as well as some melting of land-based glaciers. As a consequence, the average global sea level is expected to rise.

Although the buildup of greenhouse gases will warm the entire planet, warming due to this buildup will not be evenly distributed worldwide. Complex mathematical representations of the atmosphere, called equilibrium climate models, are used to simulate the effects of major changes in atmospheric composition. Figure 1.1 illustrates the distribution of temperature increases in a "doubled CO_2 world," as projected by several of the most advanced computer models of the atmosphere. All of these models agree that the warming will be enhanced at high latitudes and less extreme in the tropics.[5] This differential warming effect will shrink the thermal gradient between the poles and the equator, altering the historical patterns of upper atmospheric winds and oceanic currents that shape the character of local and regional weather.

Rapid Climate Change: A Threat to Global Stability?

The current generation of equilibrium climate models effectively characterizes the global impacts of climate change due to a greenhouse gas buildup. Unfortunately, these models can not accurately predict the regional implications of climate change. Nor are they capable of predicting the transient effects of global warming as the planet moves toward a new climatic equilibrium.[6]

The transient responses of the climate system are important because they determine the *rate* of global climate change in response to a given greenhouse gas buildup and the specific realization of the global change at the regional level. In many cases, it is the rate of climate change—as much as the magnitude of the change—that will determine the extent of economic dislocation experienced by human societies and the extent of disruption imposed on natural ecosystems.

Other feedback mechanisms may also affect the rate of greenhouse gas emissions or the ability of the atmosphere to cleanse itself of these pollutants. These include processes that may cause increased release of carbon dioxide by soil micro-organisms, increased release of methane trapped beneath the frozen surface of the tundra, and reduced oceanic uptake of carbon dioxide due to decreases in the rate of photosynthesis in the oceans.

Impacts of Global Warming on Human Societies

What the average person observes as the characteristic climate of his or her home region is the net effect of all these factors on several intricately linked non-linear systems. These systems (which include the atmosphere, oceans, and biota) interact in subtle and complex ways. Climate scientists try to simulate the responses of these closely coupled systems even though their individual behavior is not well understood today.

Some of the changes that result from GHG buildup will be uniform worldwide, others will be generally similar in broad bands of latitude, and some will be quite region-specific. The changes which will occur worldwide and many of those which will occur similarly in broad latitudinal bands may appear to be both gradual and smooth. Average annual surface temperatures, for example, will rise both on land and in the oceans. Although there will continue to be a significant amount of inter-annual variability, if current trends in greenhouse gas emissions continue, this generalized warming should become clearly detectable at statistically significant levels within a few decades.

Similarly, as the world warms, the global average sea level will rise by some tens of centimeters over the next century. Specific local sea-level changes will depend greatly on regional movements of Earth's tectonic plates and the presence of landed glacial ice which melts into the sea. In general, however, the large, flat, deltaic regions at the mouths of the world's great rivers will be subject to increased coastal flooding and to sea surges following large local storms. This flooding may displace significant areas of coastal zone communities and disrupt the economically important activities that take place in these zones. If, for example, sea-level rise due to global warming caused a major episode of low land inundation and storm surges at the mouth of one of the world's great river deltas (e.g., the Ganges-Brahmaputra system), the resulting flooding might create a very large number of environmental refugees in a very short period. These refugees could strain local institutions and, if they migrated across national borders, could worsen both intra-regional and inter-regional tensions.

The interaction of slow sea-level rise and the fast onset of ocean storms illustrates the fact that future interactions between atmosphere, oceans, and biotic systems can produce abrupt or discontinuous responses in the climate system. Thresholds of non-linear change may emerge in the future which scientists cannot now identify. Crossing these thresholds may cause sharp changes in regional climates—changes which could be triggered by relatively small increases in the concentration of greenhouse gases or the release of other pollutants. Some of these non-linear responses could cause the variability of weather regimes and the frequency or severity of extreme weather events to increase.

Current climate models are constrained not only in their ability to project regional impacts of climate change, they are also of little use in predicting such short duration phenomena as storms, floods, and droughts. However, scientists now have compelling geological evidence that climates have changed rapidly in the past and could do so again in the future. Some geological evidence contained in ice-cores suggests that dramatic changes have occurred during periods as short as a few decades.[7]

If global warming due to the greenhouse effect were to cause changes in the next few decades that were comparable to those observed in the long-term ice-core records, the associated changes in regional temperature, precipitation, and weather extremes could place severe stresses on human societies—societies that are already under stress from population increases, demands for economic growth, and environmental degradation. These increased climate-related stresses might materialize in a number of different ways that are difficult to link causally to the greenhouse effect. For example, if global warming caused a reduction in precipitation and available surface runoff in any one of several international river systems that form the boundary between competing riparian states (e.g., in the Jordan-Litani river system in the Middle East), the resulting shortfall could increase the likelihood of armed conflict in an already troubled region. Or if increases in the frequency of large storms (including tropical hurricanes, windstorms, typhoons, and other cyclones) cause an increase in economic damages in the developing world, the resulting displacements could release a large number of economic and environmental refugees, quickly exceeding the coping capacity of existing institutions.

But even if short-term disruptions of regional climate did not lead to weather-related disasters, these sharp changes in the timing and distribution of rainfall and snowfall could still cause significant local damage and disrupt many economically important activities. Shifts in the timing of rainfall or snowfall can leave crops parched or drowning, even if the annual average precipitation remains unchanged. Long periods of steady rain (like that which fell in the Midwestern United States in the summer of 1993) or short periods of intense precipitation can wash away tonnes of valuable topsoil and destroy much of the agricultural output in a wide area. An example of the effects of a short, hard storm occurred when a year's worth of rain fell in the Sudan during three days in 1988 following a year-long drought. Thousands of people were killed or left homeless. Most of the region's remaining topsoil was washed away.

Similarly, extended hot spells (with no rain), or hard and early freezes, can quickly destroy months of agricultural work. If the heat (or cold) comes during flowering or at other vulnerable periods in the lifecycles of certain agricultural crops, a whole year's production can be lost.

Hot spells can cause other kinds of unpredictable economic damage as well. Hot, dry weather in California from 1976 to 1978 decreased available electric power from the large hydroelectric facilities that are scattered throughout that state. The shortfall was made up with additional electricity generated at higher financial and environmental cost using thermal plants that burn fossil fuels. During the 1988 drought and hot spell in the American Midwest, eight nuclear reactors sited along major rivers were within hours of forced shutdown because the elevated water temperatures and decreased flow rates for cooling water left their managers dangerously close to violating the terms of their operating licenses. Although neither of these episodes was necessarily caused by the greenhouse effect, they are representative of the types of incidents which may occur with increasing frequency as the world warms.

Part of the difficulty in evaluating these risks occurs because these weather-related phenomena already occur naturally and will continue whether or not an increased greenhouse warming takes place in the future. Global warming may increase the frequency or heighten the severity of such episodes but scientists will remain unable to prove that any particular occurrence was *caused* by the enhanced greenhouse effect. Thus, the continuing and pervasive scientific uncertainty about future manifestations of the greenhouse effect will make it impossible to predict the value of mitigating or avoiding the regional impacts of global warming. Even if none of the anticipated events occur, the continuing uncertainty about the timing and severity of their next possible occurrence will itself impose a significant cost on human societies. Scientists will be unable in the next few decades to resolve these uncertainties in ways which will permit economists to estimate the expected future costs of global warming with precision.

Effects of Rapid Climate Change on Other Biological Systems

Rapid climate change could also significantly strain non-human components of the ecology. These effects may include changes in the composition and geographic distribution of many species. Recent analysis of ice-core data has shown that Earth's average temperature can shift up to 5°C in periods of less than one hundred years.[8] When this happens, many species must move or face extinction.

Ecological stress will be caused not only by changes in average annual temperature but also by shifts in the timing and distribution of precipitation and by changes in the frequency and distribution of extreme weather events. Changes in extreme temperatures can affect the survival and distribution of many species, particularly insects.[9] In addition, the fecundity of birds and some small mammals may be affected by changes in seasonal precipitation, snow cover, and snowmelt.[10]

The ability of species to tolerate the new regime may depend on how fast they can migrate away from inhospitable or intolerable conditions. Some species (including birds, fish, some large mammals, bacteria, and viruses) may be able to move or be carried to new areas, shifting or expanding their habitats. Other species (including many trees, land plants, and some marine organisms) may not be able to adapt their behavior rapidly enough or move to more hospitable terrain quickly enough so as to avoid large-scale die-off of local populations.

On the other hand, new ecological niches may be created by the changes in local and regional climates. Previously occupied niches may open up, ready to be filled by species that compete more successfully under the new climate regime. In some cases, pests, predators, and disease vectors may find new advantages in the changed conditions. For example, warmer winters may allow pests to overwinter in areas that were previously too cold for year-round survival. Some bacteria, fungi, and viruses may find that warmer waters allow them to spread colonies over larger regions than was possible in the past, potentially expanding the range of exposure to such diseases as cholera and schistosomiasis.

The Bottom Line

A broad consensus has emerged in the international scientific community that the greenhouse effect is real, a necessary and continuing factor in our planetary history. Human activities continue to emit large quantities of radiatively-active trace gases that amplify the natural, background greenhouse effect. Atmospheric accumulations of these gases during the last century have enhanced the greenhouse effect and turned it into the greenhouse problem. If current trends in these emissions continue unchanged for the next several decades, the atmospheric buildup of greenhouse gases will expose human societies and natural ecosystems to the serious risks of rapid climate change.

The risks of rapid climate change are closely linked to and complicated by other aspects of global environmental change. These related changes include the depletion of the stratospheric ozone layer (due to the catalytic effects of chlorofluorocarbons and halons), the destruction of tropical forests and loss of soil fertility worldwide, and the effects of land-based pollution on the health and productivity of the marine environment.

The effects of rapid climate change may include dramatic, transient shifts in regional climate regimes. There may be changes in seasonal and diurnal temperatures; shifts in the timing, distribution, and extent of precipitation; more frequent episodes of coastal inundation and destructive storm surges; and alterations in the frequency and severity of extreme weather events. The combined effects of these changes may result in shifts in the ranges and geographic distributions of species, as well as changes in the heartiness of pests, predators, and vectors of disease.

Climate change has been occurring for millennia and will continue to occur, regardless of human understanding of the climate system or the implementation of a cooperative international policy response to the risks of rapid global warming. Despite massive investments of research funds, scientists still do not know how climate will change and when. The fundamental scientific uncertainties about the inner workings of the climate system will remain unresolved for decades to come. As a result, it will be impossible to establish strict causal relationships between individual episodes of climate change and the worldwide buildup of greenhouse gases. The weather events that damage local economies can occur without any additional human intervention; they are an expected effect of the inherent variability of the climate system. No one can prove that any individual episode or weather event was caused by the greenhouse effect.

Yet given these unresolvable uncertainties, it may not be prudent public policy to wait until a monotonic increase in global temperatures signals to scientists the unambiguous onset of global warming due to the enhanced greenhouse effect. The instability and variability of regional climates seem to have increased around the globe over the last decade. More and more areas are reporting the local equivalent of hundred-year weather events, including floods, droughts, hot spells, cold snaps and windstorms. This anecdotal reporting of increased weather variability may be the best indicator we have that global climate is entering a period of unusual, and possibly rapid, transitions. Thus the high risks associated with rapid climate change and the continuing uncertainties about the workings of the climate system argue for the implementation of all cost-effective measures that can slow the rate of buildup of greenhouse gases, particularly measures which also promote prospects for sustainable economic development. Implementing such measures and policies will not only help to reduce the rate at which the risks of rapid climate change increase but, more important, will "buy time" for scientists to study the climate system, to improve our understanding of the interactions between atmosphere, oceans, and biota, and to develop adaptive responses that can minimize the damages resulting from those climate changes that can no longer be avoided. The Framework Convention on Climate Change provides a good start in this direction.

The Climate Convention in Perspective

The Earth Summit, held 1–15 June 1992, brought together the heads of state and government from more than 160 countries and international organizations. More than 10,000 people attended the meeting officially, and thousands more participated in parallel events organized by non-governmental organizations (NGOs) representing North and South, East and West. The participants in the official and unofficial activities represented democracies, dictatorships, theocracies, autarchies, religious orders, environmental groups, indigenous peoples, human rights organizations, and the international business community.

From a legal and institutional standpoint, the Earth Summit produced three important agreements. These were: (1) the Climate Convention, (2) the Biodiversity Convention, and (3) Agenda 21. In addition, the participants at UNCED agreed to a statement of principles known as the Rio Declaration. Each of these instruments established important precedents and underlined the continuing relationship between issues of economic development and environmental protection. Taken together, they will set the agenda on global environmental issues for the next several decades.

The Climate Convention and the Rio Agenda

The Framework Convention on Climate Change is the second major international legal instrument to address a problem of global environmental change before the worst impacts have been clearly documented or measured. (The first was the Vienna Convention for the Protection of the Ozone Layer with its Montreal Protocol on Substances that Deplete the Ozone Layer.) After more than ten years of intense scientific research and nearly two years of intense diplomatic negotiations, more than 150 countries agreed to cooperate on measures to reduce the risks of rapid climate change. Although the specific commitments embodied in the treaty will not reduce those risks substantially in the near term, the Convention establishes a process of continued dialogue—both scientific and diplomatic—that will focus attention on the problem, review the adequacy of the agreed measures, and deliver new and additional resources to countries with limited means to address these risks on their own. In addition, the Climate Convention institutionalizes the debate on environment and development in ways which will strongly influence North–South relations for years to come.

The overall message of the Earth Summit—and the cornerstone of the Climate Convention—is the shared recognition that all human beings have a right to economic development, but the impacts of economic activity are currently altering the planet in ways which may irreversibly damage our shared environment. With respect to the climate problem in particular, emissions from energy, agricultural, and industrial activities are changing the composition and behavior of the atmosphere in ways which scientists do not fully understand and whose impacts we cannot now predict with confidence.

The main thrust of the Climate Convention is shared by the Convention on Biodiversity, the Rio Declaration, and Agenda 21. These agreements do not provide a detailed map of humanity's best path to paradise but rather outline a process and a set of principles by which countries with very different circumstances and responsibilities can work together in the interests of sustainable development. Thus, the Climate Convention reflects a broad consensus on the seriousness of the problem and outlines a well-defined process for reaching future international agreements. It establishes a Conference of the Parties in which both developing and industrialized countries can meet regularly, review the state of atmospheric science, evaluate the adequacy of the international policy response, and adapt existing institutions to meet the new challenges. Through this cooperative mechanism, the Convention creates momentum for change. It increases public awareness of the linkages between environment and development issues, and nudges nations down the path to sustainable development. It does so without imposing unbearable burdens on any one group of countries or insisting that all Parties implement the same set of narrowly defined responses to this complex issue. In this way, the Climate Convention keeps the twin issues of environment and development on the policy agenda of governments and creates a forum in which specific future actions can be agreed upon step-by-step.

Principal Elements of the Climate Convention

The two years of negotiations leading to the signing of the Climate Convention resulted in agreement on a number of specific points. The following section summarizes the principal elements of the Convention. Although the full text of the Convention appears in this volume as an Appendix, we highlight here the most important sections of the document, in particular the sections which will be extensively referred to in the following chapters.

One of the most important elements is Article 2, the Objective of the Convention. The ultimate objective of the Convention is:

> to achieve . . . stabilization of greenhouse gas concentrations in the atmosphere at a level that would prevent dangerous anthropogenic interference with the climate system . . . Such a level should be achieved within a timeframe sufficient . . . to enable economic development to proceed in a sustainable manner.

The language of the objective is framed in terms of stabilizing *concentrations*, not emissions, of greenhouse gases. According to the First

Scientific Assessment Report of the IPCC, stabilizing atmospheric concentrations of greenhouse gases will require a reduction in the rates of emission of the long-lived gases by 60 to 80 percent relative to their 1990 levels. Thus, Article 2 sets the Parties on a path to future emission reductions that extends far beyond the commitments agreed to in the present treaty.

Furthermore, Article 2 indicates that the implementation of the Convention is to be accomplished over a time period sufficient to permit sustainable economic development to proceed, guided by the principles laid out in Article 3. These principles establish an important set of precedents for future economic and political relationships between developing and industrial countries. One significant dimension of these new relationships is the emphasis on the Precautionary Principle. The Precautionary Principle is articulated in the following terms in Article 3(1):

> The Parties should protect the climate system for the benefit of present and future generations . . . [D]eveloped country Parties should take the lead in combating climate change . . .

Additional principles outlined in Article 3 include the following:

- 3(2) "The specific needs . . . of developing country Parties . . . should be given full consideration."
- 3(4) "The Parties have a right to, and should, promote sustainable development."
- 3(5) "The Parties should cooperate to promote . . . sustainable economic growth and development in all Parties, particularly developing country Parties."

Taken together, these principles emphasize the need to protect the global environment and to maintain equity in international economic relationships. To this end, Article 4 outlines the commitments shared by all countries, reflecting their common but differentiated responsibilities. Paragraph 1 highlights the general responsibilities of all Parties. These shared commitments pertain mainly to national policies and to information that must be communicated to the Secretariat of the INC. In Article 4, Paragraph 1, all Parties agree to:

- 4(1)(a) "Develop, periodically update, [and] publish . . . national inventories of anthropogenic emissions by sources and removal by sinks of all greenhouse gases . . .";
- 4(1)(b) "Formulate, implement, publish and regularly update . . . programmes containing measures to mitigate climate change . . . and measures to facilitate adequate adaptation to climate change";
- 4(1)(c) "Promote and cooperate in the development, application and diffusion, including transfer, of technologies . . . [to] reduce

or prevent anthropogenic emissions of greenhouse gases...";
- 4(1)(d) "Promote sustainable management ... of sinks and reservoirs of all greenhouse gases...";
- 4(1)(e) "Cooperate in preparing for adaptation to the impact of climate change...";
- 4(1)(f) "Take climate change ... into account ... in their relevant social, economic and environmental policies and actions,...";
- 4(1)(g) "Promote and cooperate in scientific, technological, technical, socio-economic and other research... [and] observation ... related to the climate system..."; and
- 4(1)(j) "Communicate to the Conference of the Parties information related to implementation [of the Convention]."

Not all commitments mentioned in the Convention apply equally to all the Parties. In Paragraph 2 of Article 4, each of the developed countries takes on specific additional commitments that do not apply at present to the developing country Parties to the Convention. These include commitments by each developed country Party to begin reducing greenhouse gas emissions and strengthening the ability of natural sinks to absorb greenhouse gases. Unfortunately, the text is somewhat vague on the details of this point. However, Article 4, Paragraph 2 includes specific commitments by developed country Parties to:

- 4(2)(a) "[A]dopt national policies and take corresponding measures on the mitigation of climate change, by limiting its anthropogenic emissions of greenhouse gases and protecting and enhancing its greenhouse gas sinks and reservoirs... [which will] demonstrate that developed countries are taking the lead in modifying longer-term trends in anthropogenic emissions consistent with the objective of the Convention...";
- 4(2)(b) "[C]ommunicate, within six months of the entry into force of the Convention... and periodically thereafter... detailed information on its policies and measures referred to [above] ... with the aim of returning individually or jointly to their 1990 levels of these anthropogenic emissions of carbon dioxide and other greenhouse gases..."; [and to]
- 4(2)(d) "[R]eview the adequacy" of these commitments at the first session of the Conference of the Parties.

Article 4, Paragraph 3 is one of the most controversial sections of the Convention. This section concerns the Commitment by developed countries to provide new and additional financial resources to developing country Parties in order to cover the costs of actions taken by developing countries to achieve the objectives of the Convention. This commitment was an essential element of the compromise which

led many developing countries to sign the Convention. In Article 4,
Paragraph 3, the developed country Parties agree to:

> ... provide new and additional financial resources to meet the
> agreed full cost incurred by developing country Parties in com-
> plying with their obligations [to communicate information to the
> Secretariat and the Conference of the Parties] ... They shall also
> provide such financial resources, including for the transfer of tech-
> nology, needed by the developing country Parties to meet the agreed
> full incremental costs of implementing measures ... covered by
> Paragraph 1 of [Article 4] and that are agreed between a develop-
> ing country Party and [the Global Environment Facility].

In Paragraph 5 of this Article, the developed country Parties to the
Convention further agree to:

> ... take all practicable steps to promote, facilitate and finance, as
> appropriate, the transfer of, or access to, environmentally sound
> technologies and know-how to other Parties, particularly devel-
> oping country Parties, to enable them to implement...the Conven-
> tion.

Finally, the developed countries agree in Paragraphs 7, 8, 9, and 10
of Article 4 to give full consideration to the special circumstances
affecting the implementation of the Convention by developing countries and by countries with economies in transition. Finally, the Convention notes in Paragraph 7 that:

> The extent to which developing country Parties will effectively
> implement their commitments under the Convention will depend
> on ... [whether] developed country Parties ... [implement] their
> commitments under the Convention related to financial
> resources and transfer of technology and will take fully into
> account that economic and social development and poverty
> eradication are the first and overriding priorities of the devel-
> oping country Parties.

Articles 7 to 15 outline the institutional structures and cooperative
processes that must be developed to implement the Convention.
These institutional structures include the Conference of the Parties
(outlined in Article 7), the Secretariat (in Article 8), the Subsidiary
Body for Scientific and Technological Advice (in Article 9), the Subsidiary Body for Implementation (in Article 10) and the Financial
Mechanism (in Article 11).

Article 12 defines the commitments made by the Parties to communicate certain items of information to the Secretariat in a timely
and accurate fashion. In this article, all Parties agree: (a) to develop
national inventories of sources and sinks of greenhouse gases, (b) to

communicate "a general description of steps taken or envisaged by the Party to implement the convention," and (c) to communicate any other information considered relevant by the Party.

In addition, Article 12 requires all developed country Parties (and all other Parties included in Annex I) to communicate to the Secretariat: (a) "a detailed description of the policies and measures that it has adopted to implement its commitment under Article 4 . . ."and (b) "a specific estimate of the effects that the policies and measures referred to . . . above will have on anthropogenic emissions by its sources and removals by its sinks of greenhouse gases . . ."

Implementation of Article 12 makes it possible for each Party to review the efforts made by each of the other Parties to the Convention, thus assuring that the regime of the Convention will be as transparent as possible.

The Significance of the Climate Convention

The Climate Convention is a landmark achievement in the history of international environmental management. It is not merely another dry legal instrument, adding bulk to the international code of environmental laws. Nor is it simply a slick new employment program for legal scholars and advocates of environmental protection. It is a powerful force for change in North–South relations that will reshape the business climate for international investment and trade.

Five elements of the Convention are likely to have substantial consequences for national policies and international economic relations among the Parties. The first element—and arguably the most important—is the Objective of the Convention. The Convention sets an ambitious two-part goal for international cooperation: atmospheric stabilization of greenhouse gas concentrations and the promotion of sustainable development. In the process, the Climate Convention simultaneously endorses the Precautionary Principle as an element of international law and emphasizes the essential connection between environmental protection and economic development, giving each equal importance.

The second contribution of the Convention to international relations grows out of the treaty's emphasis on process over product. Rather than imposing a set of environmental standards or policies that must be implemented by all Parties, the Convention sets up a process of communication and negotiation that is expected to continue for decades. The diplomatic elements of this process will be informed by a systematic, long-term program of cooperative scientific research aimed at improving knowledge of the climate system, reducing scientific uncertainties, and increasing the ability of scientists to predict the future behavior of Earth's atmosphere and oceans. Regular reviews of the scientific findings will provide the inspiration and momentum for re-evaluating the adequacy of the commitments made by the Parties to reduce the risks of rapid climate change.

The third contribution of the Climate Convention grows out of the commitments agreed by developed country Parties to provide *new and additional financial resources* to support the measures undertaken by developing countries to achieve the objective of the Convention. These resources will support local priorities and provide the basis for promoting patterns of sustainable development.

Fourth, the Convention promotes joint implementation, i.e., a mechanism designed to encourage private investments made by enterprises in developed countries working cooperatively with their counterpart organizations in other countries. Joint Implementation will encourage new and expanded efforts at co-development of advanced technologies (as well as the transfer of technologies already developed for use in the North). This process will increase the technical competence of private enterprises in the poorer countries and may allow them to share equitably in the economic rents derived from these advanced technologies. In addition, this provision of the Convention may stimulate so-called "horizontal" or "reverse" transfer of technologies, introducing cooperative arrangements in which technological advances originating in private enterprises of the South are transferred—either to other enterprises in the South or to counterparts in the North. Ultimately, this process may reshape and rebalance North–South trade relationships in important new ways.

The fifth significant contribution of the Climate Convention to international relations lies in the lessons it offers concerning the dynamic process of environmental and economic negotiations. The success of the climate negotiations has led many to conclude that the secret to future success in international negotiations lies in keeping the processes as open, transparent, and participatory as possible. By welcoming the contributions from the non-governmental community (including both business and environmental NGOs), the Intergovernmental Negotiating Committee (INC) achieved a remarkable vitality and kept the climate issue high on the policy agenda of governments. The success of the climate negotiations thus re-emphasizes the importance of bringing as many "stakeholders" as possible around the negotiating table and allowing their input to enrich the process.

From Interdisciplinary Science to Prudent Public Policy:
A Drama of International Diplomacy

The Framework Convention on Climate Change is a complicated document that grew out of a complex process. Stimulated by advances in a variety of scientific fields, given momentum by increasing public concerns in many countries, the process of negotiations that produced this document stretched over almost two years. It involved more than one hundred governments and engaged thousands of individuals in different capacities—including scientists, lawyers, secretaries, ministers, economists, and diplomats. Yet aside from the substantive benefits that may accrue from the Convention, close examination of the negotiations may also reveal important lessons about the process.

For every complex chain of events, multiple interpretations are almost inevitable. Two people observing the same set of activities often have very different perceptions of the experience. Analysts revisiting a sequence of group decisions or debates may draw contrasting conclusions as to the final balance of forces or the causes of the ultimate outcome. These differences in perception are partly a reflection of cultural bias, partly due to the observer's position relative to the action, and partly a function of how the events may affect the observer's institutional and personal interests.

Such conflicting interpretations are not necessarily evidence of a subtle conspiracy to rewrite history in a more favorable light, nor the result of some personal desire on the part of participants to aggrandize their own roles. Rather, they are a natural consequence of the fact that, when related aspects of complex events take place simultaneously in different locations, no individual can see all the elements first hand. Each person is privy to some but not all of the details, and must fill in the blanks with an explanation that fits logically and consistently with the rest of his or her mental map of the larger processes at work.

We are not the first to observe the potential importance of conflicting interpretations by participants in a complex event. One graphic and accessible illustration of this phenomenon is found in the classic Japanese film *Rashomon*, directed by Akira Kurosawa. In this vivid and poignant film, three different observers reconstruct a series of events that each has witnessed. One might ordinarily expect that the visions of one crime scene as revealed by three eyewitnesses would differ only in the smallest of details. Kurosawa elegantly demonstrates that each individual's reconstruction of such events, while sharing elements with the recollections of others, traces but one thread in the tapestry. Only when all the threads are woven together can one clearly see the outlines of the larger design.

This book, like the Kurosawa movie, revisits a complicated set of events. What we present in the following chapters is not a history in the formal academic sense, but a collage of visions recounted by some of the central players in a complex drama. The differences in perception illustrate how a group of people may share a sense of the high stakes at play in a common enterprise, yet see quite different meanings in the agreed outcome and give importance to very different factors among the forces that shaped that outcome. This collection also reveals some of the common pitfalls in the international negotiating process, as well as the potential for seren-dipitous synergies to emerge among people, positions, and events.

Many who have studied the Convention have found that in both the preliminary draft documents and in the final agreed instrument, the text of the Convention supports many different interpretations. Key concepts, such as "incremental cost," "joint implementation," and "new and additional funds" are defined ambiguously, if at all. Those who have studied only the final text are often perplexed as to what was meant by some of these terms, and question how their meaning evolved as the discussions progressed.

In seeking to understand the final text, we have asked fifteen of the participants to reconstruct the key events that shaped the Convention and preceded the signing of the treaty at the Earth Summit. These key events include both the formal plenary sessions of the INC, the meetings of the Preparatory Committee for UNCED (commonly known as the PrepCom), and some of the less formal meetings, workshops, and other events during which the issues underlying the Convention were discussed. Each contributor offers a somewhat different vision of the process that led to the final agreement; each has a different interpretation of the intended meaning of the language in the treaty.

Through the Eyes of the Participants

This volume is divided into four parts. *Part I: Background* places the negotiations in a broad context. This first chapter provides an overview of the volume and an introduction to the climate debate. In Chapter 2, Daniel Bodansky discusses the recent diplomatic and legal history of the climate issue. *Part II: Views from Within the Ring* contains eight chapters (Chapters 3 to 10). These chapters offer the personal recollections of key participants who represented govern-ments in the negotiations. Some of these observers served on the

Bureau of the INC (i.e., Borione and Ripert, Djoghlaf, Dowdeswell and Kinley, and Dasgupta), others led national delegations at various stages of the negotiations (Kjellen, Goldemberg, Nitze, and Hyder). *Part III: The Outside Edges In* reflects the personal views of several observers who represented influential non-governmental organizations in the process. In Chapter 11, Faulkner provides an industry view while in Chapter 12 Rahman and Roncerel offer the perspective of environmental NGOs.

The final section, *Part IV: Prospects for the Future*, contains two chapters. In Chapter 13, Sebenius rounds out the discussion with an analysis of the strategy and dynamics of complex negotiations. In Chapter 14, we review the lessons learned in the INC process and offer recommendations for structuring the next phase of the negotiations. These recommendations are designed to expedite the preparation and acceptance of supplementary agreements that are both practical and equitable.

Each of the eyewitnesses was asked to address the same two questions:

(1) What were the key events, initiatives, issues, or decisions that shaped the dynamics of the negotiations, that caused alliances to form or coalitions of interest to dissolve?

(2) What lessons have been learned from this first phase of the climate negotiations that might be used to shape the structure of future international negotiations in ways that will lead efficiently to practical and equitable agreements?

Naturally the response of each contributor to these questions has been shaped both by his or her background and by the agenda of the constituency that he or she represented in the negotiations. It is important to emphasize, however, that each contributor has presented his or her own *personal* views. They were compiled in the months during and immediately after the Earth Summit in 1992. None of these views represents the official positions of governments or the interpretations placed on these events by the organizations with which our contributors are or have been associated.

In the remainder of this introduction, we briefly summarize the observations of the participants, highlighting the differences among them.

Bodansky: Prologue to the Climate Change Convention

In Chapter 2, Daniel Bodansky, an international lawyer and professor at the University of Washington in Seattle, outlines the events that carried the climate issue from a simple nineteenth century scientific hypothesis to the center of political attention at the Earth Summit in 1992. Bodansky brings an even-handed approach and does not pass judgment on the relationships in this historical process. Instead he focuses attention on the institutional and diplomatic initiatives leading to the final agreement in Rio de Janeiro.

As a backdrop, he describes the evolution of an international scientific consensus on the climate issue over the last four decades. Starting with the careful observational work of David Keeling and his graduate students at Mauna Loa, Hawaii, Bodansky tracks the scientific debate from the foundation laid by Revelle and Suess, through the development of general circulation models, to the completion of major multi-disciplinary reports by the international scientific community in the 1970s and early 1980s. Bodansky's concise review of the institutional and legal history of the international climate debate provides a firm foundation on which the reader may build his or her own judgement, as well as a clear and useful chronology of the events leading up to the Earth Summit.

Borione and Ripert: Exercising Common but Differentiated Responsibility

Delphine Borione and Jean Ripert of France offer a unique perspective on the INC process in Chapter 3. Within the IPCC, Ripert was the chairman of the Committee on the Special Circumstances of Developing Countries. In addition, and perhaps most relevant to his contribution here, Ripert was selected at the first INC plenary session to be the overall chair of the negotiations. Borione, a senior official in the French Foreign Ministry, was a principal architect of the French national position on climate, and an aide to Ripert.

Borione and Ripert suggest that the negotiations were initially a struggle between three contending blocs: the United States, the other members of the Organization for Economic Cooperation and Development (OECD), and the developing countries who comprise the Group of 77 and China. Economically important activities within each bloc contribute to the global buildup of greenhouse gases, but the level of contribution to current and past emissions varies considerably among the parties.

From Borione and Ripert's perspective, the United States entered the negotiations predisposed toward a weak general framework for the Convention, intent on emphasizing the need for continued scientific research and hoping to avoid any agreement to targets for emissions reductions or fixed timetables for achieving those reductions. Borione and Ripert observed that the other members of the

OECD, and in particular the European Communities (EC), took a more activist approach to the negotiations. Energized by a well-informed public that had been sensitized by NGOs and the media to the risks of rapid climate change, many of these countries pushed hard for commitments to specific targets and firm timetables for emissions reductions. In addition, the EC countries promoted common policy approaches (including the introduction of carbon and energy taxes) to stimulate changes in their domestic patterns of energy use.

The developing countries brought a sharply differing perspective. These countries appeared to Borione and Ripert to be unanimous in their initial contention that the current set of atmospheric problems had been caused by actions of the industrialized countries. Their delegations jointly conveyed to the INC the critical importance of negotiations focused on reciprocal commitments—new and additional financial resources to be provided by the developed countries in exchange for any specific commitments by the developing countries to control the rate of growth of their own emissions or to preserve the status of their existing "sinks" (e.g., forests and grasslands) for greenhouse gases. Despite this initial unanimity, the developing countries appeared to these observers to be divided over the question of how aggressive the convention should be in trying to address the issues of adaptation and abatement.

Borione and Ripert examine carefully the concept of "joint implementation," which was introduced by the Norwegian government as a vehicle to encourage efficiency and cost-effectiveness. This approach would allow an industrial country to fulfill at least some of its national obligations for emissions reductions through cooperative actions taken in developing countries or "economies in transition." Borione and Ripert point out that this approach could have significant advantages for developing countries. As originally proposed, these countries would receive additional financial transfers from the industrialized, high-emissions countries of the North as an incentive for joint implementation projects that would reduce the rate of emissions growth in the South. Although some NGO representatives viewed joint implementation as a scam orchestrated by the rich countries to avoid making emissions reductions at home, Borione and Ripert saw this concept—which was later recognized formally in the Convention—as a vehicle for promoting equity through North–South financial transfers.

Borione and Ripert challenge the reader at the outset to determine whether the ostensible success of the Climate Convention truly achieved an artful compromise by adequately balancing the national interests of competing parties, or whether the agreement of so many countries reflects an ambiguous text that describes what is, ultimately, an insignificant agreement. From their perspective, the negotiations followed an orderly progression in which widely divergent positions converged toward a mutually agreed and broadly beneficial consensus.

Djoghlaf: The Beginnings of an International Climate Law

Ahmed Djoghlaf, the vice-chair of the INC and leader of the Algerian delegation, brings to Chapter 4 the perspective of a Third World diplomat and scientist. He begins his reconstruction of the climate debate with a brief review of the scientific setting in which the negotiations began, but emphasizes that a growing, shared understanding of the science was not the sole decisive factor in the success of the INC. Rather, Djoghlaf asserts, the success of the INC was largely the result of shrewd and careful decisions made early in the process. Many of these decisions concerned the structure of the negotiating sessions and the roles of the participants. Most important among them were the measures taken to facilitate the participation of developing countries by providing financial support for travel, as well as the decision to make the deadline for agreement to the Climate Convention coincide with the convening of the Earth Summit.

Djoghlaf perceived a very different array of principal characters in the INC drama than was observed by Borione and Ripert. He saw the industrial countries acting largely as a bloc, protecting their shared interests in the status quo. From Djoghlaf's perspective, the members of this bloc were reluctant to fulfill the promises they had made to developing countries at the Second World Climate Conference held in Geneva, Switzerland, in November 1990. Furthermore, the developed countries were generally hesitant to make any new or specific commitments for additional financial resources.

In contrast to Borione and Ripert, Djoghlaf viewed the developing countries as a diverse and divided group. In place of the traditionally unified front of developing countries and non-aligned states, Djoghlaf found that the INC process promoted the formation of new coalitions among the developing countries, each sub-group defined by a narrow array of common interests. Djoghlaf observed with some regret the extent to which the debate on the climate issue catalyzed a fracturing of the Group of 77 and China (G-77) into several small groups of countries. These new small groupings included the Alliance of Small Island States (referred to as AOSIS), the Kuala Lumpur Group of countries with large forest resources, and a revitalized alliance of oil-exporting developing countries. As these new coalitions focused on their divergent interests, Djoghlaf found that the G-77 and China were able to agree on only one joint position paper concerning unresolved issues in the INC debates.

After drawing our attention to these new and strengthened sub-groups of developing countries, Djoghlaf analyses the implications of the breakdown of the former Communist bloc and highlights the new roles of these countries in the UN system. He notes that, following the dissolution of the Soviet empire, many of these countries have declined from the status of industrialized countries, have abandoned their traditional positions of support for the non-aligned states, and have begun to compete aggressively with Third World countries for Official Development Assistance (ODA) funds made available by the OECD.

Djoghlaf also observes an expanded role for another group of players on the international stage—the non-governmental organizations. From Djoghlaf's vantage point, the NGOs played a critical role in spurring the active participation of a number of governments in the international climate debate.

Djoghlaf's view of the relationships among these forces led him to draw several conclusions about the dynamics that shaped the first round of climate negotiations and that will affect the future decisions of the Conference of the Parties. Djoghlaf felt that the system of duel rotating co-chairs slowed the overall process of negotiations. Furthermore, Djoghlaf sees important differences between the role played by the INC Secretariat (i.e., the international civil servants selected by the Secretary-General to serve the Convention process, and the INC Bureau (i.e., the group of national chief delegates who acted as a steering committee for the negotiations). He suggests that the early failure of the INC Secretariat to provide an acceptable text from which to begin negotiations significantly diminished the role and importance of the Secretariat throughout the remainder of the talks. By contrast, he believed that the INC Bureau played a critical role in the success of the process, particularly in the spring of 1992 when negotiations had reached a complete impasse.

Dowdeswell and Kinley: Constructive Damage to the Status Quo

In their reconstruction of the climate negotiations, Elizabeth Dowdeswell and Richard Kinley reflect yet another perception of the balance of forces. Dowdeswell, currently the Executive Director of the United Nations Environment Programme, was the Assistant

Deputy Minister of Environment Canada and head of the Canadian delegation to the INC. She was also one of the two co-chairs of Working Group II, which dealt with the mechanisms of the Convention. Kinley, a senior official in the Atmospheric Environment Service of Canada, was a major contributor to the formation of the Canadian positions in the INC and a principal aide to Dowdeswell.

Dowdeswell and Kinley perceived the climate negotiations as divided largely along North–South lines. But in contrast to Djoghlaf, they see the process as a cooperative regime of policy development, coordination, and implementation.

Irrespective of the North–South divisions, Dowdeswell and Kinley often observed more commonality among the positions of, for example, Environment Ministries in different countries than between the Environment and Finance Ministries or the Environment and Energy Ministries in a given country. They note that this trans-national commonality of approach not only allowed fruitful informal contacts to develop in the corridors outside the plenary sessions, but also enabled colleagues in different Environment Ministries to act as a temporary bridge across which innovative ideas could pass, even before they could be formally discussed by official delegations.

Dowdeswell and Kinley reinforce an impression also developed by Djoghlaf: they observed a strong, new role emerging for NGOs in the climate negotiations. For example, they note that the most widely read source of information during the negotiations was the daily NGO publication called *ECO*. The publication of *ECO* not only served to keep the NGOs (and some delegation members) informed about the progress of the negotiations, but it also provided a sanctioned opportunity for NGO representatives to interact regularly with delegates. NGOs could advocate aggressive policy positions, provide a measure of public participation, and make participating delegations accountable to the public.

Dowdeswell and Kinley place great importance on the inventive character of the negotiations and the need for continued innovation. In contrast to Djoghlaf's view, Dowdeswell and Kinley saw great strength in the system of dual rotating co-chairs. Rather than slowing up the process, they observed that, at least in Working Group II, the presence of a developing and a developed country co-chair contributed to the step-by-step expansion of the group consensus. From Dowdeswell and Kinley's vantage point, the co-chairs were able to facilitate the working group debate. The co-chairs instigated a broader basis of understanding by framing an expanded consensus whenever the positions of the principal parties appeared to converge.

When the national positions did not converge, the co-chairs raised provocative questions. In the view of Dowdeswell and Kinley, this approach made it possible, by the close of the INC plenary in December 1991, for Working Group II to produce a fully elaborated institutional scheme and a draft text virtually free of brackets. (By UN convention, alternative phrasings are highlighted in draft texts with square brackets.) Unfortunately, they note, Working Group I was not as successful in this regard.

Dowdeswell and Kinley share the views of Ripert, Borione, and Djoghlaf on some of the critical issues in the INC debate. Dowdeswell and Kinley see the Climate Convention as "a pragmatic first step toward a new vision of international cooperation on global environmental problems." From their perspective, this new vision places equal emphasis on economic development and environmental protection. Thus, they see the INC process as a model for mutually beneficial, collaborative dialogue on sustainable development, based on sound science, and with a strong institutional foundation for continued cooperation. The INC model, in Dowdeswell and Kinley's view, can change the future shape of international environmental law as well as the character of North–South relations.

Dasgupta: The Climate Change Negotiations

Chapter 6 provides the recollections of Chandrashekhar Dasgupta, Head of the Indian delegation to the INC and a member of the INC Bureau. A skilled and seasoned diplomat, Dasgupta held the position of Additional Secretary in the Ministry of Foreign Affairs during the climate negotiations. He is now the Indian Ambassador to the People's Republic of China.

Dasgupta saw the negotiations in the context of strategic competition between North and South, East and West. His vision of the climate negotiations highlights the emerging realities of a multipolar world struggling to find a new balance of forces in the post-Cold War era. In particular, he saw a continuing struggle among the participants for control over traditional and emerging markets and for access to the types of advanced technology that would assure a competitive advantage in the future.

But rather than a world of monolithic blocs rigidly divided by political ideology, he saw sharp and persistent divisions among the positions taken by industrialized countries. These divisions were sharpened by the lack of a clear role for the industrialized countries of Eastern Europe and the former Soviet Union. Among the developing countries, Dasgupta observed a parallel breakdown of the traditional unity of the G-77 and China along lines similar to those described by Djoghlaf.

On balance, Dasgupta views the climate negotiations rather positively. But from his perspective, the negotiations that led to the Framework Convention on Climate Change, though important and constructive, fulfilled only a fraction of their potential. Initially, Dasgupta saw these negotiations as an opportunity to recognize responsibility for the damage which has been already done to the atmosphere through mankind's industrial and agricultural activities over the last two centuries. Dasgupta had hoped as well that the climate negotiations would be a vehicle to address the equity issues that have remained unresolved in North–South relations at least since the end of World War II. He recognized a potential in the climate negotiations to shape a new mechanism of compensation by which those countries that have benefitted most from the actions which caused the current level of global environmental risk could support the citizens of countries whose development strategy must now be modified to cope with the consequences.

For Dasgupta, fundamental issues of equity between states were at the heart of the climate negotiations. Thus, the debates on issues such as joint implementation, pledge and review, and the structure of the financial mechanism, were not ends in themselves but rather vehicles for addressing the deeper issues that continue to divide the interests of the parties to the Convention.

In addition to these divisions on the issues, Dasgupta draws attention to some new procedures that also affected the dynamics of the climate negotiations. This approach, called "streamlining," allowed the working group co-chairs and the Bureau to remove some bracketed text from the draft text in order to prepare a "consolidated working document" for further negotiations. By making the selection among alternative phrasings of controversial sections, this process, according to Dasgupta, substantially enhanced the role and power of the Bureau in the conduct of the negotiations.

The power of the Bureau and the chairman of the INC expanded further as a result of the convening of the Expanded Bureau and the emergence of informal "contact groups." These small groups began to meet regularly after the conclusion of the first part of the Fifth Plenary Session in New York during February 1992. In this process, negotiation moved from the full plenary to a smaller, representative group of 25–30 delegation heads. Following the April 1992 meeting of the Expanded Bureau in Paris, the chairman of the INC was authorized to prepare a final Chairman's Draft text for presentation at the resumption of the Fifth Plenary Session in New York in May.

Dasgupta concludes that the final text of the Climate Convention leaves a substantial amount of the core business of the negotiations unfinished. If the remaining questions concerning methods for calculating sources and sinks for greenhouse gases, criteria for joint implementation, and the process for reviewing the adequacy of existing commitments can be resolved, he believes that the Climate Convention can become "an effective instrument for combating climate change." From his perspective, only by addressing the equity questions in ways which lead to a "clear-cut and unambiguous, time-bound program for stabilization and reduction of greenhouse gases originating from the developed countries" will it be possible to limit the risks of global warming for all humankind.

Kjellen: A Personal Assessment

Chapter 7 presents the recollections of Ambassador Bo Kjellen, the Swedish Chief Delegate to both the INC and the Preparatory Committee (PrepCom) of UNCED. Kjellen is a seasoned foreign service officer who was chairman of the Working Group on Atmospheric Issues for the PrepCom. He has been selected as the chairman of the newly established Intergovernmental Negotiating Committee on Desertification, a role directly parallel to that of Jean Ripert in the climate negotiations.

Kjellen reconstructs the INC process principally in terms of the positions of the OECD countries and those of the G-77. As a diplomat, Kjellen sees the INC debate in terms of national positions and initiatives driven by a shared desire to achieve an overall objective, i.e., sustaining the planet in a livable form for subsequent generations. As chief of an informal negotiating group within the OECD, he was most aware of the conflicts and disagreements on specific initiatives within the OECD contingent. Perhaps more than any of the previous witnesses, Kjellen draws our attention to the finer points of disagreement among the US and other OECD member states.

From Kjellen's vantage point, some of the most difficult episodes in the INC debates occurred among the members of the OECD, as they struggled with the hard-line position of the US government. Kjellen found that addressing these differences was initially difficult because the negotiations occurred under the aegis of the UN General Assembly system, in which there is no formal mechanism available for the OECD countries to coordinate their positions.

In Kjellen's view, the G-77 and China raised several well-founded criticisms of the OECD during the fourth plenary session of the INC during December 1991. These criticisms prompted the first concrete steps to coordinate OECD positions in the climate negotiations. The process was initiated by the EC delegate, Jörgen Henningsen, who proposed the formation of a special working group and prepared an initial agenda of key issues for discussion. Kjellen became the chairman of this Special Working Group of the OECD.

Kjellen chaired meetings of the special working group during and after the first stage of the fifth plenary session of the INC in February 1992. Kjellen used these meetings to try to resolve the differences between the positions of the US, the EC, and other members of the working group. The focal point of these debates was on the proposed text of Article 4, the Convention article dealing with specific commitments of the industrialized countries. Kjellen observed an increasing level of frustration within the Special Working Group during the first part of the Fifth INC Plenary Session as the US delegation seemed "to go back on points that they had previously agreed upon." Although he knew that the differences could not be resolved in this informal group, Kjellen felt that these discussions played an important catalytic role in the negotiations.

In Kjellen's view, the success of the climate negotiations was in part due to the flexibility and spirit of compromise that many governments maintained throughout the negotiations. But, from his perspective, the personal relationships, group dynamics, and common understandings that developed between the principal negotiators may have contributed in equal measure to the ultimate success of the process.

Goldemberg: The Road to Rio

In Chapter 8, José Goldemberg provides a Brazilian perspective on the INC process and the preparations for the UNCED meeting of June 1992. During the 1980s, Goldemberg, a nuclear physicist by training, was president of the Compagnia Energetica de Saõ Paulo (CESP), the largest electric utility in the developing world. He later served as rector of the University of Saõ Paulo. During the initial stages of the INC negotiations, Goldemberg was the Minister of Science and Technology for Brazil. In the spring of 1992, he took on the additional post of Minister of Environment in the Cabinet of President Fernando Collor. He has since returned to the University of Saõ Paulo.

Goldemberg frames the evolution of Brazil's involvement in the international climate debate in terms of the interplay between domestic politics and international issues, an approach similar to that of Nitze. In 1990 and 1991, reports in the international press, focusing mainly on the burning of the Amazon forest, made Brazil appear to be an international environmental criminal, ravaging the global heritage. But reports from Brazilian scientists in the field documented a situation that was much less serious than the environmental catastrophe being pictured in the international press. These countervailing reports allowed Brazil to move away from a rigid position—one which unequivocally denied any governmental guilt— to a moderate position which acknowledged that deforestation was indeed a problem but one which could, and would, be mitigated by appropriate national policies. Goldemberg notes that Brazil offered to host the Earth Summit in part to counteract the effect of these negative reports in the international media, with the meeting providing a forum to highlight Brazil's contribution to protection of the global environment.

Like Dasgupta, Goldemberg saw the climate negotiations as an affair in which neither the industrialized countries of the North nor the developing countries of the South were able to maintain consistent positions and negotiate as monolithic blocs. In the North, the hard-line positions of the Bush Administration set the US in conflict with the Nordic countries and some of the more progressive elements of the EC. Furthermore, from Goldemberg's perspective, Great Britain and Japan appeared to take positions somewhere in between these two poles. In the South, according to Goldemberg, the pragmatic positions of China, India, and Brazil were in stark contradiction to the more ideological and rhetorical positions of some other members of the G-77 and contrasted as well with the "go-slow" approach of members of the Organization of Petroleum Exporting Countries (OPEC). Brazil, for example, sought to avoid the rhetoric of "guilt," "historical responsibility," "compensation for past deeds," and the "right" of the poor to 0.7 percent of the GNP of the rich countries in the form of ODA. Instead, Brazil sought to draw attention to the issues of excessive consumption and excessive energy use in the North. The Brazilian delegation used these as arguments for the provision of "new and additional resources" to promote sustainable patterns of development and for accelerating the transfer of less-polluting technologies from North to South on advantageous or moderately concessional terms.

From Goldemberg's vantage point, the principal tensions in the preparations for the Earth Summit derived from differences over whether the conference should focus on questions of development or environment. As he saw the situation evolving, Goldemberg felt it was naive to think that the North would allow considerations of poverty and development to take precedence over issues of global environmental degradation—since these same issues of poverty and the right to development have been the focus of confrontations between developed and developing countries for thirty years. It was significant, he felt, just to place these two issues on an equal footing.

Goldemberg observes that domestic political events in a few key countries often have a disproportionate effect on the outcome of international negotiations. He also notes that NGOs can play a critical role in international debates on the environment, focusing public attention on outstanding issues or outrageous national positions at critical moments in the process. In sum, while recognizing the important but partial successes of the Convention process, Goldemberg concludes that the North's interests in maintaining an environmentally healthy planet can only be achieved through aggressive efforts to support national economic advancement in the developing countries of the South.

Nitze: A Failure of Presidential Leadership

Chapter 9, prepared by William A. Nitze, analyzes the policy process which shaped the position of the United States in the climate negotiations and places that position in the context of several important long-term trends in US foreign and domestic policy. Nitze, formerly Deputy Assistant Secretary of State for Oceans and International Environment and Scientific Affairs and head of the US delegation to the INC, is now the president of the Washington-based non-governmental organization of industrial leaders, the Alliance to Save Energy. Until the end of 1989, Nitze was also head of the US delegation to the IPCC and represented the President of the US at most major climate meetings. He presents an insider's view of the dynamics of policy formation at the most senior level of the US government.

Like Djoghlaf and Dasgupta, Nitze sees the initial phase of the climate negotiations as a struggle between the US, other OECD countries, and the developing world. But in his reconstruction of these events, Nitze focuses on the microcosm of US decision-making rather than on the broad sweep of the overall negotiations. He draws our attention to the unique cast of characters who played out their debates in the Cabinet Room of the White House rather than in the chambers of the INC plenaries.

From Nitze's perspective, the US appeared to achieve most of its diplomatic objectives in the INC process. Many of the central tenets of its initial negotiating position—including the need to avoid binding commitments to emissions reductions at a certain date, the use of the "comprehensive approach" involving all greenhouse gases, and the desire to minimize specific commitments of new and additional resources for developing countries—are reflected in the agreed text of the Convention. Nonetheless, Nitze views the Convention process as a dramatic and unnecessary failure of US leadership, especially at the presidential level.

With control of the issue lodged safely in the White House, conservative ideologues fought successfully to keep the US from supporting any international action that might force reductions in domestic energy use per capita, raise the price of gasoline, restrict individual driving habits, or foreclose the opportunity for each person in America to buy as much gasoline or electricity as he or she can pay for. In Nitze's analysis, the critical failure lay in then-President Bush's inability to recognize the opportunities to use the climate change negotiations as a vehicle to promote increased efficiency of energy use throughout the US economy and increased competitiveness for US industry.

The domestic success of the conservative Sununu-Darman-Boskin faction of the Bush Administration imposed a very real cost on the INC process as a whole. In Nitze's view, the positions instigated by this faction made it impossible to achieve an overall international agreement on specific short-term steps for achieving emissions reductions. The success of this faction of the Bush Administration provided diplomatic "cover" for those in other countries who wanted to avoid a serious obligation to achieve future reductions in emissions or to protect existing greenhouse gas sinks.

From Nitze's vantage point first within and later outside the US government, this small group of White House advisors seemed able to hold the rest of the world hostage to their demands for strict adherence to a narrow ideological position. It appeared, from this perspective, that a minority faction of officials in one government were able to influence the overall outcome of a major international negotiation to a disproportionate extent. Without ever issuing a direct threat, this group was able to instill in the perceptions of many allied governments a fear of total impasse in the negotiating process unless the US demands were met. Further and unfortunately, in this view, it was the very success achieved by this White House clique that did the most to undermine the long-term interests of the United States and to set back international progress toward a constructive, cooperative, and sustainable solution to the problem of global climate change.

Hyder: Looking Back to See Forward

In Chapter 10, Tariq Osman Hyder shares his recollections of the INC process. Hyder is the Deputy Foreign Minister of Pakistan, responsible for the Bureau of International Economic Cooperation. Hyder led the Pakistani delegation to several of the INC plenary sessions and was the chief drafter of G-77 positions during the negotiations of the Rio Declaration.

Like Goldemberg and Nitze, Hyder sees the INC debates in political terms, a drama with three principal actors. Internationally, the protagonists were the United States, the other members of the OECD, and the developing countries (represented by the Group of 77 and China). Hyder observed firsthand the byplay among these players throughout the preparations for UNCED, including the negotiations of Agenda 21 and the Rio Declaration as well as the climate talks. Hyder sees all of these debates as part of a larger struggle among ". . . developed and developing countries to define the concept of sustainable development in a way that fits their own agendas."

Hyder believes that the developed countries placed primary emphasis on protecting the global environment from the impacts of (principally their own) economic activities. The developing countries, on the other hand, seemed to him to emphasize issues of economic development and meeting basic human needs. From Hyder's vantage point, the developed countries rejected the linkage between environment and development out of fear that the developing countries would use the North's concerns about the environment ". . . as a 'club' with which to beat additional financial resources from the recession-strapped economies of the North."

Hyder concludes that to avoid this type of coercion, the industrialized countries sought to differentiate between "global" and "local" environmental problems. For the developing countries, these two aspects of environmental degradation were inextricably linked. This difference in perception of the problem is pivotal to understanding the relative priorities which the parties placed on the issues of environmental protection and economic development. To Hyder, the developed countries' preoccupation with global effects was designed to draw developing countries into strategies that would help the industrialized world mitigate the consequences of excessive consumption. By contrast, he sees the efforts by developing countries to place equal emphasis on addressing local environmental problems as an attempt to draw global attention to the linkage between global problems and the local impacts that result from the burdens of poverty.

From Hyder's perspective, this underlying difference in approach led to the impasse reached during the first part of INC 5 in February 1992. Hyder views the chairman's decision to call the Paris meeting of the Extended Bureau in April as an effort to break this impasse. At the Paris meeting, Hyder, as the G-77 coordinator, delivered the unified position of the G-77 on the key issues. He suggested to Ripert that the text of the Rio Declaration (which he had just helped to negotiate) be adopted as the section of the Convention titled "Principles." Hyder reports that, with some modest amendments, this suggestion was accepted by the chairman (and the Extended Bureau) and consequently is now visible in the final instrument. For Hyder, with the acceptance of these principles, a basis was laid for framing a broader compromise and for reaching a general agreement on the text of the Convention.

Faulkner: Some Comments on the INC Process

Hugh Faulkner summarizes his observations of the INC process in Chapter 11. Faulkner is the Executive Director of the Business Council for Sustainable Development (BCSD), a coalition of chief executive officers and other senior officials of major multi-national corporations. Working with Stephen Schmidheiny of Switzerland, the chairman of the BCSD, Faulkner sought to bring a balance of business perspectives to the climate negotiations, the international debate on sustainable development, and the preparations for the Earth Summit.

In Chapter 11, Faulkner focuses on the emerging role of NGOs in the climate negotiations and in the preparations for UNCED. He notes that industry representatives participated both as members of official national delegations and as invited observers, representing trade associations and ad hoc coalitions. Although there were only a few opportunities for industry representatives to speak on the floor of the INC plenary sessions, Faulkner notes that there were a number of vehicles by which representatives of industry could advise and influence the delegates. Nonetheless, he observed that there was still room for improvement in the dissemination of important information and the sharing of views among various stakeholder groups.

Faulkner perceived a healthy measure of tension between the sectoral interests of the individual industries and the wider concerns of global business coalitions like the BCSD. He points out that government policy-makers should pay considerable attention to these important differences—and not assume that industry interests are uniform or monolithic.

Faulkner welcomes the opportunity for industry leaders to participate in the INC process and argues that their continued participation is crucial to achieving a broad and lasting consensus that can become the basis for future agreements. Two keys to success in the first phase of the INC process, according to Faulkner, were (1) the commitment to openness in the INC process and regular access to the INC plenaries, and (2) the encouragement of a frank and complete exchange of views with business leaders by the INC Bureau, the UNCED Secretariat and national delegations. Faulkner points out that this dialogue exposed the delegates to practical information about the impacts of various proposals under discussion that they would likely have missed without the participation of business leaders. Faulkner concludes that, with access to the views of key business leaders, governments were able to forge an international agreement that could be communicated to the general public to elicit their support in both developed and developing countries.

Rahman and Roncerel: A View from the Ground Up

Atiq Rahman and Annie Roncerel represented environmental NGOs in the climate negotiations. Rahman, a former Professor of Chemistry at Oxford University, is currently the Executive Director of the Bangladesh Centre for Advanced Studies in Dhaka, Bangladesh. Roncerel, trained in France as an international lawyer, was the founding coordinator of the Climate Network Europe. She has recently been named GEF Coordinator for Europe, North Africa, and the Middle East in the United Nations Development Programme (UNDP).

In Chapter 12, Rahman and Roncerel offer a detailed look at the evolving dynamics within the community of environmental NGOs. Their vision highlights the interplay between international NGO groups (drawn largely from the North) and national NGO groups (some based in developed countries and some in developing countries).

Rahman and Roncerel saw these NGO groups play a rich and complex role. Throughout the INC debates, the environmental NGOs sought to place the positions, proposals, and initiatives offered by the negotiators in the context of the concerns of common people "on the ground." They observe the NGOs acting as a conscience for the overall process. By publishing the daily journal, *ECO*, during the

INC plenary sessions and regional newsletters in the interim period, the environmental NGOs highlighted the knotty problems and the unresolved issues that divided North from South. Through these publications and through the official statements of an NGO representative at the close of each INC plenary, Rahman and Roncerel observed the environmental NGOs maintaining the focus of the debate on the equity issues that are the foundation of the linkages between environment and development. By using humor and passion to expose the idiosyncrasies of some of the personalities involved in the negotiations, the NGOs were able to reveal the inconsistencies between the rhetoric and the substance of national positions.

Rahman and Roncerel saw the environmental NGOs play several additional roles in the process. Drawing on independent scientific expertise, the NGOs sought to put the latest advances in climate science on the negotiating table, always trying to enrich the political context of the talks. They used this information not only to influence the plenary debate but also, back home, to push individual national governments to take a more aggressive stance in implementing the Precautionary Principle and in giving equity issues equal consideration with questions of economic efficiency.

Paralleling Faulkner's view of the international business community, Rahman and Roncerel saw the environmental NGO community as anything but a monolithic bloc. Rather it was an evolving organic whole, representing in microcosm all of the larger struggles between North and South that could be seen in the INC plenary. Rahman and Roncerel observed many Northern NGO representatives enter the negotiations with a strong grasp on the environmental science of the climate issue but with little understanding of the urgent economic pressures and equity considerations facing developing countries. In many cases, developing country NGOs initially recognized the urgent need for development assistance and improved trade relations with the North but had little background in the science of climate change and only limited understanding of the contributions of developing countries to current emissions of greenhouse gases.

Despite these deep divisions, Rahman and Roncerel observed the remarkable ability of environmental NGOs to participate in and influence the negotiations. Through countless informal consultations with members of the INC Bureau and with senior members of national delegations, the voices of environmental NGOs were heard in the INC process. Rahman and Roncerel, like Kjellen, saw the environmental NGOs as a crucial bridge for dialogue between delegations, carrying early versions of proposals from one group of principals to another, usually after debating them extensively among themselves. From Rahman and Roncerel's perspective, the environmental NGOs were a critically important component of the overall process of the climate negotiations. They conclude that continuing active participation by NGOs is an essential ingredient of successful international negotiations on environment and development issues.

Sebenius: Towards a Winning Climate Coalition

James K. Sebenius, the author of Chapter 13, is a Professor at the Harvard Business School and a specialist in complex negotiations. Though not an official delegate to the INC, Sebenius is a seasoned observer of international negotiations and was a former member of the US delegation to the Third United Nations Conference on the Law of the Sea.

Sebenius focuses on the underlying dynamics of the negotiations. He seeks to explain the outcome in terms of the factors which lead to the evolution of winning coalitions (i.e., groups of countries that agree to take concrete measures to curb climate change) and blocking coalitions (i.e., groups of countries that cooperate to block or prevent effective actions to control greenhouse gas emissions).

Recognizing that the INC process was able to achieve only a loose framework convention, Sebenius tries to understand why sustained and effective agreements on climate change have been (and will continue to be) so difficult to achieve. He notes that while the benefits of avoiding a future climate change are vague and uncertain, the costs of limiting greenhouse gas emissions are immediate and real. Furthermore, the "free rider effect" discourages participants from taking costly individual initiatives since a part of the benefits of their national sacrifices can be appropriated at no cost by every non-participant in the agreement.

Sebenius suggests that the current approach of trying to supplement the Framework Convention with specific protocols is replete with dangers. The specific opponents of any particular negotiating strategy are unknown in advance and will depend on which gases and sectors are the early targets of controls. He indicates that blocking coalitions may form based on uncertainties in the science, shared interest in a regulated or prohibited activity, a common ideology, or simply opportunism.

Sebenius' analysis suggests an alternative strategy for future negotiations. This strategy may be controversial with many of the participants in the INC process. Sebenius concludes that the limited commitments contained in the Framework Convention need not be seen as a failure of the negotiating process. Moreover, he suggests that greater success can be achieved in the long run if the next steps taken are *not* designed to reach agreement on specific protocols. Instead the next goal of the negotiations might be to tap the vast potential of actions which can be taken on a voluntary basis, even without internationally agreed emissions limits. This strategy, he believes, could lead to a flexible and expandable compact that would be less subject to obstruction and delay by coalitions of opponents. Membership in the compact could expand quickly, as the economic advantages of the participants become obvious to all.

Visions of the Past, Lessons for the Future

This volume illustrates the range of perceptions that arise when the international climate negotiations are viewed from the broad range of vantage points held by our contributors. It is not our intention to say which vision is false and which is true. Rather, we offer them here in order to alert the reader to the context in which the next and future rounds of negotiations are likely to take place. In the last chapter we provide a set of recommendations for choosing a structure for the next phase of climate negotiations and highlight some of the lessons learned in the climate debate that might be usefully applied to other international negotiations in the years ahead. Taken together, these visions offer a sense of the rich patterns and intense conflicts that must be addressed in order to reach a comprehensive agreement on future controls of greenhouse gases.

Notes

1. Houghton, J.T., G.J. Jenkins, and J.J. Ephraums, 1990. *Climate Change: The IPCC Scientific Assessment*, Cambridge University Press, Cambridge.

2. *Ibid.*

3. *Ibid.*

4. *Ibid.*

5. *Ibid.*

6. *Ibid.*

7. Oeschger, H. and I.M. Mintzer, 1992. "Lessons from the Ice Core: Rapid Climate Changes During the Last 160,000 Years," in I.M. Mintzer, ed., *Confronting Climate Change: Risks, Implications, and Responses*, Cambridge University Press, Cambridge.

8. *Ibid.*

9. Tegart, W.J., G.W. Sheldon, and D.C. Griffiths, 1990. *Climate Change: The IPCC Impacts Assessment*, The Australian Government Publishing Service, Canberra.

10. Kushlan, J.A., 1986. "Responses of Wading Birds to Seasonally Fluctuating Water Levels: Strategies and their Limits". *Colonial Waterbirds* 9, pp. 155–162.

CHAPTER 2

Prologue to the Climate Change Convention*

Daniel Bodansky
Assistant Professor
University of Washington School of Law
Seattle, Washington

The Climate Change Issue Comes of Age

Although some were disappointed with the admittedly modest achievements of the UN Framework Convention on Climate Change, what is striking about the climate change issue is not how slowly the issue has developed—but how quickly. As recently as 1985, when a major international scientific workshop was held in Villach, Austria, the US government participants went without instructions. Within three years, governments had decided to establish the Intergovernmental Panel on Climate Change (IPCC) to provide a scientific and policy assessment of the problem. And less than four years later, international negotiations on a climate change convention were completed and the UN Framework Convention on Climate Change (FCCC) was adopted and signed.

The evolution of awareness about climate change can be separated into three stages. The first stage, stretching from the mid-1950s to mid-1980s, witnessed the emergence of a broad scientific consensus. During the second stage, from about 1985 to 1988, public and political interest in the problem grew. And the final ongoing stage has involved the formulation of an international policy response.

The Development of Scientific Consensus

Although the general concept of the greenhouse effect has been understood since the 1820s, and the warming influence of increased atmospheric concentrations of carbon dioxide was predicted more than a century ago, widespread scientific concern about the problem emerged only in the last ten to twenty years.[1] This resulted from several scientific developments.[2] First, based on careful measurements made daily during the 1960s and 1970s, at a mountain-top observatory on the side of the Mauna Loa volcano in Hawaii, chemists conclusively established that the concentration of carbon dioxide was steadily increasing in the atmosphere. In the first half of this century, most scientists believed that the oceans absorbed the vast majority of anthropogenic CO_2 emissions, maintaining the balance in atmospheric composition. In 1957, the continuing validity of this assumption was questioned in a now-famous paper by Roger Revelle and Hans Suess.[3] By the early 1960s, direct measurements had unambiguously confirmed Revelle and Seuss's hypothesis that atmospheric concentrations of carbon dioxide were indeed on the rise.

As a result of this new data, scientific interest in the greenhouse warming theory began to grow in the 1960s and early 1970s. The issue was raised as a public policy issue first at a meeting of the Conservation Foundation in 1963 and then in a report prepared for the President's Science Advisory Committee in 1965. Nevertheless, there was still little sense of urgency in the scientific community about the likelihood of greenhouse warming. In the early 1970s, the Study of Critical Environmental Problems (SCEP) and the Study of Man's Impact on Climate (SMIC) identified global warming as a *potentially* serious problem, calling for more scientific study but no immediate action.

At this point, the second major factor contributing to the growth of concern about global warming kicked in—improvements in general circulation models (GCM) of the atmosphere, which form the basis of current predictions of global warming. Early attempts to model the atmosphere were constrained by limitations in computing power and observational data, and could not hope to represent the complexity of atmospheric dynamics.[4] As computing power grew, climatic models increased dramatically in sophistication and complexity. The advent of supercomputers and satellite sensing data permitted the development of more "realistic" GCMs, which represent the atmosphere in three dimensions, in greater spatial detail, and take account of some of the principal feedback mechanisms and ocean–atmosphere interactions. By the end of the 1970s, increased confidence in atmospheric models led a National Academy of Sciences panel to conclude that, if the concentration of carbon dioxide in the atmosphere continues to increase, "there is no reason to doubt that climate changes will result and no reason to believe that these changes will be negligible."[5]

Third, in the 1980s, studies of the climatological record indicated that the historical record is broadly consistent with global warming forecasts. In mid-century, such forecasts had only limited impact, given what appeared to be a cooling trend. As recently as the mid-1970s, when a series of climatological disasters—including weak monsoons in India and a drought in Europe—drew attention to the climate change issue, scientists were still split between those who believed the Earth was warming and those convinced that the planet had entered a cooling phase.[6] Some even feared the rapid onset of another ice age. Today, in contrast, a careful re-examination of the historical data, together with the fact that the 1980s were the warmest decade on record, has produced a general consensus that the earth is warming—though whether this is due to the greenhouse effect or a consequence of natural variability is not yet determinable from the climate record itself.

Finally, in the 1980s, scientists began to focus on trace gases other than CO_2 that trap heat and contribute to the greenhouse effect. Chief among these are methane, nitrous oxide and the chlorofluorocarbons (CFCs).[7] Calculations published in 1985 indicated that the global warming effect of these trace gases is roughly equal to that of carbon dioxide, making the problem twice as serious as previously believed.

From a Scientific to a Policy Issue

By 1985, these scientific developments had combined to make the theory of greenhouse warming both more convincing and more urgent. In October of that year, a scientific conference on the greenhouse problem, held in Villach, Austria, concluded that "[al]though quantitative uncertainty in model results persists, it is *highly probable* that increasing concentrations of the greenhouse gases will produce significant climatic change."[8] The conference statement observed that "the understanding of the greenhouse question is sufficiently developed that scientists and policy-makers should begin an active collaboration to explore the effectiveness of alternative policies and adjustments." It went on to recommend that the UN Environment Programme (UNEP), the World Meteorological Organization (WMO) and the International Council of Scientific Unions (ICSU) take action to "initiate, if deemed necessary, consideration of a global convention."

Whether science alone would have spurred the international community to action, however, is questionable—particularly given the scientific uncertainties that persist even now. Three additional factors catalyzed public interest in global warming and helped transform it from a scientific to a political issue.[9]

First, a number of scientists and non-governmental organizations (NGOs) acted as entrepreneurs, promoting the climate change issue through conferences, reports and personal contacts.[10] This process began in earnest in the early 1970s at the SCEP and SMIC conferences, which helped raise awareness of the climate change issue throughout the scientific community; it continued through the work of WMO's Ad Hoc Panel of Experts on Climate Change in the mid-1970s and the First World Climate Conference in 1979, and picked up momentum in the mid-1980s. In particular, the 1985 and 1987 Villach Conferences and the 1987 Bellagio Conference[11] helped consolidate the scientific consensus regarding global warming and communicate that consensus to policy-makers.

Second, the discovery of the ozone hole in 1987 dramatically demonstrated that human activities can indeed affect the global atmosphere and raised the prominence of atmospheric issues generally. Initially, public concern about global warming rode on the coat-tails of the ozone issue.

Finally, the heatwave and drought in the US during the summer of 1988 gave an enormous popular boost to greenhouse warming proponents in the United States. The testimony of James Hansen, a NASA climate modeler, to the Senate Energy Committee that "the greenhouse effect has been detected and it is changing our climate now"[12] made front page news, coming as it did during the height of the heatwave. As Senator Max Baucus noted at that Senate hearing, "I sense that we are experiencing a major shift. It's like a shift of tectonic plates." Although most scientists recognized that the causal relationship between that summer's hot weather and the greenhouse effect remained unproven, the climate change issue had emerged on the political stage, prompting President Bush's election-year pledge:

"Those who think we're powerless to do anything about the 'greenhouse effect' are forgetting about the 'White House effect.' As President I intend to do something about it."[13]

Early International Responses

Just as concern about global warming was mounting, Canada—in a follow-up to the 1987 report of the World Commission on Environment and Development (*Our Common Future*, the Brundtland Report)—sponsored an international conference on "The Changing Atmosphere: Implications for Global Security."[14] The Conference focused on climate change, and one of its principal purposes was to bridge the gap between scientists and policy-makers. It was attended by more than 340 individuals from 46 countries, including 2 heads of state, more than 100 other government officials, and numerous scientists, industry representatives and environmentalists.

The Toronto Conference Statement began with the portentous words, "Humanity is conducting an unintended, uncontrolled, globally pervasive experiment whose ultimate consequence could be second only to a global nuclear war . . . It is imperative to act now." The Statement went on to predict that, given the high rates of likely warming, "no country would benefit *in toto* from climate change." As initial actions, the Conference recommended:

- a 20 percent reduction in global CO_2 emissions from 1988 levels by the year 2005,[15]
- development of "a comprehensive global convention as a framework for protocols on the protection of the atmosphere," and
- establishment of a "World Atmosphere Fund" to be financed in part by a levy on fossil fuel consumption in industrialized countries.

In many respects, the Toronto Conference Statement was the highwater mark of policy declarations on global warming. On the one hand, although the Conference was not officially governmental in nature (the government participants all attended in their personal capacities), it had far more status and influence than other non-governmental meetings held before and after—particularly its call for a 20 percent reduction in CO_2 emissions by the year 2005. This was due in part to Canada's sponsorship and the substantial participation by high government officials, including the Prime Ministers of Canada and Norway, and in part to the fact that the Conference came at the right time: as the Conference report itself notes, it was an "event waiting to happen." On the other hand, because of its non-governmental character, the Conference Statement—including the 20 percent reduction target—was not a negotiated document and was not binding on anyone. It was drafted by a committee composed mostly of environmentalists and discussed in less than a day. Flush with the success of the Montreal Protocol, many participants did not fully appreciate the political difficulties of addressing the climate change issue. Moreover, as with many new environmental issues, environmental activists—who discovered and pushed the issue—had a head start; opponents in industry and government took longer to mobilize.

Following Toronto, the climate change issue continued to attract substantial attention. But increasingly, the discussions moved onto an inter-governmental track, where agreement proved more difficult to reach and conference statements became more carefully qualified. Indeed, as states became increasingly aware of the stakes and uncertainties involved in the climate change question, even those that had initially supported a strong policy response became more cautious. Thus, the various conference declarations following Toronto reflect a retrenchment rather than an advance.

Not surprisingly, Western industrialized countries were the first to become actively engaged in the climate change issue. They had the most influential environmentalist constituencies and, since most of the research about the greenhouse effect was done by their scientists, they were comparatively conscious of and informed about the issue. By contrast, developing countries had more immediate problems to worry about, such as poverty, drought, famine and war; to them, climate change must have seemed very remote. Thus, although industrialized countries recognized from the start that the North–South dimension of the climate change issue was important and paid lip service to the interests of developing countries, the South's perspective was not forcefully expressed in its own voice until later in the process, towards the end of 1990.

Within the West, climate change was initially viewed by most governments primarily in scientific and environmental, rather than economic terms. Scientists had long been active in climate issues, and had collaborated internationally through the WMO. As policy interest in climate change grew, these scientists were joined by environmental ministries, who tended to take the lead in developing and articulating positions at international meetings. Only later did other ministries became involved, including those of energy, transportation and finance. The United States was perhaps the only country that viewed the climate change issue from the outset through a domestic policy prism—perhaps because of its experience in the 1987 Montreal Protocol negotiations, where many in the White House felt that the State Department and the Environmental Protection Agency (EPA) had moved too aggressively, and without adequate involvement by affected domestic agencies. Following Montreal, international environmental issues were coordinated in the Reagan Administration not by the State Department or the National Security Council, but by a working group of the White House Domestic Policy Council (DPC). In this group the EPA, though represented, was increasingly outmuscled by major domestic political players such as the Departments of Energy, Interior, and Commerce, the Office of Management and Budget, and the Council of Economic Advisers. This difference in institutional perspective between the US and other Western countries produced—not surprisingly—quite different policy positions: the US emphasized the potential economic costs of response measures and argued for further research, while other Western states tended to emphasize the environmental risks, to ignore the economic dimensions of the issue and to support immediate action to curb greenhouse gas emissions.

In 1988, shortly before the Toronto Conference, governments took the first step to address the climate change issue, requesting that WMO and UNEP establish the IPCC. The IPCC was given the mandate of providing "internationally coordinated assessments of the magnitude, timing and potential environmental and socio-economic impact of climate change and realistic response strategies."[16] Earlier international assessments, such as the 1980, 1985 and 1987 Villach meetings, had been organized by the scientific and environmental communities and were viewed with suspicion in some governmental circles. Some officials still believed that the conclusions drawn at these meetings promoted environmental activism more than they reflected sound science. The IPCC, in contrast, was a governmental initiative—strongly supported by the US. It was intended in part to reassert governmental control and supervision over what was becoming an increasingly prominent political issue.

The IPCC held its first meeting in November 1988, electing Professor Bert Bolin of Sweden as chairman and establishing three working groups: Working Group I on science, chaired by the United Kingdom; Working Group II on impacts, chaired by the Soviet Union; and Working Group III on response strategies, chaired by the United States. The IPCC also adopted an expedited work schedule to allow preparation of a first assessment report by September/October 1990, in time for the 45th session of the General Assembly, the 11th World

Meteorological Congress, and the Second World Climate Conference.

The increasing governmental interest in the climate change issue was also reflected in the United Nations General Assembly, where it was raised for the first time in September 1988, when Malta requested the inclusion of an agenda item entitled, "Declaration proclaiming climate as part of the common heritage of mankind."[17] Although, as one delegate later remembered, global warming still seemed like science fiction to many people, the Maltese initiative enjoyed widespread support. On December 6, the General Assembly adopted a resolution endorsing the establishment of the IPCC, and urging governments, inter-governmental and non-governmental organizations, and scientific institutions to treat climate change as a priority issue. However, most countries felt that Malta's invocation of the "common heritage of mankind" concept—which had been applied previously to deep seabed mineral resources and the moon—was inappropriate in relation to climate and the global atmosphere. Thus, the resolution was amended to refer to climate as the "common *concern* of mankind."

Throughout 1989, discussions on the climate change issue continued to accelerate. The three IPCC working groups held frequent workshops and meetings, including the initial meeting of the Response Strategies Working Group in Washington, DC during February of that year. At the Washington meeting, James Baker made his first speech as Secretary of State, calling for "prudent steps that are already justified on grounds other than climate change." (This approach later became known as the "no regrets" policy.) In March 1989, the Netherlands, France and Norway jointly sponsored a summit in The Hague[18] on global environmental issues. This meeting was attended by representatives of twenty-four countries, including seventeen heads of state or government.[19] The Hague Conference Declaration made the radical proposal that countries develop "new institutional authority" involving non-unanimous decision-making—in effect, a partial renunciation of sovereignty. Although the Hague Conference Declaration was widely criticized and its proposal for new institutional authority quickly forgotten, the fact that climate change was now an issue for heads of state was indicative of its increased political salience. The Hague Conference was followed in July by the Paris G-7 Economic Summit, where the leaders of the seven major industrialized countries "strongly advocate[d] common efforts to limit emissions of carbon dioxide and other greenhouse gases," and urged "international organizations to encourage measures to improve energy conservation." Climate change was also discussed at the US/Soviet summit in Malta, by the Non-Aligned Movement at its September 1989 meeting in Belgrade, and in the Langkawi Declaration of the Commonwealth Heads of Government held in October 1989.

As scientists and policy-makers grew increasingly concerned that climate change was an imminent threat, not a distant possibility, a consensus gradually emerged that states should develop a legally binding convention to address the problem. This development can already be seen in the 1988 UN General Assembly resolution on the climate change issue, which called on the IPCC "to initiate action leading as soon as possible to a comprehensive review and recommendations with respect to: . . . the identification and possible strengthening of relevant existing international legal instruments having a bearing on climate; [and] elements for inclusion in a possible future international convention on climate."

Initially, two models were suggested for a convention. One approach, advocated by Canada, was to develop a general framework agreement on the "law of the atmosphere," to parallel the UN Law of the Sea Convention, and then address specific atmospheric issues such as acid rain, ozone depletion, and climate change in separate protocols.[20] The rationale for this approach was that it recognized the interdependence of global atmospheric problems. It was endorsed by the 1988 Toronto Conference Statement, which called for the development of "a comprehensive global convention as a framework for protocols on the protection of the atmosphere." In February 1989, Canada sponsored a meeting in Ottawa of legal and policy experts on the law of the atmosphere, with the hope of beginning discussions on a comprehensive atmospheric convention.[21]

The second approach was to develop a convention specifically on climate change. At the opening of the 1989 Ottawa meeting, Dr. Mostafa Tolba, the Executive Director of UNEP, strongly criticized the "law of the atmosphere" model as politically unrealistic, arguing instead for negotiation of a more narrowly focused convention on climate change.[22] Tolba's stature was then at its height, due to his recent success in the Montreal Protocol negotiations, and the Canadian proposal never regained momentum. Although the Ottawa meeting went through the motions of discussing the elements of a framework convention on the atmosphere, attention began to focus on the alternative approach.

As of early 1989, however, a governmental decision still had not been made to negotiate a climate change convention. The Toronto Conference had been a non-governmental meeting, and the 1988 UN

General Assembly resolution had spoken merely of "a *possible future* international convention on climate"; it did not actually authorize negotiation of a convention. The new US administration, and in particular the White House Chief of Staff, John Sununu, questioned the need for negotiations, arguing that more scientific study was needed.

In the spring of 1989, however, pressure intensified on the United States to accept negotiation of a climate change convention. In March, twenty-two countries, including France, Japan, Italy, and Canada, called for negotiations. Following the public flap created in early May when the White House attempted to dilute Congressional testimony of NASA scientist James Hansen, the United States finally reversed its position. On May 12, at a meeting of the IPCC Response Strategies Working Group, the US announced that it would support negotiation of a framework convention on climate change.

Several weeks later, the UNEP Governing Council, with US concurrence, adopted a resolution requesting UNEP to begin preparations for the negotiations and recommending that the negotiations begin as quickly as possible after the adoption of the IPCC's First Assessment Report. This decision was supported by the Paris G-7 Summit, which stated that the conclusion of a "framework or umbrella convention on climate change . . . is urgently required to mobilize and rationalize the efforts made by the international community." Similar conclusions were echoed by the UN General Assembly, which adopted a resolution calling on states "to prepare as a matter of urgency a framework convention on climate, and associated protocols containing concrete commitments in the light of priorities that may be authoritatively identified on the basis of sound scientific knowledge, and taking into account the specific development needs of developing countries."[23]

To begin work on the elements of a climate change convention, the IPCC Response Strategies Working Group appointed the United Kingdom, Malta and Canada as co-coordinators and scheduled a multi-disciplinary workshop for October 1989. The workshop agreed on the inadequacy of existing legal instruments and on the need for a framework convention on climate change, modeled on the Vienna Convention and framed so as to gain the adherence of the largest number of states. Some delegations proposed that the convention should include binding commitments and control measures and a mechanism for providing financial assistance, but there was no consensus on these points. Without deciding which particular elements should be included in the convention, the workshop elaborated a list of *possible* elements, which were refined and ultimately included in the final IPCC report the following year.

Perhaps the most significant meeting of 1989, however, came near the end of the year—the Noordwijk Ministerial Conference on Atmospheric Pollution and Climate Change, convened by the Netherlands on November 6 and 7. This meeting was attended by representatives of sixty-six states, roughly divided among developed and developing countries. The Noordwijk Conference was the first high-level political meeting focusing specifically on the climate change issue (the Hague summit had been more general in its orientation), and, although governments were beginning to become concerned about the economic implications of proposed response measures, most still viewed the issue primarily in environmental terms.[24] As a result, the Noordwijk Declaration was stronger than succeeding statements, which increasingly reflected the complex international and domestic politics of the subject. Forward-looking elements of the Declaration included:

- A general aim of limiting or reducing emissions and increasing sinks for greenhouse gases "to a level consistent with the natural capacity of the planet," within a timeframe sufficient to allow ecosystems to adapt naturally to climate change. (This aim was eventually adopted as the long-range objective of the UN Framework Convention on Climate Change.)
- A recommendation that states initiate actions and develop and maintain effective strategies to control, limit or reduce emissions of greenhouse gases. (The Climate Change Convention converted this recommendation into a legal obligation to formulate programs to reduce emissions and enhance sinks—perhaps its central requirement, together with reporting.)
- A recommendation that other fora, including the IPCC, explore the concept of "CO_2-equivalence" to describe the radiative effect of various greenhouse gases, which would create "a basis for negotiations in response measures for different greenhouse gases in the most cost-effective manner." (This suggestion contains the seeds of the "comprehensive approach," which the United States raised in the IPCC the following month.)
- A target for world net forest growth of 12 million hectares a year by the beginning of the next century.

In Noordwijk, although the North–South dimension of the climate issue was already recognized, the deep reluctance of the developing countries to take measures to combat climate change was not fully apparent, perhaps because they tended to be represented by officials from their ministries of environment rather than those of development or foreign affairs. The Declaration contained a number of provisions sought by developing countries, including: the principle of sovereignty; the need for international cooperation and for resolving the external debt problem; the responsibility of industrialized countries to take the lead in initiating action to combat climate change; and the need for financial and technical assistance, in part through the mobilization of additional resources. But the general assumption was that, while developing countries would need additional time and financial resources, they should be subject to targets to reduce emissions and enhance sinks. Indeed, the Declaration set a target for new forest growth which, if implemented, would have most severely impacted the South, since that is where most forests are found. The Noordwijk Declaration also recommended using existing financial institutions, including multilateral development banks, to provide assistance to developing countries—a position very much at odds with the South's subsequent positions in both the UNCED Preparatory Committee and the Intergovernmental Negotiating Committee on climate change (INC).

While the differences between North and South still lurked in the background at Noordwijk, the differences within the North were already in full view—in particular, whether to set specific, binding quantitative limits ("targets and timetables") for greenhouse gas emissions. The Noordwijk Conference recognized the need for industrialized countries to stabilize their greenhouse gas emissions "as soon as possible." But it was unable to agree on a specific date for stabilization, owing to opposition by the United States, Japan, and the Soviet Union. Nor did it establish the level of emissions at which stabilization should occur. Instead, the Declaration merely noted the view of "many"[25] industrialized countries that stabilization should be achieved "as a first step at the latest by the year 2000," and encouraged the IPCC to analyze the options for CO_2 emissions.

During the following year, little progress was made on the targets and timetables issue, and positions tended to harden. On the one side, the United States adamantly opposed establishing quantitative limitations on its greenhouse gas emissions, arguing instead for the development of national strategies, consisting of concrete policy measures. On the other side, many industrialized countries adopted targets and timetables unilaterally. In May 1990, the two sides locked horns at the Bergen Ministerial Conference on sustainable development—Europe and North America's regional conference to prepare for UNCED—with the same general result as at Noordwijk. The

Bergen Declaration called for stabilization of greenhouse gas emissions as soon as possible, and "note[d] with appreciation" the decision of some countries to stabilize CO_2 emissions at present levels by the year 2000. But it did not require or even recommend that states establish targets and timetables. It merely recommended targets as one possible option, along with the US approach of national strategies.

In May and June 1990, the IPCC Working Groups finalized their First Assessment Reports. The IPCC plenary met in Sundsvall, Sweden, in August to approve the reports and adopt an overview statement. More than 400 scientists worked on the scientific report and it quickly gained acceptance as the authoritative scientific statement on climate change. Written in clear, bold language, intended for policy-makers, the report predicted that if states continue to pursue "business as usual," the global average surface temperature will rise during the next century by an average of 0.3° C. per decade (with an uncertainty range of 0.2–0.5° C per decade)—a rate of change unprecedented in human history. Land regions will probably warm more rapidly than the oceans, and regional climate changes will likely differ from the global mean. Moreover, because of the complexity of the climate system, the report did not rule out surprises, including the possibility of more rapid warming due to changes in ocean circulation patterns and stronger than expected positive feedback effects.

The IPCC meeting in Sundsvall was followed in November 1990 by the Second World Climate Conference (SWCC) in Geneva[26]—perhaps the biggest governmental meeting focusing on environmental issues prior to the United Nations Conference on Environment and Development (UNCED). In contrast to its 1979 forerunner, the SWCC included a ministerial as well as a scientific component, reflecting the heightened political interest in climate issues. By this time, momentum was building towards the climate change negotiations, which were expected to begin early the following year. As a result, the negotiations on the SWCC Ministerial Declaration became, in essence, a dress rehearsal for the INC, with countries already jockeying for position.

Like the Noordwijk and Bergen Declarations, the SWCC Ministerial Declaration stressed the need to stabilize emissions of greenhouse gases, but was unable to establish when or at what level stabilization should be achieved. Shortly before the Conference began, at a meeting of Environment Ministers held in Luxembourg, the European Community (EC) had agreed on a Community-wide goal of stabilizing carbon dioxide emissions at 1990 levels by the year 2000. The SWCC Ministerial Declaration welcomed this decision, as well as decisions and commitments by other states "to take actions aimed at stabilizing their emissions of CO_2, or CO_2 and other greenhouse gases not controlled by the Montreal Protocol, by the year 2000 in general at 1990 levels." But it did not recommend that states adopt this target; instead, it merely urged developed states "to establish targets *and/or feasible national programmes or strategies.*" (Italics added.)[27]

While there was little movement on the targets and timetables issue, the SWCC was marked by one important development: for the first time, developing states participated as equal partners in the discussions and made clear, to a much greater extent than previously, that North–South issues would play a prominent role in any future convention negotiations. This development was part of a larger transformation of international environmental politics in the late 1980s, in which North–South issues began to move to center stage. In the ozone negotiations, developing countries had begun to assert themselves following the adoption of the Montreal Protocol in 1987. They demanded "new and additional" financial resources and technology as the price for becoming parties to the Protocol. At the London Conference in June 1990, the developing countries successfully pressed to establish a special fund to provide assistance in the implementation of the Montreal Protocol. Similarly, in the UN General Assembly debate in December 1989 over the authorizing resolution for the UN Conference on Environment and Development, developing countries insisted that the Conference give equal billing to environment *and development*.

Developing countries took somewhat longer to become fully engaged on the climate change issue—in part because it was new and comparatively unfamiliar. Thus, the IPCC was generally dominated by developed countries, which held most of the leadership positions. Even before the SWCC, however, this had begun to change. The introduction of the climate change issue in the UN General Assembly in 1988, and its consideration again the following year, began to familiarize developing countries with the climate change issue. In 1989, some of the bigger developing countries, in particular Brazil and Mexico, expressed dissatisfaction with the lack of developing country representation in the IPCC, leading to the creation of a "Special Committee on the Participation of Developing Countries," chaired by Jean Ripert of France (later the chairman of the INC). At the August 1990 IPCC meeting in Sundsvall, Brazil was very vocal and almost prevented adoption of the First Assessment Report—providing a preview of the later North–South debates in the SWCC and INC.[28]

By October 1990, when the SWCC convened, many developing countries had become fully aware of the significance of the climate change issue. The preparations for UNCED were by then in full swing, and the UNCED Preparatory Committee had just held its first meeting in Nairobi. At the SWCC, the large developing countries knew that negotiations on a climate change convention were just around the corner and the results were expected to feed into the UNCED process.

The emergence of developing countries at the SWCC had two effects. First, it became obvious that the climate change negotiation would not be simply about the environment, but about the linked issues of environment *and* development. Second, at the SWCC the differences among developing countries began to show. On the one hand, the oil-producing states emerged as a major voice in the discussions, questioning the science of climate change and arguing for a "go slow" approach. On the other hand, small island and low-lying coastal states—who feared the adverse effects of sea-level rise—argued for strong response measures, and organized themselves into the Alliance of Small Island States (AOSIS), which eventually included thirty-five countries.

Both the IPCC and the SWCC called for negotiations on a climate change convention to begin as quickly as possible. To prepare for these negotiations, UNEP and WMO convened an open-ended ad hoc working group of government representatives, which met in Geneva in September 1990.[29] The ad hoc group considered the ways, means and modalities for the negotiations, including the structure of working groups, the venue for the negotiating sessions, the duration of the sessions, rules of procedure, the structure of the bureau, the secretariat, and whether protocols should be negotiated simultaneously with the convention. The group made twenty recommendations; the most important was that a single negotiating process should be established to discuss both policy issues and legal instruments. The ad hoc group, however, could not agree on two important questions:

(1) Should protocols be negotiated simultaneously with the convention?

(2) Who should organize and conduct the negotiations?

With respect to the latter issue, the meeting identified two options:

(a) a negotiating committee under the auspices of WMO and UNEP,

in essence carrying forward the IPCC process, or
(b) a special conference under the authority of the UN General
    Assembly.
Western countries tended to support the former option, while many
developing countries, who felt excluded from the IPCC, preferred
the second option.

The question of who should conduct the negotiations had been
looming in the background for some time. In 1989 and early 1990,
most observers assumed that the negotiations, like the IPCC, would
be conducted under the joint auspices of UNEP and WMO. Accordingly, these two organizations took the lead in the early preparations. As developing countries became more actively involved in
the climate issue, however, the equation changed. Developing countries tended to see the climate change issue in developmental as well
as environmental terms, focusing on its implications for industry,
transportation and agriculture. Moreover, they argued that climate
change is a political and not merely a technical issue. For both reasons, they felt that it should be addressed under the auspices of a
political body—namely the UN General Assembly—rather than by
more technical agencies such as UNEP and WMO.

On December 21, 1990, the UN General Assembly adopted Resolution 45/212, establishing the INC as "a single intergovernmental
negotiating process under the auspices of the General Assembly."
With the mandate to negotiate a convention containing "appropriate commitments," Resolution 45/212 directed the INC to "take into
account" the work of the IPCC, and invited UNEP and WMO to
make "appropriate contributions" to the negotiation process, but
called for the establishment of an ad hoc secretariat. In early February 1991, during the first meeting of the INC, the Secretary-General of
the UN appointed Michael Zammit Cutajar, a UN staff member previously with the UN Conference on Trade and Development (UNCTAD),
to serve as Executive Secretary of the Committee; staff were assigned
to the Secretariat by UNCTAD as well as WMO and UNEP.

Negotiating History of the INC

Between February 1991 and May 1992, the Intergovernmental
Negotiating Committee for a Framework Convention on Climate
Change (INC/FCCC also referred to simply as the INC) held five
sessions, the last divided into two parts. Under the terms of UN
General Assembly Resolution 45/212, the maximum duration of each
negotiating session was two weeks and the INC was to complete its

work in time for the Convention to be signed at UNCED in Rio de Janeiro. Inter-sessional work was forbidden by the INC's rules of procedure, although this rule broke down towards the end of the negotiations, as time ran short and pressure mounted to conclude the agreement.

The discussions in the INC followed a pattern common to international negotiations. At first, little progress was apparent. During most of 1991, states debated procedural issues and enunciated and reiterated their positions instead of bridging their differences by finding compromise formulations or engaging in tradeoffs or deals. Where they disagreed about a provision, participants simply bracketed the language in question (to indicate the disagreement) or added alternative formulations. At the end of the year, much if not most of the text remained bracketed—sometimes in brackets within brackets within brackets.

This sparring process, although frustrating to those seeking rapid progress, played a necessary role by giving states an opportunity to voice their views and concerns. They learned about and gauged the strength of other states' views. They sent up trial balloons and explored possible areas of compromise. Indeed, without this mutual learning process, it is hard to imagine that agreement would have been possible.

Real negotiations, however, began only in the final months before UNCED. Given the public visibility of the UNCED process, most delegations wished to have a convention to sign in Rio. As one critic quipped, "The INC was doomed to success." Thus, when it became clear in the spring of 1992 that the United States would not accept definitive targets and timetables, that Western states would insist on some role for the Global Environment Facility (GEF), and that developing countries would not accept strong commitments or implementation machinery, delegations finally got down to the hard work of crafting compromise language and produced a text that the INC adopted on the final night of the negotiations.

INC 1: Chantilly, VA, February 4–14, 1991[30]

At the invitation of the United States, the INC held its first meeting in February 1991, at the Westfields Conference Center in Chantilly, Virginia. The Committee elected Jean Ripert of France as chairman along with vice-chairs from Algeria, Romania, Argentina and India. The INC also established two working groups, adopted rules of procedure, and heard general statements by delegations at this session. Owing to the extreme shortness of time between the passage of the General Assembly resolution establishing the INC and the Chantilly session, the INC secretariat had still not been formed when the session began and only limited financial assistance could be provided for attendance by delegations from developing countries. The absence of such financial support led to far less participation by developing countries than at later sessions—despite the close proximity to New York and Washington where most developing countries have missions.

Going into the Chantilly session, many delegations expected to develop a negotiating text. Draft texts had been in circulation for more than a year, prepared by NGOs and in some cases governments.[31] At the beginning of the Chantilly session, the United Kingdom circulated a complete draft convention in its capacity as joint coordinator of the IPCC legal measures group, based on the consensus elements of the group's report.[32] But given the extreme sensitivity of many delegations, extending even to concerns of how to organize the INC's work, these expectations proved overly optimistic. It was not until December 1991—ten months and three sessions later—that the INC finally agreed on a negotiating text.

Most of the Chantilly session was occupied by the question of what working groups to establish. Ultimately, the INC decided to create two working groups—one on commitments, the other on mechanisms. As set forth in Decision 1/1, the mandate of Working Group I (WG I) extended to all types of commitments, including appropriate commitments to:

- limit and reduce emissions of carbon dioxide and other greenhouse gases, "taking into account that contributions should be equitably differentiated according to countries' responsibilities and their level of development;"[33]
- protect, enhance and increase sinks and reservoirs;
- provide "adequate and additional" financial resources to enable developing countries to meet incremental costs required to fulfill their commitments under the convention;
- facilitate the transfer of technology on a "fair and most favorable" basis; and
- address the special situation of developing countries, in particular, those specially vulnerable to the adverse effects of climate change.

The mandate of Working Group II (WG II) was to develop legal and institutional mechanisms to implement the convention, including mechanisms related to scientific co-operation, monitoring and information; compliance; assessment and review; transfer of financial resources and technology; and entry into force and withdrawal.

While many delegations expressed frustration with the seemingly procedural nature of the Chantilly session, the discussions on the structure and mandates of the working groups in fact represented the initial skirmishes on the main substantive issues before the INC, and served as a proxy for negotiations on a convention text. For example, EC–US differences regarding targets and timetables were reflected in Chantilly in the dispute over whether to establish different working groups to address the sources and the sinks of greenhouse gases, or a single working group dealing with both together. The EC wanted to establish different working groups,[34] reflecting its desire to negotiate concrete commitments on energy and forests. The United States pressed for a single working group, because it favored more generic commitments that addressed sources and sinks comprehensively.[35] Ultimately, they reached a compromise: one working group would address both sources and sinks, but its mandate would include commitments aimed at "limiting and reducing emissions of carbon dioxide and other greenhouse gases," a formulation reportedly suggested by the US.

Similarly, the debate about which working group should address financial and technology transfer issues was, in effect, about the nature of the *quid pro quo* between developing and developed country commitments. Developing countries generally wanted financial and technology issues assigned to the same working group as sources and sinks, to keep all of these issues tightly linked.[36] The US, in contrast, proposed dealing with them in separate working groups. Here, the developing country position ultimately prevailed.

More obviously substantive in nature was India's proposal that the "Guidelines for the Negotiation" refer to the need for "new and additional funding" to help developing countries implement measures to combat global warming. In effect, the Indian proposal would have implied an up-front commitment by developed countries on financial resources, prejudging the outcome of the negotiations. Following intense negotiations, compromise language was adopted directing Working Group I to prepare text "related to . . . appropriate commitments on adequate and additional financial resources"— short of India's language but an important indication that the framework convention, in contrast to its predecessors on acid rain and ozone, would contain substantive financial commitments.

Finally, Decision 1/1 of the INC Plenary contained two "procedural points" that greatly affected the INC's work: (1) a prohibition on holding more than two meetings at any one time during a session;

and (2) a prohibition on inter-sessional meetings. Both rules were intended to promote transparency as well as full participation by developing countries—particularly those with small delegations. But, by restricting the opportunities to negotiate, they made reaching an agreement more difficult and, toward the end of the negotiations, both were informally abandoned.

INC 2: Geneva, June 19–28, 1991[37]

Both during and after the Chantilly meeting, many states submitted papers providing their views on the convention, including general position statements, specific elements for inclusion in the convention and, in some cases, entire draft texts. These were compiled by the Secretariat into an ongoing documents series, the so-called "Misc.1" papers,[38] which helped to clarify national positions and identify options.

Despite this preparatory work, substantive discussions were again delayed at the beginning of INC 2 by disputes among the various regional groups over who should chair the two working groups. INC 1 had been unable to resolve this question, and inter-sessional consultations by the INC chairman had also proven inconclusive. In order to allow more regions to be represented and to avoid a divisive vote, the INC finally agreed to appoint two co-chairs for each working group along with one vice-chair. Reflecting the increasingly prominent role of island states in the negotiations, Ambassador Robert Van Lierop of Vanuatu was appointed as one of the WG II co-chairs—apparently the first time the island states as a group had received a leadership position in a UN forum. Mexico and Japan were elected as co-chairs of WG I, Mauritania as vice-chair of WG I, Canada as the other co-chair of WG II, and Poland as vice-chair of WG II.

Substantive discussions finally got underway during the second week of INC 2. In both working groups, the discussions proceeded in a general manner. WG I extensively discussed a proposed article on general principles, which had not been mentioned in Decision 1/1 but was introduced by China and supported by other developing countries. The group made no progress on the issue of specific limitations on greenhouse gas emissions. The United Kingdom had proposed a compromise formulation (subsequently called "the phased comprehensive approach"). Under this approach, countries would receive credit in calculations of their emission levels both for cuts in sources of greenhouse gases other than CO_2 and for enhancement of sinks. The US, however, maintained its opposition to any targets and timetables.

WG II began its discussions of legal and institutional mechanisms, unanimously agreeing that science was the basis upon which the convention should rest. In contrast to WG I, WG II gave its co-chairs the mandate to prepare a single text for the third session.

The main source of controversy at INC 2 concerned the concept of "pledge and review"—a concept introduced by Japan as a potential compromise on the targets and timetables issue. Under the Japanese proposal, states would be required to make unilateral pledges, consisting of national strategies and response measures to limit their greenhouse gas emissions, together with an estimate of the resulting emissions. These pledges would be periodically reviewed by an international team of experts, which would publicly evaluate the pledges and make recommendations. According to proponents, pledge and review would serve two purposes: the unilateral pledges would be a one-way ratchet towards stricter commitments by parties; and the international review process would promote transparency and accountability. Although the United Kingdom and France made similar proposals, most European Community members expressed reservations about substituting "pledge and review" for internationally defined commitments. Environmental NGOs also sharply criticized pledge and review, dubbing it "hedge and retreat." Following INC 2, pledge and review was discussed in detail at a symposium organized by the Royal Institute of International Affairs in London, attended in an informal capacity by a number of governmental representatives.[39]

INC 3: Nairobi, September 9–20, 1991[40]

Progress continued to be slow at the third session. Discussions were held on virtually every aspect of the proposed convention, based primarily on consolidated texts prepared during the inter-sessional period by the working group co-chairs. WG II made some progress in elaborating the convention's implementation mechanisms and institutions. But there was little movement in WG I on the substantive provisions of the convention, including general principles, commitments on sources and sinks, financial resources, or technology transfer. Rather than narrowing alternatives in order to move towards consensus, WG I instead produced ever longer compilations of alternative proposals. Some appeared satisfied with this approach, arguing that the session should be devoted to a full airing of views;

others were frustrated by the lack of progress towards consensus. In any event, by the end of the session, this process had played itself out, and WG I finally gave its co-chairs the mandate to prepare a consolidated text for INC 4.

Near the beginning of the session, the plenary held an extensive informal discussion of "pledge and review," at which proponents of internationally defined targets and timetables succeeded in turning the pledge-and-review concept on its head, insisting that pledge and review should be in addition to, rather than a substitute for, international commitments to limit greenhouse gas emissions. The EC in particular criticized the pledge and review concept, characterizing it as "the twin ghosts [that] have been haunting" the session. At the end of the session, most delegates thought that the pledge and review concept had been safely buried. But, as it turned out, the reports of its death were premature. While the term "pledge and review" was abandoned, the substantive concept continued to be discussed and was at the heart of the final compromise package.

INC 4: Geneva, December 9–20, 1991[41]

The fourth session continued the pattern established at INC 3. Rather than negotiate, states tended to reiterate their previously enunciated positions, reintroducing proposals and language that had been omitted from the co-chairs' consolidated draft texts.

Perhaps the most notable aspect of INC 4 was the breakdown of unity in the G-77. At INC 3, the G-77 had played a relatively low-key role, but, early in the December session, it began to meet on a regular basis in an attempt to prepare alternative proposals on the core provisions of the convention. Initially, the G-77 initiative had some success when, after lengthy meetings, it agreed on a "general principles" text, which was in large part accepted by other countries as a basis for discussion. But, during the second week of INC 4, the G-77 foundered on the question of what commitments to support. The small island states supported strong commitments; Saudi Arabia opposed such commitments even for developed countries; and the large industrializing countries such as China and India were somewhere in the middle. Ultimately, the G-77 chair announced in plenary that the G-77 would no longer meet at the session as a group, leaving G-77 member states free to submit proposals on their own. The G-77 draft, as it stood when discussions collapsed, was immediately introduced by a G-77 rump (initially dubbed the G-24 but ultimately including forty-three states). The small island states submitted an alternative proposal containing farther-reaching commitments, including a CO_2 stabilization target at 1990 levels by the year 1995 for developed countries and reductions thereafter on a timetable to be agreed by the conference of the parties.

Despite these problems, the INC made some progress in Geneva—at least from a cosmetic standpoint. At the end of the session, the INC combined the various texts from the two working groups into a "Consolidated Working Document,"[42] allowing participants to claim that they finally had succeeded in preparing a negotiating text. In fact, the Consolidated Working Document was no different stylistically than the consolidated texts prepared for the session by the working group co-chairs, which the groups had agreed at the outset of the session to use as negotiating texts. The main difference was that, by assembling the texts from the two working groups into a single document, the result looked more like a complete convention.

INC 5: New York, February and May 1992

Originally, the INC planned to hold only a single session in 1992. However, towards the end of 1991, the INC was still so far from completing its work that some delegations suggested the need for two sessions in 1992, one in February and another in April. Still, no final decision was made, for fear that, if delegations knew that there would be an additional session, they would merely continue to postpone making compromises. Ultimately, a resumed fifth session did prove necessary.

First Phase of INC 5: February 18–28, 1992[43]

At the February session, intensive negotiations finally began. To expedite the negotiations, the working group co-chairs encouraged contact groups to form on various issues (financial resources, technology transfer, implementation committee, and so forth), which met on a frequent basis. While the more focused work style represented a substantial change from previous sessions, differences among delegations remained wide on many of the core issues, and the INC succeeded in bridging few gaps in the text that emerged at the end of the meeting, the Revised Text under Negotiation. States still seemed engaged in a battle of nerves, hoping that, with the Rio Summit fast approaching, the other side would blink first.

Following its disarray at the December session, the G-77 resumed meeting in New York, primarily to press its views on financial resources and technology transfer. Early in the session, OECD countries also began to caucus almost continuously, under the chairmanship of Sweden, to try to narrow their differences on specific commitments to limit greenhouse gas emissions—in particular, targets and timetables. Despite the hopes by some observers that the removal of John Sununu as White House Chief of Staff would lead to a softening of the US position, the United States remained firm in its opposition to targets and timetables. Moreover, other OECD countries continued to disagree about the exact terms of the proposed targets and timetables. As a result, even after days of negotiation, the OECD text still contained numerous brackets and alternatives. When, towards the end of the session, it was finally introduced in Working Group I, developing countries expressed dismay and proposed numerous alternative formulations of their own, further complicating the Revised Text.

OECD and Extended Bureau Meetings, Paris, April 1992

At the end of the February meeting, wide differences remained between delegations on the core issues of targets and timetables, financial resources, and the financial mechanism. Basic differences together with the sheer volume of work remaining on the rest of the convention—which was politically less sensitive but still heavily bracketed—made many participants doubtful that the INC would be able to complete the text of a convention at the resumed fifth session. The chairman therefore scheduled a meeting of the so-called "Extended Bureau"—a group of the 20–25 key delegations—in mid-April in Paris.

Immediately before the Extended Bureau meeting, the OECD countries met as a Common Interest Group in a final effort to resolve the targets and timetables issue. Despite substantial pressure, however, the US maintained its opposition to targets and timetables and no compromise could be reached.

The Extended Bureau met from 15 to 17 April and discussed most aspects of the convention, making progress on a few issues. More importantly, however, the Bureau unanimously urged the chairman to develop his own compromise text for the final meeting, so that the Committee would not have to work its way through the many brackets and alternatives remaining in the Revised Text under Negotiation. Somewhat reluctantly, the chairman agreed—a move that most participants believed was critical to the ultimate success of the negotiations.[44] The chairman declined to introduce a text on targets and timetables, however, because of its political sensitivity. This issue continued to be discussed at a high political level in late April, including at a meeting in Washington between the EC President, Jacques Delors, and President Bush. The deadlock was finally broken at the very end of the month, just as the fifth session resumed, when the US and the UK worked out the compromise language that now appears more or less unchanged in Article 4(2) of the Framework Convention.

Second Phase of INC 5: April 30–May 9, 1992[45]

When the fifth session resumed on 30 April in New York, the INC decided to abandon its working group structure and divide instead into three groups to consider various clusters of articles. The core articles on commitments (Article 4), financial mechanism (Article 11), and reporting (restyled in Article 12 as "communication of information" to make it more politically palatable) were considered in a group chaired by the chairman of the INC. The preamble and the objective and principles articles (Articles 2 and 3) were assigned to a group chaired by Dr. Ahmed Djoghlaf, the Algerian vice-chair. Finally, the articles on institutions, dispute settlement, and final clauses were considered under the chairmanship of Ambassador Raoul Estrada, the Argentine vice-chair. Towards the end of the session, this latter group also served as a legal drafting group and considered articles from the other clusters.

The core group met initially in an open-ended meeting, at which the US–UK compromise text on specific commitments was heavily criticized, particularly by developing countries led by India. After giving all delegations the opportunity to express their views, the chairman then reconstituted the Extended Bureau, which met practically around the clock during the final days of the session to hammer out compromises on the outstanding issues. The provision on specific commitments—as drafted by the United States and United Kingdom—although questioned, was amended only slightly after a strictly limited discussion. In the end, the major sticking point was Article 11 ("Financial Mechanism"), with OECD countries seeking to designate the Global Environment Facility (GEF) as the mechanism, and developing countries pressing either to create a separate fund or to leave the issue to the conference of the parties. Ultimately, a compromise was reached that the GEF would operate the financial mechanism on an interim basis.

Even after this compromise was reached, however, it was still unclear on the final day of the session whether some states would object in plenary to the compromise text on the ground that they had been excluded from the Extended Bureau meetings. Iran, in particular, reportedly threatened to open up the package by reintroducing a previously rejected principle on the right to development. Although the INC's rules of procedure permitted votes, the INC had up until then been able to operate on the basis of consensus, and few delegations—if any—wanted a vote now. Following a general discussion and some minor amendments, the Convention was finally adopted by acclamation on the evening of 9 May and prepared for signature at the Earth Summit in Rio de Janeiro.[46]

Notes

*   Portions of this article appeared in: Daniel Bodansky, "The United Nations Framework Convention on Climate Change: A Commentary," *Yale Journal of International Law,* 18: 458-75, 481-92 (1993).

1.   For general discussions of the history of greenhouse warming science, see Jesse Ausubel, "Historical Note," *Changing Climate*, Annex 2, p. 488 (Washington, DC: National Research Council, 1983); Melinda L. Cain, "Carbon Dioxide and the Climate: Monitoring and a Search for Understanding," in David Kay & Harold K. Jacobson, eds., *Environmental Protection: The International Dimension* (1983); William W. Kellogg, "Mankind's Impact on Climate: The Evolution of an Awareness," *Climatic Change* 10:115 (1987); Roger Revelle, "Introduction: The Scientific History of Carbon Dioxide," in E.T. Sundquist and W.S. Broecker, eds., *The Carbon Cycle and Atmospheric CO_2* (1985); Jonathan Weiner, *The Next One Hundred Years: Shaping the Fate of Our Living Earth* (1990).

2.   The development of a scientific consensus on climate change, as reflected in the IPCC report, also depended heavily on the international character of the climatology community, dating back to the Global Atmospheric Research Program (GARP) in the 1960s.

3.   Roger Revelle and Hans E. Suess, "Carbon Dioxide Exchange Between Atmosphere and Ocean and the Question of an Increase of Atmospheric CO_2 during the Past Decades," *Tellus* 9:18 (1957).

4.   The first attempt to calculate the effects of a CO_2 doubling using a three-dimensional GCM was performed in 1975. R.T. Wetherald and S. Manabe, "The Effects of Doubling the CO_2 Concentration on the Climate of a General Circulation Model," *Journal of Atmospheric Sciences*, 32:3 (1975).

5. National Research Council, *Carbon Dioxide and Climate: A Scientific Assessment*, p. viii (Washington, DC: NRC, 1979).

6. See National Defense University, *Climate Change to the Year 2000: A Survey of Expert Opinion* (Washington, DC: NDU, 1978).

7. More recent evidence suggests that CFCs may not be a net greenhouse gas, since their heat-trapping effect is counterbalanced by their depletion of stratospheric ozone, which also is a greenhouse gas.

8. WMO/UNEP/ICSU, *Report of the International Conference on the Assessment of the Role of Carbon Dioxide and of Other Greenhouse Gases in Climate Variations and Associated Impacts, Villach, Austria, Oct. 9–15, 1985,* WMO Doc. No. 661 (1986) (emphasis added).

9. See generally Rafe Pomerance, "The Dangers from Climate Warming: A Public Awakening," in Dean E. Abrahamson, ed., *The Challenge of Global Warming,* p. 259 (1989).

10. See Peter Haas, "Banning Chlorofluorocarbons: Epistemic Community Efforts to Protect Stratospheric Ozone," *International Organization* 46:187 (1992).

11. *Developing Policies for Responding to Climatic Change, A Summary of the Discussions and Recommendations of the Workshops Held in Villach, Sept. 28–Oct. 2, 1987, and Bellagio, Nov. 9–13, 1987,* WMO Doc. WMO/TD-No. 225 (April 1988).

12. *Greenhouse Effect and Global Climate Change, Hearings before the Senate Comm. on Energy and Natural Resources,* 100th Cong., 2nd Sess. 40 (1988).

13. At about the same time, Prime Minister Margaret Thatcher of Britain gave a speech to the Royal Society, in which she expressed concern about the greenhouse effect, warning that we may have "unwittingly begun a massive experiment with the system of this planet itself."

14. Proceedings of the World Conference on the Changing Atmosphere: Implications for Global Security, Toronto, June 27–30, 1988, WMO Doc. 710 (1989).

15. It is often overlooked that the 20 percent reduction target was a global goal. If, as the Conference Statement says, developing countries will need to increase their energy use "significantly," then industrialized states would need to reduce their emissions by more than 20 percent to offset these increased emissions.

16. UN General Assembly Res. 43/53 (1988).

17. The characterization of climate as "the common heritage of mankind" echoed Malta's introduction of the "common heritage" concept in 1967 to describe deep seabed mineral resources. See D.P. O'Connell, *The International Law of the Sea,* vol. 1, p. 25 (I.A. Shearer ed., 1982).

18. Declaration adopted at The Hague, March, 1989, reprinted in UN Doc. A/44/340-E/1989/120, annex.

19. Neither the United States nor the Soviet Union were invited to the meeting initially, and neither attended.

20. For a discussion of the "law of the atmosphere" model, see Durwood Zaelke and James Cameron, "Global Warming and Climate Change—an Overview of the International Legal Process," *American University Journal of International Law and Policy*, 5:249, 276-78 (1990).

21. Meeting Statement, *Report of the International Meeting of Legal and Policy Experts on the Protection of the Atmosphere*, Ottawa, Feb. 20–22, 1989.

22. The law of the atmosphere model was also questioned by Sir Crispin Tickell, the British Ambassador to the United Nations, in a talk to the Second North American Conference on Preparing for Climate Change, Dec. 6–8, 1988. Climate Institute, *Coping with Climate Change*, p. 4 (Washington, DC: 1988).

23. UN General Assembly Res. 44/207 (1989).

24. The Noordwijk Conference was organized by the Dutch Ministry of Housing, Physical Planning and Environment, which issued invitations to other environment ministries. The statement by Saudi Arabia at the Noordwijk Conference is a revealing illustration of the environmental orientation of many delegations. In this statement, Saudi Arabia characterized global warming as "a life or death issue for considerable areas of the earth," acknowledged that there is "no argument" that the "main culprit" for global warming is carbon dioxide, and recognized the need to move to non-greenhouse emission energy production and consumption systems and to stabilize and reduce emissions of greenhouse gases. In contrast, by the end of 1990, Saudi Arabia was stressing the uncertainty of climate change and strongly opposed establishing targets and timetables to reduce carbon dioxide emissions.

25. "Many" was a compromise between "most" (supported by the European Community) and "some" (supported by the US).

26. Jill Jager and Howard L. Ferguson, eds., *Climate Change: Science, Impacts and Policy. Proceedings of the Second World Climate Conference.* (Stockholm and Cambridge: Stockholm Environment Institute and Cambridge University Press, 1991).

27. The Declaration recognized "the difference in approach and in starting point in the formulation of targets." In addition, at the insistence of the United States a phrase was added "acknowledg[ing] the initiatives of some other countries which will have positive effects on limiting emissions of greenhouse gases."

28. At the insistence of Brazil, the Preface to the Overview of the IPCC First Assessment Report contains the significant qualification: "It should be noted that the Report reflects the technical assessment of experts rather than government positions, particularly those governments that could not participate in all Working Groups of IPCC."

29. *Report of the Ad Hoc Working Group of Government Representatives to Prepare for Negotiations on a Framework Convention on Climate Change*, UNEP/WMO Doc. Prep./FCCC/L.1/REPORT.

30. *Report of the First Session of the INC/FCCC*, UN Doc. A/AC.237/6 (March 8, 1991).

31. Both the United Kingdom and the United States had prepared drafts of a climate convention in 1989, based on the Vienna Ozone Convention. An NGO draft was prepared in early 1990 under the Climate Institute's sponsorship.

32. *Draft Framework Convention on Climate Change prepared and circulated by the United Kingdom at the first session of the INC*, UN Doc. A/AC.237/ Misc.1/Add.1 (May 22, 1991). Ordinarily the Secretariat would have prepared an initial discussion text, but because Res. 45/212 mandated the creation of an ad hoc secretariat, which had not yet been created as of February 1992, the UK stepped into the void.

33. According to Decision 1/1, para. 6(a), the commitments to limit and reduce net emissions of CO_2 and other greenhouse gases were to go "beyond those required by existing agreements"—a reference to the US attempt to receive credit for the CFC reductions it was already required to make under the Montreal Protocol.

34. Initially, the EC proposed establishing two groups on: (1) greenhouse gas emissions, in particular CO_2; and (2) sinks and reservoirs, focusing initially on forests. In the IPCC, sub-groups had been established in the Response Strategies Working Group on a sectoral basis, including on energy and industry, agriculture and forestry, and coastal areas.

35. The United States also proposed establishing working groups on scientific and economic research and monitoring, implementation, and legal issues. Forest countries such as Malaysia joined the US in opposing the establishment of a separate working group on greenhouse gas sinks.

36. For the same reason, developing countries supported dealing with commitments on sources and sinks and commitments on financial resources and technology transfer in a single commitments article, rather than in separate articles as is typical in other conventions.

37. *Report of the 2nd Session of the INC/FCCC*, UN Doc. A/AC.237/9 (Aug. 19, 1991).

38. *Set of informal papers provided by delegations, related to the preparation of a framework convention on climate change*, UN Doc. A/AC.237/Misc.1/ Add.1–15.

39. Michael Grubb and Nicola Steen, *Pledge and Review Processes: Possible Components of a Climate Convention* (London: Royal Institute of International Affairs, 1991).

40. *Report of the 3rd Session of the INC/FCCC*, UN Doc. A/AC.237/12.

41. *Report of the 4th Session of the INC/FCCC*, UN Doc. A/AC.237/15 (Jan. 29, 1992).

42. *Consolidated Working Document*, UN Doc. A/AC.237/15 Annex II (Jan. 29, 1992).

43. *Report of the First Part of the 5th Session of the INC/FCCC*, UN Doc. A/AC.237/18 (Part I) (March 10, 1992).

44. *Working Papers by the Chairman*, UN Doc. A/AC.237/CRP.1/Add.1–8.

45. *Report of the Second Part of the 5th Session of the INC/FCCC*, UN Doc. A/AC.237/18 (Part II).

The author's research was supported by an International Affairs Fellowship from the Council on Foreign Relations.

PART II

Views from Within the Ring

Exercising Common but Differentiated Responsibility

Delphine Borione
Ministry of Foreign Affairs
Paris, France

and

Jean Ripert
Chairman
Intergovernmental Negotiating Committee

In February 1991, at the first plenary session of the Intergovernmental Negotiating Committee, the prospect of successfully negotiating a Framework Convention on Climate Change seemed problematical at best. The complexity of the issues, the continuing scientific uncertainty about the causes and the impacts of the historically observed climate change, the wide range of countries' diverging views as to what action (if any) should be taken, the decision by many southern countries that the responsibility for action to reduce the risks of climate change should fall to the industrialized nations—each of these aspects of the situation in which negotiations began seemed to present insurmountable obstacles to the early conclusion of a worldwide agreement. They certainly made it seem unlikely that any agreement could be reached which would include significant commitments to emissions reductions or legal constraints on economically important activities. To complicate the situation further, economic crises and changing political priorities focused public interest on other challenges. For those at the highest levels of government, in the industrial countries in particular, the collapse of the Berlin Wall, the Gulf war, and the upheaval in Eastern Europe emerged in turn to divert political attention away from the climate issue.

The result obtained one and a half years later—156 signatures to the Convention gathered during fifteen days at the United Nations Conference on the Environment and Development (UNCED, also called the Earth Summit) in Rio de Janeiro—must be assessed in light of these considerations. Nonetheless, we pause here to ask: Is the universality of this endorsement a response to the legal void created by agreement to a text of no significance or to a correct balance of competing interests given full realization of the serious threats facing the planet?

The authors of this chapter may not be the best qualified to provide an objective answer to this question. Their role throughout the negotiation was to work obstinately for a compromise that could be supported by the greatest possible number of participants. This chapter will underscore some of the reasons why the final agreement seems to us to have a sound basis and gives the international community an effective tool for future cooperation on issues of common concern.

It is appropriate here to underline the fact that the opinions expressed are our personal views and are not to be considered as those of the French government.

Prologue to the Negotiations

Political concerns about the risks of rapid climate change arose abruptly on the international scene following the Toronto Conference on the Changing Atmosphere in June, 1988. Scientists participating in this conference voiced their alarm and called for early action to reduce the risks. They recommended a 20 percent reduction of carbon dioxide emissions by 2005. Despite an emerging consensus in the international scientific community, these scientists were nevertheless still isolated from the mainstream of international political debate. Most leaders of governments, with a few notable exceptions, were only beginning to discover the existence of a distinct and important threat to national economic development strategies.

One must consequently welcome the speed with which the United Nations Environment Programme (UNEP) and the World Meteorological Organization (WMO), both of which had already taken the initiative of launching major research programs on climate in the 1970s, decided to propose that their governing bodies establish an intergovernmental panel to review the scientific aspects of the climate problem.

A few months later, the United Nations General Assembly adopted a resolution on climate change at the initiative of the government of Malta. This resolution set in motion a veritable explosion of political initiatives. UNEP and WMO initiated an assessment program that would coordinate the efforts of over 300 experts from more than fifty countries. A new body, called the Intergovernmental Panel on Climate Change (IPCC), was asked to proceed with a scientific and technical assessment of atmospheric changes, their impact, and possible response strategies, based on all available data. The IPCC was invited to present a preliminary report of its findings at the Second World Climate Conference to be held in the fall of 1990.

In March 1989, a Conference was convened in The Hague, at the initiative of the governments of the Netherlands, Norway and France. This meeting brought together twenty-four heads of state and government. A declaration negotiated at this meeting was adopted by the participants and signed subsequently by the representatives of more than forty countries. This statement highlighted the "vital, urgent and worldwide" nature of the response that should be given to the greenhouse problem. It called for the establishment of a world authority to protect the atmosphere and argued for the payment of compensation to those developing countries that would be required to take on "abnormal" burdens as a result of the climate problem or the measures needed to address it. For many, the institutional aspects of this initiative seemed too revolutionary. It nevertheless succeeded in mobilizing attention and gave a strong impetus to discussions already underway, most notably in the IPCC.

The economic summit of the Group of Seven (G-7) in July 1989 in Paris also played a key role in moving the climate problem onto the international political agenda. The G-7 Summit Communiqué identified threats to the atmosphere as a top priority in environmental matters, explicitly called for the urgent adoption of a framework convention on climate change and argued forcefully for protocols containing concrete commitments to limit emissions of carbon dioxide and other greenhouse gases. The Communiqué bore witness to the involvement of the most important leaders in the Western world on the climate issue and strengthened the credibility of the efforts underway to address the problem in a diplomatic context. Later summits confirmed the continuity of this concern.

The Ministerial Conference on Atmospheric Pollution and Climatic Change held in Noordwijk, the Netherlands, on 6–7 November 1989 made this awareness more concrete. At this conference, nearly sixty ministers affirmed the need to stabilize CO_2 emissions as a first step toward a cooperative international response strategy. The concentration level at which this stabilization was to occur and the reference dates for achieving it were proposed as subjects for evaluation by the Intergovernmental Panel on Climate Change (IPCC) and review by the Second World Climate Conference when it convened in November 1990. It should nevertheless be noted that the United States delegation clearly stated at the Noordwijk conference that it was not ready to accept any binding commitment concerning the level of emissions.

The year 1990 witnessed further advances in the scientific work on the greenhouse problem. The first Assessment Report of the IPCC was adopted in August in Sundsvall, Sweden. The Second World Climate Conference established a broad scientific consensus on the nature and potential severity of the greenhouse problem when it convened in Geneva in November. From that point, the negotiation of a convention could be based on two pillars: a "scientific" pillar, the IPCC report, and a "political" pillar, the Ministerial Declaration adopted in Geneva. Still, the framework of the negotiation remained to be defined.

Key Issues in the Negotiations

In its first Assessment Report, the experts of the Intergovernmental Panel on Climate Change were clear: humans are causing the climate problem. Emissions from human activities are enhancing the natural greenhouse effect. If present trends continue, the IPCC estimates that by the year 2050, the average global surface temperature will rise by between 1.5° C and 4.5° C and the average sea level will increase by between 30cm and 50cm.

Although much remains to be learned, there are several aspects of the greenhouse problem on which nearly all agree. Among these are the universality of the greenhouse problem, continued uncertainty with regard to the science, and the linkage between environment and development problems.

Universality

The greenhouse problem has worldwide implications: the problem concerns all economic sectors, and its impacts will affect all countries. Energy, industry, transport, agriculture, forests—virtually every economic activity in a given country produces or absorbs greenhouse gases. Energy activities are the principal source of greenhouse gas emissions: they are responsible for more than 50 percent of anthropogenic emissions. This situation necessarily focuses attention on the question of development and, for some countries, the vital interests of single-product economies. The universality of contributions heralded tough negotiations which would deal not in accessory interests, but in the vital interests of countries and economic sectors. It was therefore natural for powerful interest groups to organize and mobilize, more especially so because scientists did not mask their sense of the remaining uncertainties.

Continuing Uncertainty

Despite the broad consensus on the causes of the enhanced greenhouse effect, the IPCC Assessment emphasizes that certain phenomena are not fully understood. At the present time, it cannot be ruled out that the observed historical warming has stemmed from natural causes. Furthermore, major uncertainties persist concerning such important aspects of the dynamic climate system as the role of the clouds and of the oceans. As a consequence of these uncertainties, the regional impacts of climate change cannot now be predicted with confidence. Although the precise distribution of regional impacts is unknown today, the IPCC Assessment concludes that some countries are particularly at risk of suffering serious adverse effects as a direct consequence of climate change. This is clearly the case for low-lying coastal zones (such as the small Pacific islands or the coast of Bangladesh).

A Problem of Environment and Development

By analogy with the Basel Convention on Waste and the Vienna Convention for the Protection of the Ozone Layer, most of the industrialized countries were of the view that UNEP and WMO should be entrusted with organizing the climate negotiations. Brazil and India, however—with support from most of the other developing nations—argued in favor of a broader base in the UN system. Other countries quite speedily rallied to this second approach.

The reasons why this broader approach seemed so persuasive are simple. The problem of climate change is not only an environmental problem but is, as has already been said, a development problem. Its implications affect numerous sectors of national economies—e.g., energy, industry, forests, agriculture—that range far beyond the competence of an institution responsible solely for environmental matters. In fact, for completeness, all the specialized institutions and programs of the United Nations system would have to be considered in and associated with this process. Only the General Assembly of the United Nations had the necessary authority to ensure this participation.

Other considerations argued in the same direction. The developing countries were particularly concerned with ensuring respect for certain principles of procedure—including transparency and universality of representation—that would enable them all to be present, to benefit from the weight of numbers and to ensure political monitoring of the process. The UN General Assembly format gave them this guarantee. Experience was to prove the judicious nature of this choice, which, in practice, did not reduce the effectiveness of the negotiating mechanism.

Initial Negotiating Positions

At the time it began, the climate negotiations could be presented in simplified fashion as representing a test of strength among three groups of participants. The United States adopted certain clear-cut positions even before the start of the negotiation and was not afraid to appear isolated. Defending the concept of an extremely general Framework Convention—a convention similar to the Vienna Convention for the Protection of the Ozone Layer—it argued that there was an overriding need for further scientific research and a better assessment of the economic costs and benefits of the different possible options (including adaptation) before discussing the adoption of quantitative targets and timetables. The US was slow in lending support to the idea of a Framework Convention with a solid institutional mechanism. It was only in the final phase that the US agreed to play an active part in formulating the objectives of the industrialized nations and in drawing up other key provisions. The US was joined by Saudi Arabia and Kuwait which sought to avoid any binding commitments to specific quantitative reductions in carbon emissions by a fixed date.

The other industrialized countries, pushed by public opinion that is generally well informed about threats to the environment, entered the negotiations with rather ambitious objectives. This was particularly true for many EC Member States and the European Commission. At the Council of Energy and Environment Ministers on October 29, 1990, the EC adopted the objective of stabilizing regional CO_2 emissions in the year 2000 at the 1990 level. This objective was presented as a first phase target en route to subsequent reductions. During the Convention negotiations, the EC decided to consider a tax on CO_2 and energy.

From the outset, Germany, the Netherlands, Denmark, Italy, and France promoted an ambitious concept of the Convention and insisted on the need to begin the negotiation of protocols without awaiting the outcome of the negotiation of a Framework Convention. In contrast, the countries of southern Europe maintained very cautious positions for reasons similar to those of many developing countries. While the government of the United Kingdom came round to these positions, its negotiators were always careful not to compromise the chances for a final text that could be agreed to by the United States. The UK was certainly pivotal to reaching agreement in the final phase of negotiation. The European Free Trade Association countries coordinated their positions closely with those of the EC. Japan and other OECD member countries defended comparable positions, though taking care for reasons which were not unrelated with the ongoing General Agreement on Tariffs and Trade (GATT) negotiations, not to give the impression that they were taking part in what sometimes appeared to be a conflict between the United States and the EC.

Despite their relatively minor economic weight, from the outset a small number of developing countries played a very large role in the negotiations. The small island states, the sub-Saharan African countries, and some of the countries with large, exposed deltaic plains, pressed aggressively for an action-oriented convention that would address both adaptation and abatement issues.

Other developing countries feared that they would be adversely affected by measures taken to reduce greenhouse gas emissions. Countries dependent on the production and export of fossil fuels (oil-producing countries, in particular) feared the consequences of the introduction of energy taxes. Representatives of these countries consequently emphasized the scientific uncertainty about climate change and the flaws in existing economic analyses. They pointed out the risks that preventive measures could pose to world economic growth, in particular for developing countries. They went to great lengths to minimize any specific commitments to emissions reductions, avoid any reference to energy and generally to delay the conclusion of an agreement.

The developing countries were unanimous in putting forward the "historical" responsibility of the industrialized nations for exacerbating the greenhouse effect and refused to envisage any specific commitment to emissions reductions of their own. Nevertheless, based on the principle of "common but differentiated responsibility," they agreed to participate in the negotiation process on condition that their development priorities were recognized and that they received guarantees of financial and technological aid. They emphasized the need for reciprocal commitments. They sought binding commitments ensuring the mobilization of new and additional financial resources by the industrialized nations to correspond to specific commitments by the developing countries. In the absence of financial support, they would be bound only by "moral" commitments.

More generally, many countries, both industrialized and developing, voiced concern about possible economic impacts that could result, for example, from the introduction of an eco-tax by the EC. This being the case, the positions of the EC Member States were regarded warily. It is true that a tacit agreement was reached quite early on among the delegations that recognized the impossibility of negotiating anything other than a Framework Convention in the time available. Consequently, no detailed debate was held concerning implementation measures for the agreed commitments. The elaboration of any implementation strategies was left to fall within the competence of the national governments.

Negotiations Begin with a Lengthy Prelude

Given these conditions, and the very tight deadlines, the general skepticism at the start of the negotiations was understandable. The slowness of initial progress reinforced this view. The final success cannot be explained without a reference to the efforts, imagination and tenacity of the negotiators, in particular those who agreed with the Bureau members and the leaders of the negotiating groups to make the process truly a team effort. The negotiators had two major assets at the outset. First, the quality and rigor of the work done by the IPCC, under the wise chairmanship of Professor B. Bolin, gave the diplomats a firm scientific foundation that was never once seriously challenged during the negotiations. The IPCC's deliberations on the possible policy responses to future global warming led, in fact, to a preliminary "roughing-out" of possible options. Second, the process benefited from a strong political impetus in the period preceding the negotiation. Public opinion was put on the alert by the scientists; then non-governmental organizations (NGOs) and the public media kept the spotlight on the climate problem.

UN General Assembly Resolution 45/212 formally launched the negotiation, giving the negotiators the charge of adopting a framework convention open to signature in June 1992, i.e., ready for signature at the Earth Summit in Rio de Janeiro. It was hoped by many that this text would be matched with some "related legal instruments." But the negotiation of the Convention itself was to prove so complex that the negotiation of these protocols was by tacit agreement postponed to a later date.

The negotiation of the Convention can be broken down into several phases. The first session, held in February 1991, at Chantilly, VA, near Washington, DC, was disappointing for most participants. To most outside observers, the very limited progress at this meeting seemed to augur badly for the process as a whole. Ten days turned out to be barely enough to elect a bureau, set up two negotiating groups and prepare their briefs. The chairmen of these groups were not appointed until four months later, at the start of the second session. In fact, however, this time was not wasted. Preparing the briefs of Working Groups I and II was tantamount to defining the scope and content of the Convention. The importance of financial elements, the sectoral approach, the special concerns of low-lying small island states and other developing countries—all these elements had to be included from the outset in the Working Group briefs. This was necessary in order to give the affected countries a guarantee that their concerns would be taken into consideration.

The main objective at this stage was to bring all the participants fully into the process—a process which many initially entered with great misgivings. The negotiators needed time, however, to gradually familiarize themselves with the subject and to get a clearer idea of the concerns of their interlocutors. This start-up phase continued between March and August 1991. In this period, many countries forwarded either complete draft conventions or more limited contributions that they wished to see considered for inclusion in the text.

What was very encouraging to the members of the Bureau was that such proposals emanated not only from industrialized delegations, but also from developing countries such as India, Malaysia, Vanuatu and China. The submissions from these developing countries confirmed the message issued by more than forty developing country governments at the ministerial meeting that was held in Beijing on June 18–19, 1991. On certain crucial points, the positions of these countries were very far from those of the industrialized nations, but they jointly asserted the will to conclude an effective action program to fight the greenhouse effect. China added an insistent reference to the need to adopt a pragmatic attitude.

From the positions voiced at the first two sessions, as well as from the reading of the drafts submitted, it was clear that an agreement could not be reached based solely on the traditional concepts of international conventions. The Framework Convention would have to go beyond the usual structures that systematically include uniform commitments applicable to all parties and that are controlled through a procedure of conformity reports. Many participants nevertheless rejected what appeared to others as a reasonable solution: the adoption of general provisions, including the formulation of commitments, and the postponement of detailed, specific provisions to a subsequent stage of negotiations—a stage that would be focused on the text of the companion protocols. This approach, inspired by the process adopted to fight deterioration of the ozone layer, was not acceptable to those wishing to see action right away—especially those arguing for action on the basis of explicit commitments to emission reductions.

New Concepts Emerge

A new concept was proposed by Japan as a way to move beyond the growing impasse. The new concept was called "pledge and review." These terms were hard to translate into other languages, and we found, in the end, that they did not simplify the debate. "Pledge" refers to a voluntary commitment, or an "announcement." The term is used frequently in the area of finance. Its "binding" nature remained ambiguous, which was precisely its virtue for some participants. The concept of "review" was not much more precise. It definitely did not imply "monitoring of compliance with an obligation" in the legal sense.

Through extensive discussions, elements of clarification began to emerge. The Japanese delegation suggested that the "pledge" could take on a more binding nature if the means a given government planned to use to fulfill the objective of the convention were made explicit and if the contracting parties agreed to "make every effort" to achieve the desired result. It was suggested that a periodic public examination procedure ("a review") of government reports could be binding on national political authorities or at least facilitate the acceptance of certain collective disciplines.

Given such contradictory interpretations, the concept of "pledge and review" unfortunately engendered a good deal of confusion. This new concept was finally rejected by the developing countries, which viewed it as a one-sided attempt by the industrialized countries to impose a new set of obligations upon the poor which were likely to affect their autonomy in development matters. They were prepared to communicate information about their "emissions" or "sinks," but they were not willing to commit to formal external reviews of their national policies. These debates were held less often in plenary meetings but frequently in informal meetings. In our view, in spite of everything, these debates in the negotiating groups favored the progression of ideas that moved forward elements of the final consensus. Discussion of "specific" commitments was gradually made less alarming. The idea progressed that the Framework Convention could become as much a process for further dialogue as a formal instrument chiseled in stone.

Similarly, discussion of the proposals tabled by the Norwegian delegation contributed to the progression of ideas. Highlighting the need to ensure cost effectiveness and economic efficiency in all the measures to be taken, the Norwegian government introduced a proposal concerning the joint implementation of national commitments to emissions reductions. This proposal would allow an industrial country to achieve some of its commitments to emissions reductions through cooperative projects with developing countries. Recognizing that the available room for action and the estimated cost of reducing greenhouse gas emissions vary widely between industrialized and developing countries, and, indeed, even among industrialized countries, many delegates observed that large and immediate gains in limiting the greenhouse effect could result from programs to reduce emissions and increase sinks in the developing countries. But developing countries have no obligations and few incentives to take actions in the short term. Thus, it seemed reasonable that if the rich countries were to provide a financial incentive by paying for such measures, perhaps the credit for such efforts at joint implementation could be shared among the parties participating in a program or project.

In an effort to convince their interlocutors of the validity of this approach, the Norwegian delegation had advanced this idea through bilateral contacts and at least one workshop in France. For developing countries, one advantage would be an increase in the transfer of resources from North to South. Many other countries, however, expressed their fear that such a possibility would divert an interested country from fulfilling its own obligations domestically and would therefore increase the differences of emissions between countries.

The concept of "joint implementation" has finally been introduced in the text of the convention Article 4(2)(b), but its concrete modalities will need to be further defined by the Conference of the Parties. Reflecting these concerns, no clear consensus emerged about how to implement this approach.

Another controversial concept which has taken much time to be clarified and accepted is the idea of "net" emissions. The debate on this notion has been tense because it reflected different approaches to national obligations under the convention. Many delegations feared indeed that it was a way for some countries to prevent a clear identification of the specific actions to be taken on individual gases— and in particular, on CO_2.

It is in fact quite impossible, in the present state of scientific knowledge, to aggregate on a common basis the figures of reduction of emissions from the different greenhouse gases, given the uncertainties concerning the "global warming potential" of each gas. It is also difficult at this stage, to identify and to quantify all the sinks. Commitments on "net emissions" could therefore be considered impossible to implement.

A similar debate has taken place when discussing the merit of "comprehensive" actions, instead of more focused attention on individual gases. The final compromise speaks about action on "greenhouse gases" but there is also a specific reference to CO_2.

Despite the untiring efforts of the co-chairs of both working groups, the sessions in September and December 1991 only succeeded in including each party's ideas in the texts, which all recognized could not yet form the basis of conclusive negotiations. The resulting texts were crammed with brackets. The texts were entirely illegible to an uninitiated reader, and impossible to submit for assessment by government ministers who are always pressed for time. At the February 1992 session the tension resulting from dealing with all the bracketed text was crystallized. There seemed to be insufficient time left to reach a final agreement.

The Climax Approaches

With the agreement of the Expanded Bureau of the INC, the chairman decided to hold detailed consultations among a representative set of countries. These consultations were held with a view to determining the bases on which an agreement could be concluded, which the main participants publicly affirmed they wanted and privately told him they thought possible. These consultations were not held solely on a bilateral basis; the chairman also organized an unofficial consultation session from 15 to 17 April in Paris. The members of the principal negotiating groups were invited to this session on a personal basis, along with the members of the Expanded Bureau and the representatives of some of the countries most active in the negotiation. The meeting convened with around twenty-five delegates in all. The convening of such a group during earlier phases, even when only involving organizational or procedural problems, had demonstrated the progressive establishment of a climate of mutual respect among the key participants.

The Paris meeting did not reach formal agreement on an integrated text. Such was not its objective, in any event, but these consultations allowed the shape of a possible convergence to appear. For example, concerning reporting obligations, it appeared possible to differentiate between the obligations of developed countries and those of developing countries. The developed countries which had met earlier among themselves (following a suggestion by the chairman of the INC) were, unfortunately, unable to reach agreement on the formulation of specific commitments to be accepted by them.

It was not until the resumption of the Fifth Session in New York in May 1992 that an agreement could be reached on an integrated text. Earlier, the chairman of the INC, at the request of the participants in the Paris meeting, had prepared a draft Convention, free of brackets, that was likely, in his view, to constitute a basis for agreement. In preparing this text, the chairman benefited from the suggestions of the co-chairmen and the solid backing of the excellent Secretariat. The main merit of this text was no doubt that it confirmed the intuition of all those who thought it still possible to reach agreement on a unified text.

It was no doubt these initiatives and the spirit of compromise shown by the delegations—especially those representatives who met with the chairman for several days and nights in succession—that made it possible to conclude an agreement during the final debate of the "last chance" session in May. We must not underestimate the importance of the many discussions held between governments during this period, including at the highest levels. They led in particular to the wording allowing the United States and its Western partners to agree on a compromise text for Article 4(2)(a) and (b), the text concerning the specific commitments of the developed countries which the hairman of the INC was able to include in his proposal.

The Time for Action has Come

The Framework Convention on Climate Change was signed by almost all United Nations member countries. At least fifty countries must ratify it to enable its entry into force. This figure may seem high, more than twice the threshold set for the Conventions on

Protection of the Ozone Layer or Biological Diversity. It is the will to ensure a large and representative participation which dictates this choice. The objective of this Framework Convention is to launch a process which will be effective only if it is universal and if it includes, in solidarity with each other, the countries which are responsible for the main greenhouse gas emissions and those which can be more dramatically affected by climate change.

Getting Off to a "Prompt Start"

Because of the need for fifty countries to ratify the convention, legal entry into force will occur at the earliest in late 1994. Many countries are interested in taking action right away. Whilst awaiting the Convention's entry into force, these countries seek a quick start— even before legal entry into force—in order not to lose the momentum engendered by the negotiation and the broad political agreement registered in Rio.

This is the object of the "prompt start," a concept which proved very popular in the final days of the negotiations. A resolution, discussed at the initiative of the Canadian delegation and several others, was approved by the INC following adoption of the draft Convention. It encourages the governments to take anticipatory measures. Interim provisions are included in the Convention itself (Article 21) concerning research, financial mechanisms and the continuation of the work of the Secretariat.

The "prompt start" concept responds to the sense of *urgency* expressed on several occasions by the governments of the industrialized countries. The "prompt start" should enable them to prepare their first national programs and to organize a minimal level of consultation in this regard. It thus provides a response to those who regret the "vagueness" of the specific commitments in the Convention. Financial aid will be granted to developing countries for this purpose. Under the auspices of the Global Environment Fund, detailed country studies will be developed. Indeed, some country studies have already been started; they have been initiated with a view to specifying and harmonizing assessment methods or to developing sustainable development strategies.

Many Legal Issues Remain to be Resolved

The implementation of the Convention also requires intense legal preparation. Under the pressure of time, the negotiators postponed decisions on many procedural questions until the "first session of the Conference of the Parties." Some saw this postponement as an evasion of responsibility in the face of seemingly insurmountable difficulties. But there was no other way to meet the Rio deadline. Debate on these legal technicalities is expected to grow less impassioned with the passage of time. Many observers expect that national positions will become more flexible once political agreement has been achieved and the need for action has become more pressing.

One of the questions on which the negotiators will focus their attention deserves mention here. It is whether new legal instruments should be urgently negotiated to complement the Framework Convention. As mentioned above, the negotiation of the famous "protocols," called "related legal instruments" in Resolution 45/212, was proposed at the outset by a large group of industrialized nations. Only the United States, among the OECD countries, remained in opposition to this proposal.

The Noordwijk Ministerial Statement and the Geneva Declarations of the Second World Climate Conference each recommended the negotiation of protocols on energy and forests. The advisability of protocols on each gas, such as CO_2 or methane, was also discussed. The difficulty involved cannot be concealed. It is not a question, as for the Montreal Protocol for example, of regulating a few easily defined substances such as CFCs, that are produced by a limited number of companies in restricted quantities, and for which substitutes exist. The climate issue will affect entire areas of economic activity in extremely varied circumstances and emissions under widely varying conditions. It is consequently not conceivable to commit to a program of uniform limitation. If such a regime were proposed, each party would emphasize its unique position, the unavoidable external constraints that limit its own choices and many other circumstances necessitating an exemption from a "common law." Even in the European Community, which is relatively uniform as a whole, considerable difficulties are being experienced in adopting a set of consistent targets for emissions stabilization.

Key Conclusions and Outcomes

This observation suggests a simple conclusion: it is premature at the present time to seek to negotiate specific obligations as regards means or results. It would be better to concentrate attention and efforts on effectively implementing contractual obligations and setting up the Convention's institutions. The formulation of national policy and the publication of national reports have the specific objective of beginning action in the industrialized countries. Review of these reports and comparison of these policies will exert useful pressure.

However, discussions among industrialized countries need to be pushed forward. They should, for example, explore ways to better harmonize national economic policies for combating the greenhouse effect. This concerns, in particular, the proper use of economic instruments—e.g. taxes—which require coordination among countries in order to avoid distortions and unbalanced effects on the competitiveness of national economies.

Meanwhile, scientific and technical work should continue. The insufficient number of economic analyses, particularly concerning the use of economic instruments and the impact of policy measures, should not continue to be used as a justification for lack of action.

New Impetus is Given to the IPCC

From this point of view, the work of the IPCC will be decisive. One should welcome therefore, the initiative of the chairman, Professor Bert Bolin, and the decision of the committee itself concerning its program of work and its structure. We attach great importance to the economic dimension that, according to these decisions, will be incorporated into the next phase of this group's analyses. In the coming phase of their work, economic and social analyses must be made at microeconomic and macroeconomic level. They should be integrated into all of the IPCC's other work on prevention and adaptation.

One point will need to be clarified quickly in terms of cooperation with IPCC: the question of the difference between the composition of the IPCC, which is open to all countries, and the composition of the Parties to the Convention. That seems to exclude for a certain period of time, the possibility of the IPCC being recognized as the scientific organ of the Convention. Nevertheless, every effort should be made to avoid duplication in work and conflicts between the respective schedules.

Other fora of scientific discussion may also be used to advance the consideration of these issues. The OECD, for example, constitutes an important forum for debate and allows an open exchange of ideas and experiences. The OECD, in liaison with the IEA, has already undertaken specific studies. Doing away with the link between discussion and negotiation enables the development of genuine expertise without "polluting" the scientific debate with political considerations, which should become prominent at the time of decision-making.

The respective roles of the different bodies should be clearly distinguished: the INC today—and the Conference of the Parties tomorrow—should have control respectively of the negotiations and political decisions. The IPCC must have responsibility for scientific and technical assessment, which is given legitimacy by the quality of its expertise and facilitated by the adjustment of its structures that is now underway. The OECD can constitute a forum for debate and further work by the industrialized countries, which are the main parties concerned with emission reductions in the coming years. Developing countries could have the possibility of being associated with these discussions if this option interests them. The development of regional or sub-regional research centers in other parts of the world should also be encouraged.

Just as the "Atmosphere" chapter of Agenda 21 was adopted in the framework of the UNCED, the question of climate change falls within the scope of the future Commission on Sustainable Development. This Commission should play a valued role of general supervision without intervening in the implementation of the Convention.

In particular, the Commission on Sustainable Development will have a role to play in ensuring that the development constraints of southern countries are taken into consideration. Although concern for sustainable development should be at the center of actions and programs implementing the Framework Convention on Climate Change, the Convention does not provide the instruments needed for development concerns to be taken fully into account. It is to be hoped that this might be done in this new body.

The Challenge of the Financial Mechanism

In contrast, the definition of the financial mechanism falls within the competence of the Conference of the Parties to the Climate Convention. More precisely, the definition of "the policies, the priorities of its programme and the approval criteria" for projects should be reviewed, evaluated, and, if need be, readjusted by the Conference of the Parties at its first meeting. (See Article 11 of the Convention.) In particular, the role of the GEF after the interim period will be decided by the Conference of the Parties.

One of the crucial points in this regard will be the definition of "agreed full incremental cost" used in the financing of Global Environmental Facility (GEF) projects. This notion, contained in Article 4 of the Convention, is one of the provisions adapted from the Amendments to the Montreal Protocol, creating a multilateral fund. While this notion seems well defined in the case of replacing a CFC substance with another less-dangerous compound, or in the event of the substitution of a production process, the question is considerably more complex in the case of the climate. Serious, practical case studies in a large number of countries and different sectors of activity are needed to clarify the debate.

From our point of view, this type of debate must in the end be decided by the Conference of the Parties. It will require, however, without delay, the attention of the national delegations. It should be prepared by discussions and analyses to be developed with a maximum of transparency in the GEF as well as in the INC. Informal discussions among members of these two bodies could be very beneficial.

Implementation of the Framework Convention Should Mobilize Many Participants

Responsibility for action on the climate problem does not fall exclusively on governments. Admittedly, governments must take legislative or regulatory measures, adopt economic instruments or incentives or dissuasive measures for economic operators. Their role is of strategic importance, but they are not the ones who will apply these measures: private enterprises, producers, and consumers will be the real participants in the battle against the greenhouse effect, through consumer choices and production techniques. The struggle to limit the harmful effects of climate change will fail if these groups are not mobilized.

It is vital, for example, to develop technical alternatives in the field of energy. To accomplish this, basic scientific research as well as applied research on new technologies must be expanded. Advanced designs for low-risk nuclear electric systems and for different types of renewable energy systems must be completed and commercialized because these systems produce minimal amounts of greenhouse gases. But, above all, we need safe, cost-competitive technologies which can be disseminated in specific end-use applications. A good deal of progress remains to be made to ensure broader use of these technologies. In the case of nuclear energy, developing a safe, reliable method of long-term waste management and reducing the risks of weapons-related proliferation of fissile material are crucial public policy goals.

Thanks to technological progress, other forms of energy or production processes may be found which also produce low levels of greenhouse gas emissions. Biomass is another source of energy which could be developed in a cost-effective and sustainable way. Waste recycling technology will allow currently uncontrolled methane sources to be tapped. There is still much progress to be made to find substitutes for CFCs that have lower global warming potential; transitional HCFCs still seem to have dangerously high warming potential.

Industry must be directly involved in fighting global warming, as it was in the search for ozone-friendly substitutes for the most dangerous CFCs. Technical innovation must be encouraged in all countries according to arrangements responding to their market structure. We note here a positive sign: the announcement by the group of the world's seven largest electricity companies of plans to develop technical assistance programs to aid electric companies in developing countries in their efforts to improve efficiency and to mobilize financial resources to this effect.

Another group of essential participants is made up of non-governmental organizations (NGOs). As groups exerting pressure on governments, they will have increasing influence as the number of so-called "green" voters increases. These voters tend to favor more "environmental" positions. Environmental NGOs express the views of these voters, serving to transmit opinions to decision-makers. These groups also play an active role in educating the population and contributing to necessary changes of behavior. As in the past, they should be allowed to attend the sessions of the INC and to have the possibility of expressing their views. The established modalities for their participation in the process seem to have been satisfactory.

The participation of other experts is certainly desirable. One should nevertheless be careful about the modalities in order to avoid mixing competence. Specific workshops can help to elaborate more difficult concepts, to clarify uncertainties and to achieve progress on the definition of adequate actions.

Other problems as complex as those we have just enumerated must be solved to ensure effective implementation of the Convention. The industrial countries have accepted responsibility for the majority of greenhouse gas emissions; each of them must make an effort in accordance with their contributions to global warming and their economic capacities. There is a long road ahead. Imagination and political will must continue to be mobilized to serve constructive action. The spirit of solidarity underlying the negotiation of the Framework Convention must not be diminished.

Having a target date for this next phase of negotiation will certainly be helpful to press the governments toward decisions and actions. The convention itself, in Article 4(2), calls for two specific meetings of the negotiators: an initial review of the specific commitments of the industrialized countries at the first Conference of the Parties; and a second review not later than December 31, 1998. Other deadlines will be given by countries willing to accelerate the process.

Epilogue

The final result of the negotiation falls short of many of our initial hopes. The recommendations made in Toronto in 1988 and in the Noordwijk Declaration of November 1989, signed by sixty Ministers of the Environment and recommending the stabilization of CO_2 in the year 2000 at 1990 levels, were ignored. The implementation of a tax on CO_2 and energy in the European Community is proving difficult to achieve! It is understandable that so many are expressing disappointment with the international response to the climate problem.

The result obtained by the Convention is nevertheless of major significance. The Convention comprises commitments which are sufficiently specific to play a "pedagogical" role in defining internal policies and enabling action to begin. The question of climate change, which appeared on the international scene less than four years ago and was originally raised by a few northern countries, has managed in this very short time to unite the entire international community.

The discussion of Agenda 21 at UNCED was enlightening from this point of view: more specific elements were able to be included, such as favoring energy savings and promoting more rational use of energy. Agenda 21 encourages the use of economic and regulatory instruments for all sectors concerned. Admittedly, these are only "recommendations," but they would not even have been conceivable only five or ten years ago.

This type of negotiation in a United Nations body must pay the cost of its universality. The rule of consensus, breached only in very rare cases, requires that we can reach agreement at the level of the least common denominator. The challenge is to place this " least common denominator" at the highest level possible. This is a very heavy burden but its outcome will determine the success of a response to a worldwide phenomenon, for which the only effective response must necessarily be global. Only if the industrialized countries accept their historic responsibility and preach by example is there any hope of our future success.

CHAPTER 4

The Beginnings of an International Climate Law

Ahmed Djoghlaf
Vice Chairman
Intergovernmental Negotiating Committee

Prologue

The twenty-fold increase in world industrial activity since the start of this century has had many negative repercussions for the environment. Pollutants emitted from these activities have seriously altered long-standing balances in the chemical composition of the Earth's atmosphere. During the 1980s, the annual increase in world industrial output was approximately equivalent to the total industrial production of Europe during the 1930s. Annual emissions of carbon dioxide in 1992 are estimated to be more than 20 billion tonnes, a figure that is expected to double before the year 2030. The effects of these steadily increasing emissions are reflected in today's elevated atmospheric concentrations of CO_2 and other greenhouse gases. The concentrations of several greenhouse gases will remain high for decades. This is particularly true for carbon dioxide, which has a residence time in the atmosphere of up to 200 years.

Moreover, while mankind has succeeded in considerably increasing the sources of greenhouse gas emissions, we have at the same time significantly reduced the natural sinks (i.e., biological reservoirs that store carbon), especially through the processes leading to deforestation and desertification. More than one-third of the earth's original forest cover has already been lost due to the frenzied pace of deforestation. The global rate of forest loss today exceeds 17 million hectares per year. In some regions, twenty-nine trees are cut for each one that is planted. Over the last 100 years, we have cleared more land for cultivation than in the entire previous history of humankind.

To make matters worse, in many developed countries more than 30 percent of the remaining temperate forest suffers from extensive damage due to air pollution. The damage is caused by pollutant emissions emanating from both local and cross-border sources. But the problem goes beyond the direct effects of pollution and human-induced deforestation. Forest fires due to natural causes further reduce the Earth's carbon-capturing capabilities. In the Mediterranean area alone, more than 500,000 hectares of forest and woodland are damaged by the 30,000 to 40,000 forest fires occurring each year.

The cumulative effects of the increase in greenhouse gas sources and the decline of greenhouse gas sinks form the basis of projections suggesting a possible global warming of 2–5 degrees Centigrade (C) over the next century. A warming of this magnitude would likely be accompanied by a rise of up to 1 meter in average global sea level, thus threatening the existence of certain island countries and of low-altitude coastal regions. These considerations, together with concerns about stratospheric ozone layer depletion, form the foundation for international efforts to prohibit the production of chlorofluorocarbons (CFCs), a family of man-made chemicals which account for 25 percent of greenhouse gas emissions. The efforts to control CFCs have met with considerable success, notably the adoption of the 1985 Vienna Convention for the Protection of the Ozone Layer, and the 1987 Montreal Protocol on CFCs and its subsequent amendments.

These same considerations led participants in the Tenth World Meteorological Congress, held in May 1987, to call on the Executive Committee of the World Meteorological Organization to set up an appropriate mechanism to address the technical aspects of the greenhouse issue. As a result, the following year, the Intergovernmental Panel on Climate Change (IPCC) was set up to study climate change. The IPCC was assigned the following specific tasks: (1) to assess the science and impacts of climate change; and (2) to evaluate possible policy response strategies that could be implemented at the national level.

In 1990, the IPCC finished its first Scientific Assessment Report, and presented it for review at the Second World Climate Conference

(SWCC), held in Geneva in November. This first IPCC Assessment Report shaped the statement adopted at the Ministerial Conference on Climate Change (a diplomatic meeting held in Geneva immediately following the SWCC). The Ministerial Statement following the SWCC recommended to the 45th session of the General Assembly of the United Nations that an intergovernmental negotiating committee be formed and given responsibility for drawing up an international treaty on climate change. Consequently, at its 45th session, the United Nations General Assembly (UNGA), adopted Resolution 45/212 entitled "Protection of the World Climate for Present and Future Generations." This resolution established an Intergovernmental Negotiating Committee (INC) responsible "for formulating a framework convention on climate change which includes appropriate commitments and for preparing any other related instrument that may be agreed."

A Fast-track Process Yields a Unique Achievement

On 9 May 1992, at the conclusion of five working sessions, the Intergovernmental Negotiating Committee completed a draft treaty. The Framework Convention on Climate Change based on that draft was signed by 154 countries at the United Nations Conference on Environment and Development (UNCED) in Rio de Janeiro one month later. Barely sixteen months elapsed between the first meeting of the INC in Chantilly, VA, in February 1991, and the ceremony for the signature of the convention held at the Earth Summit.

What factors made possible this remarkable achievement? What conclusions can be drawn from the experience acquired by the Intergovernmental Negotiating Committee? What has been learned in the INC process that could help when setting up the mechanisms for implementation of the Framework Convention on Climate Change? This chapter will attempt to make a small contribution to answering these questions.

The INC Faces Major Challenges

The INC was required to proceed simultaneously with a dual task—both drafting and negotiating a convention. Generally speaking, the negotiation of any internationally binding legal instrument is preceded by protracted consultation culminating in a preliminary draft. Due to time constraints and to the pressure of both national and international public opinion, the usual drafting procedure could not occur before the beginning of formal negotiations by the INC. Despite the complexity of the issue, the scope of possible impacts and the controversies within the negotiations, the INC was compelled both to draft and to negotiate the various provisions of the Convention on Climate Change in a comparatively short time.

Although the Intergovernmental Negotiating Committee did not have the benefit of a preliminary legal drafting process at the time its work began, it did receive substantial input from a number of high-level international meetings on the scientific, technical, and political aspects of the climate change issue. In this regard, it is important to highlight the contribution of the Ministerial Declaration of the Second World Climate Conference and the related declarations made by ministers assembled in Bergen, Noordwijk, Langkawi, Ontario, Toronto, Tokyo, and The Hague. In addition, the work of the Intergovernmental Panel on Climate Change provided a critically important part of the foundation underpinning the negotiations of the INC. While these scientific and technical meetings certainly provided support for the diplomatic negotiations and helped to establish its legal argumentation, science was not the sole decisive factor in the success of the INC process.

Through these various scientific, technical, and diplomatic meetings, substantial debate preceded the political and legal negotiation of the Intergovernmental Negotiating Committee. Recapturing this advantage in future climate negotiations will place particular responsibility on the Subsidiary Body on Scientific and Technical Advice that will be established under Article 9 of the Convention on Climate Change.

The exceptional conditions forming the foundation of this success are not likely to be duplicated in the future. What remains of the negotiations on climate change is a process leading to the first Conference of the Parties (COP). It is the duty of the Conference of the Parties to resolve the many unfinished issues from the Convention process. The COP must therefore precede the negotiation of any additional legal instruments related to the Convention. In this next phase, there must be a careful drafting process before any formal negotiations begin. This separation of preliminary drafting from direct negotiation is as necessary as it is laborious.

Keys to Success in the Climate Negotiations

The experience of the Intergovernmental Panel on Climate Change, in particular, merits further study. The work of the IPCC was accorded much greater international legitimacy following the decision at its second session to set up a Special Committee on the Problems of Developing Countries. As a consequence, the number of developing country participants rose from eleven to fifty-four in 1990.

The decision adopted by the UN General Assembly Nations in Resolution 45/212, binding the negotiations to the June 1992 convening of the Earth Summit (i.e., UNCED), was critical to the success of the INC. This decision prevented a breakdown of the negotiating process, as no country or group of countries could accept responsibility for the failure of the negotiations without risking grave public relations consequences.

Moreover, the specific nature of the process leading to UNCED was unique in the annals of multilateral cooperation. It featured the simultaneous organization of negotiations on three separate tracks converging towards a single event in June 1992. This inexorable deadline was a healthy stimulus to all three processes of negotiation, promoting a spirit of mutual cooperation among the delegates to the Intergovernmental Negotiating Committee on Climate Change, the committee set up for negotiation of the Convention on Biological Diversity, and the Preparatory Committee of the Conference itself.

The Preparatory Committee of UNCED (known as the PrepCom), like the INC, met in New York, Geneva and Nairobi. This use of multiple meeting sites and the proliferation of preparatory bodies heavily handicapped Third World representatives, making it difficult for these delegates to maintain continuity of representation at all these meetings. These difficulties had repercussions for the negotiating process. The continual and irregular changes of government representatives at the negotiating table made it more difficult for the representatives of Third World countries to develop personal relationships with their colleagues from the developed world.

Critical Moments

Many factors contributed to the success of the first two years of the INC process. But at several critical moments, the entire outcome seemed to hinge on the actions of a few key players. The following paragraphs highlight two such episodes.

The Paris Meeting of April 1992

In February 1992, at the conclusion of the first part of the INC's fifth session, the consolidated negotiating draft was saturated with bracketed text, denoting scores of conflicting positions strongly held by the Parties. It resembled a simple compilation of contradictory positions more than a recognizable legal instrument. The text included no fewer than 800 brackets and over 300 alternative formulations for key sections of the text.

An informal consultation of the Enlarged Bureau of the INC was held in Paris from 15 to 17 April 1992, on the eve of the final negotiating session, including representatives of the twenty-five key delegations. As Vice-Chairman of the INC, I pointed out at this meeting that if the INC, during its final negotiating session, devoted a mere five minutes to examining each of the 800 outstanding brackets in the consolidated working draft, it would take seventy-seven hours to debate the text. But the final session of the INC included only forty-eight hours. This argument helped to persuade the chairman of the INC that he should submit his own complete negotiating text with no remaining brackets. Indeed, had this meeting of the Enlarged Bureau not been held, and had the chairman not been assigned the task of submitting his own negotiating text, the stalemate that stymied the INC at the end of February 1992 would doubtless have continued.

INC Chairman and of the Bureau Overcome Obstacles to Practical Agreements

The leadership role played by the Bureau of the INC, and especially by its chairman, extended beyond the preparations for the INC meetings. This leadership continued during INC plenary sessions as well. It was especially important during the final phase of negotiations. These negotiations included some thirty delegations during marathon working sessions held by the Bureau at the United Nations headquarters. In particular, the draft text submitted for adoption by the INC plenary on 9 May 1992 had been negotiated extensively within the Enlarged Bureau during the second part of the fifth INC session.

The critical role played by the Bureau of the INC could not be accurately foreseen at the outset of the negotiating process. In part, its successful contribution was made possible by the unusual decision to select co-chairs to head two working groups, one related to commitments and the other to the mechanisms of the convention.

Nonetheless, the decision to have dual co-chairs had a negative influence on the negotiations. The selection of co-chairs reflected the underlying impasse between North and South that would persist throughout the INC negotiations. The INC Bureau elected the representatives from Japan and Mexico to chair the working group on commitments, and the representatives from Canada and Vanuatu to lead the working group on mechanisms.

As a result of the division of labor among the working groups, each working group was assigned several chapters of the draft agreement. The dynamic links between the key aspects of the negotiations were thus largely ignored; instead topics were approached by the working groups in separate and narrow ways. Furthermore, the formal rotation of the co-chairs between working group sessions tended to limit the continuity and to obstruct the global vision needed for the success of these complex negotiations. In hindsight, perhaps it would have been better to have had only one permanent chair for each working group.

Fortunately, this artificial division of responsibility, which could have dealt a fatal blow to the negotiating process, was healed at the final negotiating session. The leadership of the INC decided to handle the two issues of commitments and mechanisms in a simultaneous and integrated fashion at its last plenary session. During the second part of the fifth INC session, three new working groups were created directly under the auspices of the plenary. This was a constructive move. The remaining work was completed in a number of intensive meetings of the Enlarged Bureau. The output from these meetings formed the foundation of the final package offered by the chairman as a draft text for the Convention. This experience suggests that sometimes it is better to suggest a compromise at the end of a negotiation, rather than at the beginning. For the final phase of negotiations, the merit of the chairman's text resulted not only from the principles embodied in its submission, but also from the judicious timing of its presentation. Introduced following months of tedious negotiation, at a point which appeared to represent a complete impasse, this text was accepted almost immediately by all participants as the basis the final draft.

If an initiative of this kind had been taken prematurely by the chairman, the Bureau would have been compelled to establish some sort of an impartial arbitration procedure, in order to find a "new" compromise. As a consequence, the negotiations would have dragged on and the impasse would have remained unresolved.

The Role of the INC Secretariat

Disregarding the basic rule of right timing in multilateral negotiating tactics early on, the ad hoc Secretariat of the INC was prevented, due to the suspicion of developing countries, from playing the role it should have. At its first organizational meeting the ad hoc Secretariat presented the INC with a preliminary negotiating text, with the intention of facilitating and activating the negotiating process. Unfortunately, this commendable initiative was not appreciated because it was taken prematurely. In the minds of many delegations, this action disqualified the Secretariat as a leading influence in the negotiations.

Because the INC rejected this first initiative by the Secretariat in Chantilly, the Secretariat was forced to confine itself to the task of tallying the positions expressed during the remainder of the negotiating process. The very fact that the document to be adopted by the Committee would be legally binding on the signatory states contributed to the decision to reject this initiative.

In contrast, the Preparatory Committee (PrepCom) of the Rio de Janeiro Conference, assigned a leading role to the UNCED Secretariat, both through the action program outlined in Agenda 21 and in other documents prepared by the PrepCom. Indeed, the negotiation of Agenda 21 took place on the exclusive basis of the documents submitted by the Secretariat of the Conference. This occurred at the express request of many of the participating member states. Thus, in the case of the preparations for the UNCED meeting itself, the basic negotiating documents were presented—for the first time in the annals of North–South conferences—not by regional groups of states, but by the Conference Secretariat.

The Outline of a New Order

UNCED was the first major international conference to be held in the post-Cold War era. This meeting helped to shape new procedures for the conduct of future international meetings. The deliberations of the INC marked the eclipse of regional negotiating groups founded on ideological considerations. Such groups have historically set the tone for many UN negotiations. The implosion of the USSR has led to a permanent dissolution of the former Communist bloc, eliminating the group (but not the individual states) from the negotiating arena of the UN system. This transformation of the former Communist bloc has eliminated what was traditionally a strong ally of the Group of 77 developing countries (G-77). The two groups now have separate interests and the former Communist countries have become serious competitors with members of the G-77 for official development assistance. Many of the former Communist countries have also become formidable economic adversaries, due to the enthusiasm of the new converts to the market economy. Unfortunately, this heightened competition comes at a time when aid flows themselves are stagnant or perhaps even declining. This situation, compounded by the serious economic problems that face many developing countries, has subsequently led to a notable withdrawal of the G-77 from international negotiations.

Most developing countries (with the notable exception of the Alliance of Small Island States, the AOSIS group) were reluctant to become actively involved in the negotiations of the climate convention. This reticence was aggravated by the lack of concrete financial commitments from the industrialized countries to pay for the efforts taken by developing countries to reduce the risks of rapid climate change. The unwillingness on the part of the industrialized nations seemed directly to contravene the promises they had made to the developing countries at the Second World Climate Conference, promises for additional financial aid which were contained in UN Resolution 45/212.

It is significant to note in this regard that, over the course of the negotiations, the G-77 presented only one unified position paper. This paper broke no new ground; it simply restated the general considerations which have historically guided the actions of the G-77. Unfortunately, owing to contradictory positions and differing approaches by its member states, the Group of 77 was unable to form a consensus on the important issue of specific commitments to reduce greenhouse gas emissions.

Neither the G-77 nor the Movement of Non-Aligned Countries managed to convene a high-level coordination meeting in preparation for the Earth Summit. This task had been assigned explicitly at the fourteenth Ministerial Meeting of the G-77. Moreover, a parallel and complementary task was assigned at the Ninth Summit of the Non-Aligned Countries held in Belgrade in September 1989.

New Actors Emerge to Play Significant Roles in the Drama

This disorganized situation resulted in the emergence of a multitude of small groups of developing countries as actors on the negotiating scene. These ad hoc groups were founded and cemented by their shared interests.

The most important group of developing countries which emerged as an actor in the INC negotiations was unquestionably the Alliance of Small Island States (referred to as OASIS in French or AOSIS in English), chaired by Vanuatu. The importance of the OASIS group was quickly recognized by the INC leadership and Vanuatu was officially asked to join the INC Bureau. The representative of Vanuatu was named deputy-chair of the Working Group on Mechanisms, not in order to represent the Asian region but to reflect his position as a member of AOSIS.

This was followed by the emergence of the Kuala Lumpur Group, made up of forty-two developing countries. During the fourth INC session, the Kuala Lumpur Group presented an influential document concerning proposed national commitments under the Climate Convention. This timely move compensated somewhat for the inability of the G-77 to reach consensus on this critical issue.

Concerns of another kind provided the initial impetus for the mobilization of the Kuala Lumpur group within the INC. This group was composed of the large developing countries that hold important forest resources. Many of these countries viewed the negotiation of a Convention on Climate Change as an attempt by Western countries to dictate the use of their forest resources. They believed that they detected a maneuver by some participants in the negotiations to transform the provisions of the Convention on Climate Change into a legal instrument on forests. Throughout the UNCED preparatory process, these countries rejected such a move as a violation of their national sovereignty.[1]

Furthermore, the work of the Intergovernmental Negotiating Committee gave OPEC member states a chance to renew the tradition of holding periodic consultations in conjunction with major international events. This useful process had been suspended since the early 1980s. In addition to convening regular consultative meetings in conjunction with the sessions of the INC, the OPEC member countries held three separate seminars in Vienna to harmonize their positions in regard to the major provisions of the draft Convention on Climate Change. The reactivation of this tradition of consultation by OPEC member countries parallel to UN meetings was motivated by the concerns of the principal oil-exporting developing countries about the potential impact that the Convention on Climate Change might have on the future of the international oil market. Especially worrisome to these countries were the variety of proposals for an international tax on carbon emissions that would apply to all sales of petroleum products.

Thus, the disappearance of some and the withdrawal of other traditional protagonists led to the emergence of new coalitions of participants in these multilateral negotiations. The cohort of new players included many non-governmental organizations (NGOs). The participation of NGOs in the preparatory process for Rio '92, including within the INC, was indisputably an important factor in achieving our ultimate success. Indeed, history may yet mark the emergence of an effective NGO presence in international negotiations as a critical milestone on the path to international cooperation on environment and development. With the emergence of these new groups of actors, the United Nations and its subsidiary bodies will never again be the kind of private diplomatic club it once was.

Unresolved Issues

The environmental effects of unbridled development in industrialized countries are primarily responsible for upsetting the essential ecological balances of our planet. The lack of political will on the part of most of the industrial countries to control these negative impacts was clearly evident during the negotiations of what were referred to as the Specific Commitments section of the Convention. The same attitudes emerged during the discussions of "new and additional" financial resources and during the debate on the transfer of environmentally safe and secure technologies. The acrimony was most visible when the developing countries requested that any new and additional resources be provided under preferential and concessional terms.

These issues are all of vital significance to developing countries. The poor results on these questions left many participants with a feeling of a deep of frustration. This threatens to lead many of these countries to delay the process of ratification of the Convention.

Lessons Learned in the INC Process

Participating countries learned many important procedural lessons from the INC process. Because the composition of the Bureau of the INC was limited to only five members, representation was by regional groups. Although many delegations viewed the decision initially as unnecessary, the arrangement proved to be quite practical.

By contrast, the Bureau of the UNCED PrepCom was composed of forty-two members. This large, often unwieldy collection of delegates played no significant role, meeting only three times during the negotiating process. Their failure was due principally to unresolved differences on issues related more to form than to substance.

Faced with an unproductive Bureau, the Chairman of the UNCED

PrepCom took the initiative of setting up a parallel mechanism composed of the chairmen of the three negotiating groups and the coordinators of the different items on the agenda of the plenary. Each participant was appointed on the basis of their personal competence. This group, which met daily during the sessions of the PrepCom with the members of the Secretariat, ensured the necessary coordination of the work. Here again, however, considerations related to the delegate's desire for the prestige that accrued from their honorary membership of the Bureau took precedence over the principle of efficacy. I feel compelled to point out that during the negotiations on the composition of the Bureau of the PrepCom, the Algerian delegation steadfastly recommended that the UNCED Bureau be established in a way that reflected the traditional composition of such bodies within the United Nations. No other delegations maintained this position.

Given this situation, the Bureau of the UNCED PrepCom held no meetings during the inter-session periods. The Bureau of the INC on the other hand, held three enlarged meetings during its inter-sessions. The meetings of the Enlarged Bureau and, in particular, the Paris meeting in April 1992, played a decisive role in the success of the Intergovernmental Negotiating Committee.

The specificity of the process helped the participants to avoid repetition in the negotiation. Related issues could be grouped together and referred to one or another of the working groups. Many of the procedural agreements reached by the INC were later adopted in other UN fora. For example, the provisions of the voting process outlined in the Committee's rules of procedure and adopted at the first meeting in Chantilly, VA, were incorporated by the PrepCom into the draft rules of procedure for UNCED, which were adopted during the third session of the PrepCom.

Looking Ahead

The first Conference of the Parties (COP) to the Convention on Climate Change will continue the negotiating process and will have responsibility for following the course already traced. The COP should take account of the lessons learned from this experience as they make plans for the future process of negotiations. It should, early-on, set a deadline for finalization of the negotiations on the related legal instruments.

It would be desirable for such a timetable to coincide with the convening of meetings of a universal nature or of a major summit conference. The celebration of the fiftieth anniversary of the founding of the United Nations, which is certain to be a high point in UN history, could be one such occasion. The proposal submitted by the Tenth Conference to the Summit of Heads of State and Government of the Non-aligned Countries in September 1992, calling for, by 1997 at the latest, a second Earth Summit to assess the results of the Rio de Janeiro Conference, offers a similar target. Given the relatively long periods of time necessary for the preparation of the legal instruments related to the Convention, the COP should quickly set up negotiating groups equipped with a supporting bureau.

The Climate Convention will enter into force following its ratification by fifty countries. Ratification by developing countries may be delayed owing to cumbersome national administrative procedures and the lack of follow-up and coordination by the agencies of these governments. If such delays were to occur, it could lead to a flagrant imbalance in the composition of membership at the first COP. Such an imbalance in representation at the first COP would likely be to the advantage of the industrialized nations.

This imbalance at the outset would have repercussions on the composition of the subsidiary bodies of the Convention that are due to be set up at the first session of the COP. The result could seriously undermine the future work of the INC and the credibility of the Convention on Climate Change. But this outcome might still be avoided if the industrialized nations make the effort to fully accept their responsibilities, in particular as regards the granting of new and additional funds and the transfer of technology to developing countries.

With adoption of the Convention on Climate Change, the Intergovernmental Negotiating Committee laid the foundation for the emergence of a new field of law: international climate law. It also laid the groundwork for a new UN institution: The Conference of the Parties. The Conference of the Parties will negotiate all the related legal instruments necessary to achieve the objectives of the Convention, in particular, any required additional protocols.

It will fall to the first COP to establish a definitive mechanism to finance the activities envisaged by the Convention. This permanent financial mechanism should also be organized in such a way as to guarantee implementation of the obligations contracted by the industrialized nations. Only in this way can we give full effect to the principle of joint but differentiated responsibility of the states.

The first COP must avoid any superficial and simplistic attempts to hide the controversy sure to surround the choice of a permanent financial mechanism. Any attempt to institutionalize the provisional funding mechanism which maintains a dichotomy between the decisions of the COP and the financing criteria of the World Bank would inevitably lead to a disaffection by developing countries with this new institution. The result would be similar to the skeptical attitude that predominates in many developing countries with respect to certain United Nations bodies, especially its Economic and Social Council.

Epilogue

In the emerging world order, trade, aid, and environmental protection are inescapably linked to global economic relationships. The countries most likely to be negatively affected by rapid climate change include the low-lying island countries and those developing countries most vulnerable to desertification and drought. A differentiated ecological regulation that is more favorable to these developing countries should reinforce the differentiated and more favorable commercial treatment set up for them under Chapter Four of the General Agreement on Tariffs and Trade (GATT). Only in this way, will the future of humanity truly be "our common future," shared equitably by all.

We can only hope that this new field of international climate law will in turn give rise to a new ethic in North–South relations—an ethic based on the principles of equity. This new ethic will, and indeed must, recognize the ever-increasing interdependence of nations and the interwoven destiny of humanity. It is therefore important to reconcile man with his tools of production, reinvent his relationship with nature and rethink the relations of men and nations. To this effect, the exclusive and resolute objective of seeking short-term mercantile profit must give way to a reverent commitment to the Precautionary Principle, avoiding potentially large negative impacts in the future by erring on the side of caution rather than profit. A commitment to equity must be realized through a concern for the survival of the human species and this planet, incorporating a vision of the long-term interests of present and future generations. Our shared stake in the global commons can and must be viewed without any form of exclusivity or pretense of ownership.

Notes

1. This state of affairs explains the deep suspicions that surrounded the negotiations of the Convention on Climate Change. The proposal on forests submitted to the Summit of Heads of State in The Hague during March 1989 helped keep this suspicion alive. The Hague proposal called for the creation of an international authority endowed with the power to impose economic penalties. If implemented as originally conceived, this authority would have been set up outside the framework of the United Nations. A similar proposal was presented by the United Kingdom at the 1989 spring session of the Economic and Social Council of the United Nations. The 1989 UK proposal would have added the issue of environmental protection to the agenda of the Security Council of the United Nations. If the UK proposal had been adopted as proposed, national environmental protection policies would have become subject to the penalty provisions of Chapter Seven of the San Francisco Charter.

2. The recollections of the author were compiled into this chapter in the Fall of 1992.

CHAPTER 5

Constructive Damage to the Status Quo

Elizabeth Dowdeswell
Assistant Deputy Minister
Environment Canada
Ottawa, Canada

and

Richard J. Kinley
Atmospheric Environment Service
Environment Canada
Ottawa, Canada

The Framework Convention on Climate Change probably qualifies as the first global "sustainable development convention." It integrates the basic human quests for economic security and for a safe and healthy environment. In this chapter we hope to do four things. First, we want to review how the nature of the climate change problem, and of its solutions, influenced the negotiations and resulted in a "sustainable development convention." Second, the chapter will discuss how the process evolved over time—from science to negotiations to agreement—and attempt to draw some lessons from this experience. Third, we will review why we think the Convention's mechanisms and processes are significant. Fourth, and finally, we would like to highlight the importance of a "prompt start" to follow-up on the negotiations. We emphasize that the views expressed here are our personal views and are not to be considered as the positions of either the United Nations Environment Programme or the Government of Canada.

Climate change is a particularly challenging policy question. It is, first of all, a policy problem of unprecedented scope and complexity. The enhanced greenhouse effect is the unanticipated result of industrialization, land use and technological changes, modern lifestyles and our dependence on energy. In fact, someone has described it as the result of normal, not aberrant, behaviour. There is no particular villain, for we are all responsible. It is the product of innumerable decisions that are taken by individuals, by industry and business in virtually every sector, and by governments—every day, every hour, and all around the world. It is this scope that distinguishes it from most of the environmental problems we have faced to date. As a result, there is no single-sector solution, no simple "end of the pipe fix." Rather, the solution lies in a fundamental shift in the composition of economic activity—a shift to sustainability in both the North and the South.

In Canada, we have thought of sustainability as integrating environment and economics in government, business and individuals' decision-making. This is clearly valid for climate change where continued growth—albeit of a different sort than in the past—is essential to finance the adjustments that are required. This is, however, doubly true on a global scale. The dominant dynamic throughout the climate change negotiations has been the tension between pressures for environmental protection and forces promoting economic development. Any successful resolution of the problem of global climate change must respect the development imperative of the South. All this argues for the proposition that the prospect of global climate change demands a fundamental rethinking of the way we approach economic development in both the North and the South.

The second reason why climate change as a policy question, and the negotiations of the Framework Convention on Climate Change in particular, have been challenging is that decisions had to be made, and must continue to be made, in the face of persistent uncertainty. This is an area where we simply have to take some actions now— even before we find "the smoking gun" that would identify the principal culprit. We know that the uncertainties exist and that they are significant, contentious and slow to be reduced. Few of our climate modelers, for example, expect to be able to improve the precision of their temperature and rainfall forecasts within the next decade. We know that it will be even longer before we can predict things like regional variations and a very long time before we can predict precise local impacts with any degree of confidence. But, with the best advice of the world's leading scientists and the knowledge that current trends are unsustainable, it is clear that some precautionary actions are warranted. The debate has turned on how much must be done now and by whom.

The third characteristic that defines climate change is global inter-dependence—both environmental and economic. In the early 1960s, we considered environmental problems to be purely local in nature—as are the effects of urban air pollution or sewage discharges. In the 1970s, we came to understand that some problems can be continen-tal in scale, with the impacts felt thousands of kilometers away—as are the effects of acid rain. In the late 1980s, we have entered the age of *global* environmental consciousness—scientists know that CFCs released in Europe affect the ozone layer over Antarctica, that a car driven to work in Toronto or a forest cleared in the tropics affects the interests of all of us living today and of many future generations as well. Just as all countries contribute, or will contribute, to the prob-lem of climate change, all countries will suffer the consequences of inaction.

The actions that must be taken to address climate change will have important economic consequences. Some will improve efficiency and competitiveness while others will have net costs. If there is to be a concerted response to the problem, as there must be, it is preferable that countries act together cooperatively and thus minimize the potential for the economic and social distortions that could follow unilateral actions.

Because of this global interdependence, both environmental and economic, it was, and remains, crucial that as many countries as possible be involved in the solution to the climate problem. This applies to large and small emitters as well as to current and future emitters. Up to 140 countries participated in the INC and almost 160 countries have now signed the Convention. This experience is encouraging—and indeed unprecedented for an "environmental" agreement.

In an interesting twist on the global character of the climate prob-lem, we saw throughout the negotiations, at least in the corridors, that there was often a greater affinity of interests between represen-tatives from similar departments in different governments than between representatives of different departments on the same del-egation. Thus, departments of the environment, or of energy, or of finance (at least from developed countries) tended to have similar perspectives that crossed traditional national borders. These diver-gent departmental views were consolidated within delegations in different ways, depending on the relative influence of the depart-mental actors involved.

The Evolution of the Process

In the case of the ozone depletion problem, it took over a decade to move from the emergence of a scientific concern to the conclusion of a meaningful international agreement. Despite the fact that climate change is a more complex issue, we were able to agree on a Framework Convention in about five years. No doubt, some will say that we had no other choice.

A Brief Chronology of Events

The chronology of the climate issue is instructive, and may provide parallels with other issues. A clear scientific concern emerged from meetings at Villach, Austria, in 1985 and in Villach and Bellagio, Italy, in 1987. The June 1988 Toronto Conference on the Changing Atmosphere sought to bridge the science–policy gap by bringing this concern to the attention of decision-makers. The Toronto Conference, combined with the heatwave of 1988, sparked an explosion of public interest in the climate issue. The political pressure that emerged in developed countries in 1988 continues to this day. The next step was to establish more firmly the basis of scientific concern and to assess possible impacts and policy responses from a technical perspective. Consequently, the World Meteorological Organization and the United Nations Environment Programme established the Intergovernmental Panel on Climate Change in late 1988. In 1989, the Ottawa meeting of legal and policy experts began consideration of the elements of a possible future convention.

Climate Change Moves Up on the International Political Agenda

In the summer of 1989, the political agenda began to move forward. At the July 1989 Paris Economic Summit, the leaders of the Group of Seven called for the development of a climate change convention. In November, ministers from about eighty countries, meeting at Noordwijk in the Netherlands, put the stabilization of greenhouse gas emissions on the international policy agenda. The discussions resumed and the benchmarks were moved forward slightly at the May 1990 Bergen Ministerial Conference.

In August 1990, the IPCC issued its first Scientific Assessment Report confirming what we did and did not know about climate change. This Assessment, with only slight adjustments in early 1992, reflected the scientific consensus on which the negotiations have proceeded to the present day.

In November 1990, the Second World Climate Conference provided the political and policy backdrop, and the United Nations General Assembly provided the legal mandate, for the negotiation of a "framework convention" containing "appropriate commitments." The UN Conference on Environment and Development (UNCED, sometimes called Rio '92 or the Earth Summit) provided a climax and a deadline for the negotiators—June 1992 in Rio de Janeiro, Brazil.

While some would argue that we had entered a negotiating mode some months earlier, the negotiations only began officially in February 1991. The usual preparatory period was dispensed with, resulting in a rather slow and difficult start. The first session, held in Chantilly, VA, caused even the most optimistic observers to despair. Initial text for the Convention did not begin to emerge until August 1991 and there was no working document even vaguely resembling a convention until December of that year. Nonetheless, and despite the complexities discussed above, an agreement was concluded in the space of fifteen months. This was an international negotiation conducted at unprecedented speed; it must surely go down in history as a minor miracle.

Over this five-year period, climate change became a compelling issue that caught the attention of scientists, policy-makers, environmentalists, industrialists, world leaders and the general public. The growing spotlight illuminated the process and caused the international debate to evolve. At the start, it was purely a scientific exercise with very few countries involved. As the IPCC finalized its first scientific assessment, the dynamics changed abruptly. Countries realized that climate change was not simply an issue of meteorology; it was an issue of energy and economic policy. Things were never again the same in the climate debate.

At the same time, developing countries strengthened their call for equality of membership in the international process. They thought they had been shut out of the earlier phase of the debate. The formal international process had become a dialogue among scientists, run largely by representatives from the North. It became increasingly clear that the climate issue was a priority for the North, but of less urgency in the South, where governments were driven to address more urgent problems of development and basic human needs. Developing country representatives sensed that climate change was an issue that could be used in the long-standing North–South debate. Pictures of winners and losers started to emerge from the IPCC Assessment. It became evident that some national delegations could not distinguish between their own country's interpretations and an objective, technical analysis of the global problem. For these countries, it was impossible to separate the self-interest of their own country's initial negotiating positions from an informed sense of the common good.

It was clearly time to move into formal negotiations. When the United Nations entered the scene, the full weight of that system was brought to bear. But at this point, the momentum of the IPCC process, which had brought the issue to the "take-off point," was very nearly lost. Responsibilities shifted from specialized agencies to political organizations, from scientists to professional diplomats. In the process, the rules of the game changed dramatically.

From the first negotiating session, all delegations recognized that the "stakes" in the climate change negotiations were very high. This raised the level of political interest and politicized the process. Compared with other "environmental" negotiations, participation by economic and other interests was much higher. This meant that the environment and development discussions that were evolving in the context of preparations for the Earth Summit spilled over into the climate change negotiations. The Intergovernmental Negotiating Committee (INC) for a Framework Convention on Climate Change (FCCC) sometimes found itself out in front on these issues, forced to move forward by a tighter deadline and the need to reach a legally binding agreement.

As with any negotiations, there were frustrations: countless hours of discussion on process and rules of procedure; endless days of country position statements; a formal prohibition on inter-sessional work; undue influence from certain capitals (and not enough from others); the additional complexity of dealing with the international media and with the political agendas of dozens of non-governmental organizations. Nonetheless, an agreement emerged at the conclusion of the fifth INC negotiating session which, while not perfect, was the best that could be expected at that moment in time.

Five Key Factors Contributed to INC's Success

Many factors contributed to this result, of which five are worthy of mention here.

Chairman Ripert Prepares an Integrated Text

First, the text prepared by the chairman for the last session was of supreme importance. Its appearance and its quality enabled the negotiations to conclude successfully. Drawing on the months of negotiations, the chairman's text described a consensus position which countries had been unable to negotiate. The magic compromise contained enough for those who wanted "more" but not too much for those who wanted "less." It therefore could serve as the basis for the concluding session, and the first real "negotiations."

The Structure of the Process

Second, the structure of the negotiating process contributed significantly to the outcome. The fact that over 100 countries were attempting to negotiate a complex international agreement in plenary, without sufficient preparatory negotiations, meant that the process was slow. Furthermore, as all the issues were interlinked, it was virtually impossible to proceed in a step-by-step fashion. Thus, we entered the second half of the fifth session in May 1992 with virtually none of the principal issues resolved. This meant that the only possible solution, given the looming UNCED deadline, was for the chairman to issue a text which could not be altered in any significant way and for key issues to be negotiated in private by a small group of representative countries.

The lesson from this experience is that the traditional United Nations model of active participation by all states, in an international community that is approaching 200 members, may no longer be the best way to develop international law. Just as one *cannot* negotiate in a forum of over 120 countries, one *should not* negotiate in a forum of thirty hand-picked men and women. Some middle ground is required—although it will clearly be a challenge to break this new ground. For example, some system of representative constituency groupings might be appropriate so that all countries would in fact be *participating*, although not necessarily directly, while the number of interlocutors would be more manageable. There are many existing groupings that could serve as the basis for such a system (e.g., EC, CANZ, EFTA, AOSIS, ASEAN, etc.).

There continues to be an important role for a vigorous Bureau—both in the negotiations and in the Conference of the Parties. An eleven-member Bureau, including the officers of the working groups, is sufficiently representative to take the steps necessary to promote serious negotiations. But for this approach to be successful, the Bureau must display a willingness to take responsibility for advancing proposals that will facilitate the negotiations.

NGO Participation

Third, not only were unprecedented numbers of countries involved, but there was unprecedented participation by non-governmental organizations (NGOs) and, especially, by business and industrial interests. (See Chapter 11 and Chapter 12, eds.) The climate change negotiations were in the forefront of opening UN processes to participation by non-governmental representatives. Virtually all of the sessions, including those of the working groups, were open. Several countries (e.g., Australia and Canada) included environmental and business representatives as members of their delegations. And the most widely read source of information on the negotiations was *ECO*, the daily publication of the environmental non-governmental organizations. This was an important element of the negotiations—it meant that the non-government sector was fully informed of developments, had a better understanding of the negotiation process, had an opportunity to interact with delegations, and could serve as a mechanism to assure accountability and a source of public pressure. One can only hope that the openness and transparency demonstrated in the INC have set a precedent for future international negotiations on environmental issues.

While all of these factors—political interest, economic implications, breadth of participation, number of countries—made, and will continue to make, the negotiations more difficult, the breadth of involvement also means that any deal will be more likely to be effective. With so many actors involved from so many sectors and countries, and with many of those at senior levels, there should be a high level of confidence in, and a strong commitment to the result. This bodes well for the future and is an important benchmark for future negotiations.

The Importance of the Science

Fourth, the INC process and the resulting Framework Convention demonstrate the importance of having a credible scientific base for international negotiations. Fora like the INC should not negotiate on the science but they should be informed by it. The efforts of the IPCC to develop its 1990 Scientific Assessment Report, and the 1992 supplement thereto, were fundamental to the success of the negotiations. An independent technical assessment which draws on the views of the top scientific experts in each field is an important precursor to any international convention designed to address a global environmental problem. The IPCC process also demonstrated the need to involve specialists from developing countries in the scientific assessment from the outset. Future negotiations demand that greater efforts be made to build the scientific and research capacities in developing countries so that they can be full players in both the scientific and the diplomatic phases of any future international negotiations.

Innovation is Essential

Fifth, the INC process demonstrates the importance of *not* being bound by precedent—in terms of either process or substance. Just because something has always been done a certain way, or because something has never been done before, should not preclude change. No one doubts that there are problems with the current international system. That is why the international community is striving to design a new one—one that is sustainable. This demands innovation, creativity, vision and new ways of doing business.

Agents of Change: The Convention's Mechanisms and Processes

The Convention's institutions, and the processes that underlie them, may well emerge as its strongest legacy. They will have the most lasting effect because, in the absence of targets and schedules, it is the mechanisms and processes that give vigor to the Convention. They will lead (or perhaps push) the international community to more rigorous commitments in the years ahead.

Working Group II was the scene of some of the most interesting, provocative and innovative discussions during the negotiations, particularly in the early days when the focus was on concepts rather than specific wording. Although given a mandate to develop a text relatively early in the process, the co-chairs adopted a strategy of incrementalism. When there was similarity between the various informal texts (often called "non-papers") that had been presented by delegations, it was possible to draft consensus positions. However, there were some significant areas that were not addressed by the non-papers or where original and ground-breaking work was necessary. Consequently, the co-chairs' text produced for the Nairobi session was a combination of draft legal text and questions from the co-chairs. These questions were designed to elicit the views of delegations and thereby, it was hoped, lead to a subsequent consensus on additional text.

This approach worked well. In addition to provoking some stimulating debate, it enabled the co-chairs to produce a coherent, virtually unbracketed text on mechanisms in time for the fourth negotiating session in December 1991. The text described an array of Convention institutions and processes which were activist in nature and which, in some instances, broke new international legal ground. Thus, Working Group II considered a text during the December 1991 session that was virtually a fully elaborated institutional scheme. Unfortunately, by contrast, in Working Group I, there was a continuing deadlock on "commitments." Delegations, when confronted with this deadlock, put the brakes on. Given the linkages and the required tradeoffs among issues—including developed country commitments, funding obligations, the scope, guidelines, and role of the financial mechanism, as well as the requirements for national reporting and review—this should not have come as a surprise. It was, nonetheless, rather disheartening. While progress was made on a number of institutional elements in the subsequent negotiations, the key issues were stalemated until the chairman released his final text.

The Subsidiary Bodies for Implementation and for Scientific and Technological Advice

The final Convention provisions are, however, a rather remarkable achievement. They frame the Convention in terms that are both flexible and adaptable. Such flexibility will be necessary to successfully address the climate change problem in the years ahead. The key provisions are those that require the Conference of the Parties:
(1) to review progress in climate science,
(2) to evaluate the implementation of the Convention, and
(3) to examine periodically the national obligations and the institutional arrangements codified in the Convention.
The Convention provides the ability and the incentive to strengthen national obligations in the future. We note that similar provisions for periodic review that were included in the Montreal Protocol have resulted in accelerating the phaseout of CFCs at each available opportunity.

Driving this ongoing re-evaluation process will be new scientific knowledge and updated reports on the status of national programs to implement the Convention. The implementation process will be documented by the national "communications" (i.e., the reports) that are required under the Convention.

With regard to science, the Convention's Subsidiary Body for Scientific and Technological Advice (SBSTA) must ensure that the Conference of the Parties (COP) has the information necessary to carry out its mandate. While it is clear that the SBSTA will have to work closely with the IPCC, and that the IPCC will conduct scientific assessments, the exact division of responsibilities between the two bodies is not yet clear.

Their respective roles will have to be defined in the period leading up to the first session of the COP. It may be possible that the establishment of the SBSTA, as the interface between the COP and the IPCC, may allow the IPCC to avoid becoming any more politicized than it already is and, thus, to carry out its "technical" role more effectively. This is not to say that the SBSTA should be overly political. However, as a Convention organ it is likely to have a somewhat more political character than the IPCC has had. As the functions and mandate of the GEF's Scientific and Technical Advisory Panel (STAP) evolve, its role and mandate will have to be coordinated with the roles of the IPCC and SBSTA in order to ensure that there is no duplication of effort.

The regular scientific assessments that are required under the Convention need not be too frequent. To be so would place an undue burden on the international scientific community. Moreover, it is unlikely that new scientific data will emerge with breathtaking speed. In due course, the scientific assessments could be coordinated with the cycle of future negotiations. It is important to recognize, however, that the scientific assessment and the negotiation processes need to be kept separate. Science must not be politicized, lest the quality of the intellectual enterprise suffer irreparable damage.

The second element in this re-evaluation equation is the information communicated by the Parties and its consideration by the institutions of the Convention. The dynamics of the negotiations made it impossible to elaborate fully on the working details of a "reporting and review" process—even the words describing the process proved too controversial. The reporting system and the Subsidiary Body for Implementation (SBI) that were outlined in the Convention have the potential to act as agents of change, but only if the Parties allow it. The goals of the reporting and review process are:
(1) to promote public and international accountability,
(2) to facilitate the sharing of practical experiences, and
(3) to build new and more complete databases.
The process should be facilitative, open and transparent and should be coordinated with other similar international reporting and planning exercises to avoid duplication.

The Subsidiary Body for Implementation will require considerable attention to ensure that its potential is not squandered. Because it is an open-ended body that will have to digest vast amounts of information, both from the Parties and from other sources, it will require the support of a strong and expert Secretariat which possesses a solid analytic capacity and has access to adequate resources. If the SBI is to be effective, strong leadership from its Bureau will also be important because the Bureau has "political legitimacy" that the Secretariat can never have.

Promoting Innovative Ideas in the INC

In Working Group II, we strove to promote innovative ideas. Three particularly important ideas related to implementation of the Convention are presented below. In some cases, it proved impossible to agree on anything beyond a concept. Our objective was then to ensure that no options were prematurely closed.

The first example relates to the financial mechanism. The designation of the Global Environment Facility (GEF) as the interim operator of the financial mechanism was the last issue to be resolved in the negotiations. The strength of feeling among key donor countries on this issue was crucial in determining the outcome. In a spirit of compromise, they were prepared to accept the "interim" designation, but it was clear that they would not accept a Convention-specific funding arrangement. Similarly, developing countries would not accept the GEF as the permanent financial mechanism. Thus, some compromise on the financial mechanism was a prerequisite to successful conclusion of the negotiations.

The Convention set an important precedent in entrusting the funding responsibility to an institution somewhat separate from the Convention and the United Nations. This arrangement, if it works as envisioned, will marry the GEF's strengths of efficiency, higher resource flows and cross-issue linkages with the COP's expertise on climate change policy and its commitment to action. It is an innovative approach which will present many challenges because, in many cases, entirely new ground will need to be broken. Developing the required operational "arrangements" between the Convention and the GEF should be one of the highest priorities for the interim period.

In looking ahead to the arrangements between the Convention and the GEF, it obviously will be crucial that the GEF restructure itself to meet the political requirements of the Convention. This should be achieved within a reasonably short timeframe. Getting agreement on the funding that must be "necessary and available" for the implementation of the Convention will be the most challenging issue. The INC should expand and improve its linkages with the

GEF so that the expected replenishment of the GEF takes the Convention fully into account. Looking further down the road, it might be appropriate if the negotiations on GEF replenishment and the periodic renegotiation of the Convention (or the negotiation of any necessary additional protocols) were to proceed in parallel. In this way the level of developing country commitments and the amount of funds available to support their implementation could be coordinated.

The second example of innovative ideas to emerge from the discussion of mechanisms was the concept of "joint implementation." The notion that one country could participate in emissions reducing actions taken in another country may be logical given the special characteristics of the climate problem. Climate change is a global problem which cannot be resolved by phasing out a particular product or activity, but can be resolved only by altering the pattern of economic and social development. By most standards, cost-effective emissions reductions may be cheaper to achieve in one country than in another. Why not let two countries share the costs and the credits for these cost-effective investments? This is a rather revolutionary concept—but new ways of thinking are needed if we are to resolve the climate change problem.

Joint implementation is about choosing the most cost-effective opportunities to reduce greenhouse gas emissions, regardless of where they occur. In a time of limited resources, efficient use of scarce capital must be a high priority. Joint implementation cannot be about shifting responsibilities or commitments. If joint implementation is to succeed, it must have a private sector orientation. It cannot be just another official development assistance flow. Joint implementation therefore introduces the possibility of tapping a new source of revenue and technology namely, the private sector.

Much work remains to be done to elaborate the concept of joint implementation and to flesh out the operational details, including who can participate. However, the willingness to consider new approaches is an encouraging hallmark of the climate change negotiations. It would be useful if some countries could move ahead and demonstrate how a regime of joint implementation could work in practice. In the absence of internationally agreed terms and conditions, and in the absence of binding targets and schedules, any "credit" earned would be political rather than legal. In the process of learning from experience, however, quantities of physical emissions could be reduced and technologies could be transferred.

The third example relates to the issue referred to as "resolution of questions." This initiative, which came from the co-chairs, provoked both strong support and high anxiety among delegations. It was designed to buttress the idea that the Convention should be a facilitative rather than a punitive document. Traditional, bilateral dispute settlement procedures are highly legalistic and confrontational. To rely on these procedures would have been insufficient and inappropriate for a complex global problem like climate change. The concept of resolution of questions was conceived as a way to avoid the traditional sort of confrontation. A Party that had a concern about the implementation efforts of another Party, or about how the Convention was being interpreted, could bring that concern to the Conference of the Parties. The COP would review the issue and possibly establish a facilitative rather than a judicial body to find a solution. The Convention leaves the door open to further consideration of how such a process might be carried out.

It would be difficult to conclude a discussion of "innovative processes" without mentioning "pledge and review." In the end, this concept amounted to little more than a diversion on the way to our final conclusions. It stands as living testament to the dangers of launching ill-defined and ambiguous ideas at inappropriate times in a negotiation. Designed as a mechanism to avoid targets and timetables, it had no legitimacy in the debates and consequently did not survive the third session. When, in the final days of the negotiations, it was obvious that agreement on legally binding targets and timetables would prove elusive, another concept had to be found. What emerged was the vaguely formulated goal now included in Article 4.

Prompt Start

As the negotiations approached their final phase, the prospects for finalizing all of the detailed provisions of the Convention in the time remaining grew slim indeed. The problem was how to ensure that the momentum generated by the negotiations, and the ideas raised through the five sessions, would not be lost until the first Conference of the Parties. A number of countries and non-governmental organizations, particularly a group of academics at the Massachusetts Institute of Technology, promoted a new concept, which became known as "prompt start." "Prompt start" was used to describe the process for making interim arrangements that could be implemented between the date of the signing of the Convention and its legal entry into force. It was picked up by the Bureau of the INC and considered at an informal workshop hosted by the Rockefeller

Foundation in Bellagio, Italy in January 1992.

At the final negotiating session, an informal group was established to give life to the "prompt start" idea. The discussions proved unexpectedly difficult but were successfully concluded in the INC's resolution on "Interim Arrangements." While the name "prompt start," and the detail of what such a process might entail, did not survive the final session, the crucial concept of completing unfinished business in preparation for the first session of the Conference of the Parties was embraced (with more enthusiasm by some than by others), enabling the negotiating process to continue.

"Prompt start" encompasses one more crucial element of successful negotiations—a *deadline*. It can only be hoped that the pressure to prepare for the first COP will be as "inspirational" as was the pressure to complete the negotiations before UNCED.

Although it is not without precedent in recent international environmental negotiations, the prompt start concept had its own special characteristics in the climate negotiations. While the original purpose was to ensure that countries kept talking to one another, the prompt start approach evolved to support other objectives as well. Related aspects of the prompt start included the promotion of early ratification and early reporting of national programs, moves designed to encourage the GEF to fund relevant activities without waiting for the Convention to enter into force, and the development of an effective interim Secretariat. Many countries have taken steps to contribute to this "prompt start" both domestically and internationally. The ultimate success of the concept, however, can be measured only when the Conference of the Parties meets for the first time. The criteria for measuring success include the speed of entry into force, the decisions the COP will be able to take based on the preparatory work already completed, and the funding provided by the GEF.

Epilogue

The Framework Convention on Climate Change is a pragmatic first step toward a new vision of international cooperation on global environmental problems. Its objective is to promote economic development while minimizing the irreversible long-term damage to the global environment. It is oriented toward action—action to reduce dangerous emissions, to retain and enhance natural sinks, to transfer financial resources and technology, and to promote cooperation in science and education. But all of this is founded on a solid institutional base that will enable the Convention to evolve as our knowledge improves.

We owe it to ourselves and to future generations to make the Convention work! The early signs are most heartening: political commitments at the highest level, an unprecedented number of signatures, ratification processes in full swing, high public interest, and national reports already in preparation.

The 140 or so governments that participated in the climate change negotiations were engaged in a collective effort for change. We had varying priorities but were united in a common commitment to find an instrument that would work to meet our many needs. We were trying hard to change the shape of environmental law—and it appears that we succeeded. But the work we began in the last two years must now be completed.

CHAPTER 6

The Climate Change Negotiations

Chandrashekhar Dasgupta

Ministry of External Affairs
New Delhi, India

and

Indian Ambassador to China
Beijing, China

Prologue

Few issues have acquired priority in the international agenda in as short a time as global warming. Up to the early 1970s many scientists believed that the Earth might be moving towards a new ice age and that increasing levels of emissions of greenhouse gases (GHGs) from human activities would not alter this course and might even be helpful in slowing down this movement.[1] It was only towards the close of the decade that scientific opinion tentatively endorsed the view that on average global temperatures might be rising. The first World Climate Conference, organized by the World Meteorological Organization (WMO) in Geneva in 1979, cautiously concluded that:

It can be said with some confidence that the burning of fossil fuels, deforestation and changes of land use have increased the amount of carbon dioxide in the atmosphere by about 15 per cent during the last century and that it is at present increasing by about 0.4 per cent per year . . . it appears plausible that an increased amount of

129

carbon dioxide in the atmosphere can contribute to a gradual warming of the lower atmosphere, especially at high latitudes.[2]

Despite continuing uncertainties in many of its aspects, this hypothesis made rapid headway in scientific circles in the 1980s. Workshops held in Villach (Austria) in October 1985 and in October 1987, and in Bellagio (Italy) in November 1987 gave impetus to this development. Apart from accumulating further evidence, these workshops focused attention on the need for a political response to the problem and for a global convention to combat climate change.

The transition to the international political agenda took place in June 1988 when, at the initiative of Canada, a conference of scientists and policy-makers was held in Toronto. The Toronto Conference called for an Action Plan for the Protection of the Atmosphere, including an international Framework Convention. This was followed in rapid succession by the World Congress on Climate and Development held in Hamburg (November 1988), the Hague Conference organized by France, Netherlands and Norway (March 1989) and a Ministerial Conference on Atmospheric Pollution and Climatic Change (Noordwijk, the Netherlands, November 1989). By 1989 the question of global warming had achieved a degree of prominence which assured it attention in international summit meetings such as those of the Group of Seven (Paris, July), the Non-Aligned Movement (Belgrade, September) and the Commonwealth (Kuala Lumpur, October).

Meanwhile, at the instance of Malta, climate change was also brought on to the agenda of the UN General Assembly. In December 1988 the General Assembly adopted a resolution in support of the climate-related activities of the United Nations Environment Programme (UNEP) and WMO and endorsed their joint establishment of the Intergovernmental Panel on Climate Change (IPCC).[3] The findings of the IPCC in turn provided further momentum for consideration of the question at the political level. Influenced by the first Assessment Reports of the IPCC, the Second World Climate Conference (held in Geneva in November, 1990) called for early negotiations on climate change. Finally, in December 1990, the UN General Assembly established the Intergovernmental Negotiating Committee on Climate Change.[4]

Thus, within barely a decade, the conventional wisdom in the scientific community had shifted from speculation about the planet's gradual movement towards a new ice age to concern about unacceptable levels of global warming in the near future, albeit qualified by a recognition of continuing uncertainties over several aspects of the phenomenon. The depth of scientific concern had prompted a political decision to commence multilateral negotiations on a Climate Change Convention.

Because of the nature of global warming, a convention to contain the phenomenon could have wide-ranging economic implications. Anthropogenic emissions of the main "greenhouse gas," carbon dioxide, arise from consumption of coal and other fossil fuels. A climate change convention would therefore necessarily affect levels and patterns of energy use. It would have implications for energy generation, transmission and utilization, for industry and transportation. Moreover, since forests act as carbon dioxide sinks, a convention would also have a major impact on land-use patterns. Efforts to control emissions of other greenhouse gases would also have similar implications, though on a less extensive scale. Methane emissions, for example, are associated with agriculture and animal husbandry. In short, a climate convention could constitute a major multilateral economic agreement. The sharing of costs and benefits implied in the convention could significantly alter the economic destinies of individual countries.

It is remarkable that a framework convention with such vast potential implications was concluded in a mere fifteen months. The first session of the Intergovernmental Negotiating Committee (INC) was held in Washington in February 1991 and the last session in New York concluded in May 1992. The negotiations were driven by a political decision to conclude a convention in time for the UN Conference on Environment and Development (UNCED) Summit in Rio de Janeiro (June 1992). By way of comparison, the Law of the Sea negotiations covered a full decade; although the economic implications of the climate change convention are arguably more far-reaching, at least in their immediate impact, the negotiations of the climate convention were concluded in less than two years.

As we shall see, this tight timeframe influenced the nature of the INC negotiating process and, eventually, its outcome. In certain respects, the negotiations were not truly completed; the convention itself specifically deferred resolution of some important questions to the first meeting of the Conference of Parties.

The INC Takes Shape

The first session of the INC was held in Chantilly, near Washington, DC, on 4–14 February 1991. The main organizational questions were resolved in this session. A Bureau was elected with the distinguished former international civil servant, M. Jean Ripert of France, as its chairman. Two working groups were established to deal respectively with "commitments" and "mechanisms," and Rules of Procedure were drawn up reflecting a balance between the requirements of consensus and majority voting.

Consensus procedures in large multilateral conferences tend in practice to confer something approaching a veto only on the most influential states; procedures based on voting have obvious attractions for the less influential, where they are in a majority. Not surprisingly, therefore, many of the developed countries urged that a convention would have little meaning unless it were to be adopted by consensus, in contrast to the Group of 77 which argued that, while the best endeavor should be made for a consensus, deviating from standard UN General Assembly procedures was undesirable.

It was finally agreed that the Committee should "make its best endeavor" to ensure general agreement but if, on any substantive issue, "no agreement appears to be attainable" despite the best efforts of the Committee and its chairman, the "Committee shall decide on the steps to be taken," in accordance with the rules of procedure of the UN General Assembly (i.e., voting).[5] In the event, as negotiations unfolded in this and subsequent sessions, the chairman was able, often through strenuous efforts, to ensure consensual decisions. There was not a single instance of a substantive decision taken on the basis of a vote. The possibility of recourse to the voting procedure itself tended to act as a restraint on influential parties and this obviated the need for actually employing the procedure.

In establishing the INC, the UN General Assembly had specified that participation would be open not only to UN members but also members of UN Specialized Agencies. This paved the way for inclusion of a number of small island states which would have otherwise been unable to take part in the negotiating process. The decision had a major impact on the negotiations. An Alliance of Small Island States (AOSIS) was created on the basis of common concern over the consequences of sea-level rise induced by global warming. Represented in many cases by able British and American lawyers or academicians, AOSIS members vigorously argued the case for a strong convention.

The General Assembly had also provided for non-governmental organization (NGO) participation. As in the parallel case of the UNCED Preparatory Committee, this provision was generously interpreted to enable participation not only of NGOs in consultative status with ECOSOC but also a wide range of other bodies, including groups representing powerful industrial and commercial interests.

At the first session itself, as many as seventy-seven NGOs were represented.[6] NGO representatives were permitted to participate in debates and were only excluded from informal negotiating sessions. They significantly affected the course of negotiations through their powerful advocacy role both within and outside of the conference halls.

Initial Negotiating Positions Were Far Apart

At the Washington session itself, a number of delegations had outlined their views on the contents of a framework convention. The second session in Geneva saw a further elaboration of positions on substantive issues and the initiation of detailed discussions. By June 22, delegations had submitted as many as twenty-five written papers containing substantive proposals or comments.[7] These included the full text of a possible draft framework convention, circulated as a "non-paper" by India.[8] These papers, and the initial discussions, showed a wide divergence in approach between developed and developing countries as well as significant differences within each of these groups.

In general, developing countries maintained that since developed countries were mainly responsible for causing climate change, it should also be their responsibility to take measures for a solution. Thus, the Chinese non-paper included in the preamble, a paragraph stating that the "emission of greenhouse gases affecting the atmosphere has hitherto originated mainly from developed countries, which should therefore have the main responsibility in addressing the problem." The Indian delegate elaborated the following approach:

> The problem of global warming is caused not by emissions of greenhouse gases as such but by *excessive* levels of per capita emissions of these gases. If per capita emissions of all countries had been on the same levels as that of the developing countries, the world would not today have faced the threat of global warming. It follows, therefore, that developed countries with high per capita emission levels of greenhouse gases are responsible for incremental global warming.

> In these negotiations, the principle of equity should be the touchstone for judging any proposal. Those responsible for environmental degradation should also be responsible for taking corrective measures. Since developed countries with high per capita emissions of greenhouse gases are responsible for incremental global warming, it follows that they have a corresponding obligation to take

corrective action. Moreover, these are also the countries which have the greatest capacity to bear the burden. It is they who posses the financial resources and the technology needed for corrective action. This further reinforces their obligations regarding corrective action.[9]

In contrast, the United States took the view that participation of countries in the international response to climate change should be "in accordance with the means at their disposal and their capabilities,"[10] without any reference to the extent of their responsibility for contributing to global warming. In general, developed countries tended either to ignore or to de-emphasize the link between responsibility for contributing to climate change and responsibility for taking corrective action. Germany, at least, was willing to face the issue: a German "non-paper" recognized that developed countries have a special responsibility and obligation to take appropriate measures and strategies "since these countries have been the main sources of the increase in atmospheric concentrations of climate-relevant gases." However, the document called upon developing countries also to accept commitments since "it is only with their broad participation that the global challenge can be met effectively."[11]

These differences were reflected in divergent views on the commitments which countries should accept in regard to emissions and sinks. In general, developed countries maintained that all signatories should accept certain binding commitments, while acknowledging their need to "assist" developing countries to fulfill their commitments. They were also prepared to accept "differentiated" commitments, although there were wide differences between them on specifics. Many West European countries were ready to accept time-bound stabilization targets. Thus, the Netherlands recalled the European Community (EC) position that "EC Member States and other industrialized countries should take urgent action to stabilize or reduce their carbon dioxide and other GHG emissions. Stabilization of carbon dioxide emissions should in general be achieved by the year 2000 at 1990 level."[12] Norway proposed that OECD countries should stabilize GHGs emissions not covered by the Montreal Protocol at 1989 levels by the year 2000.[13] France signalled its readiness to limit emissions below an absolute level of 2 tonnes of carbon equivalent per capita by the end of the decade, provided other industrialized countries accepted a similar obligation.[14]

The United States, which had signalled its willingness to accept a stabilization target in the first session, changed its stand in the next session. It now maintained that "specific commitments for emissions reductions should not be included in the framework convention, because of the need for flexibility in nations' choices of their own measures."[15] Britain took a midway position between the EC target and the US by proposing stabilization "as soon as possible."[16]

Sweden proposed that all countries should make the best effort to limit greenhouse gas emissions on the basis of best available technology and practices. A series of specific energy efficiency measures were prescribed for countries grouped into categories on the basis of their ability to take on commitments (apparently without reference to the extent of their contribution to causing climate change).[17]

Japan proposed a simple "pledge and review" arrangement under which each country "makes public a pledge, consisting of its past performance strategies" to limit greenhouse gas emissions and targets or estimates for such emissions as the result of the strategies; these would then be subject to periodic reviews by experts and a review committee.[18]

A common feature of these approaches was that they did not directly link commitments with responsibility for inducing climate change. Rather these proposals sought binding commitments even from countries which were potential victims of a phenomenon induced by others.

A widely held view among developing countries was that the Convention should not impose binding obligations on them since they were not responsible for inducing climate change. They were, however, prepared to accept commitments of a conditional or contractual nature. Such measures would be conditional upon provision of "new and additional" financial resources to cover the full incremental costs involved in implementing these measures, as well as transfer of the required technology on non-commercial terms.

With the exception of the US, developed countries generally acknowledged the need for providing additional funds to developing countries. But their description of this as "assistance," did not conform to the position of the developing countries regarding the conditional nature of their commitments. With a few exceptions (e.g., Norway), the developed countries avoided any specific commitment to the net increase in financial flows for development that was implied by the term "new and additional" financial resources; moreover they referred to "agreed" rather than "full" incremental costs. On technology transfer, they rejected the demand for non-commercial terms on the ground that this was incompatible with protection of Intellectual Property Rights. They maintained this position despite efforts by representatives of the South to demonstrate that this was not necessarily the case.

Thus there were basic differences between the North and the South in their approach to the Convention. But there were also significant differences within each of these groups.

As we have seen, the EC and the US positions on specific commitments were divergent. This applied to both stabilization of emissions as well as provision of financial resources to developing countries.

Within the South, there were major differences between the AOSIS countries, which wanted a Convention with strong commitments, and countries whose economies were heavily dependent upon oil exports and which, therefore, were concerned over the implications of the Convention for the petroleum market. Thus on the question of commitments to emissions stabilization that would be binding on developed countries, many of the oil-exporters took a position similar to that of the US. Meanwhile, the AOSIS countries (together with some other developing countries such as India) held positions which went beyond that of the EC in calling upon developed countries to commit themselves to a time-bound program that went beyond stabilization.

Building New Coalitions

Against this background, efforts were made within both the North and South to forge negotiating coalitions. The first such initiative among countries of the North came from the EC which took on board the Japanese proposal for a "pledge and review" arrangement. On June 28, 1991, the concluding day of the second INC session, the EC declared that the Convention "should include what has come to be called a pledge and review proposal... The need for flexibility stems from the nature of our task, which extends over a long time span, well into the next century at least."[19] In accepting the pledge and review formula, the EC reiterated its commitment to a time-bound program of stabilization of carbon dioxide emissions by industrialized countries but in language which suggested flexibility. The EC expressed as its "view" that the stabilization target was "an example of a commitment that should preferably be embodied in a protocol."

To many delegates, this position appeared to be an effort on the part of the EC to formulate a common position with Japan and also to soften somewhat its differences with the US. If this was indeed the intention, it proved unsuccessful. In an early demonstration of the influence which NGOs could exercise on the negotiations, the NGO newsletter distributed daily during the negotiations, *ECO*, dismissed the proposal as amounting to "hedge and retreat." The proposal also drew criticism from developing countries on several grounds: (1) it was vague; (2) it diluted the specific commitments of the North; (3) it inequitably imposed binding obligations on developing countries; and (4) absent agreed criteria or guidelines, a review exercise would necessarily be arbitrary in nature.

Heeding these criticisms, the EC modified its position at the next INC session. The next EC statement noted that:

> the concept of Pledge and Review has caused a great deal of confusion. We are quite ready to admit that this was also the case among the Member States of the European Community. The Group of 77 was, therefore, completely right when it stated, through its Chairman, that the concept of Pledge and Review lacked precision and transparency.[20]

The EC concluded that the phrase "pledge and review" had "generated too much heat" and that it would be preferable, therefore, to "concentrate on the kind of structure we envisage for this Convention and the process for its implementation." In eschewing the phrase the EC did not, however, give up its position on certain elements of the concept, particularly review of national policies. This remained a bone of contention with the South until the final phase of the negotiations.

Further efforts at evolving a common position for the North proceeded on the basis of the EC stabilization formula. There was some progress toward consensus among the EC, Japan, Canada, Australia and other OECD countries but the differences with the US remained very wide. In informal conversations, delegates from EC Member States remained resolutely optimistic about the prospects of a change in the US stand. Nonetheless, an attempt to forge a common OECD position proved a failure even at the end of the first part of the final session of the INC negotiations in February 1992.

After arduous negotiations amongst themselves, the OECD countries finally presented a common formulation on stabilization on February 26, but even this contained no fewer than ten brackets and four alternative formulae! The basic divergences between OECD delegations were fully reflected in the heavily bracketed text. Portugal, on behalf of the EC, made an anguished statement expressing "regret that the positions reflected in the document are as far apart as is the case. We find it particularly regrettable that not even the very general texts agreed at Noordwijk, Bergen, the Second World Climate Conference, and the last OECD Ministerial Conference can be found without brackets in the present document."[21]

The OECD formulation was silent on reduction targets, despite the readiness of some of its members to accept such provisions. Thus, an amendment proposed by India, which required all developed countries to accept reduction targets within a stipulated timeframe, received full support from Germany, but was rebuffed by other members of the OECD.

Efforts at consensus-building in the South were rather more successful. From the outset of the negotiations, China participated fully in the work of the Group of 77, whose chairman spoke on behalf of "the Group of 77 and China." China's role enhanced the effectiveness of the South in the negotiations. At the third session, as we have seen, the G-77 and China took a common position on the pledge and review proposal. Even at a fairly early stage, the Group of 77 and China were able to forge a common position on the section on "Principles."[22] On a range of other issues as well, the Group was able to arrive at a consensus. These included consensus positions on:

- "adequate, new and additional resources";
- technology transfer on "favorable concessionary and preferential terms";
- the requirement that the "characteristics of any potential delivery mechanisms for financial resources and technology transfers [should be such] . . . that their structure, governance, eligibility and project selection criteria, administration, implementation, assured and adequate replenishment, are defined and ruled by the Conference of Parties";
- "Commitments that might be entered into by developing countries under this Convention are contractually dependent on the fulfillment of the financial and technology transfer obligations that must be entered into by developed countries who are in the main responsible for the urgency of the present situation";
- maintaining that "broad national reviews cannot be used as a means for institutionalizing a new form of conditionality."[23]

Nevertheless, as we saw earlier, there were significant differences within the South. From the outset of the negotiations, AOSIS functioned both as a participant in the larger "Group of 77 and China" as well as an autonomous coalition presenting its group position on a variety of issues. There were wide differences between AOSIS and the oil-exporters, in particular on the crucial section on "commitments." There were divergent views also in regard to the admissibility of any obligation for review of national policies. As a result of these differences, efforts to evolve a comprehensive consensus text covering the Group's position on commitments proved unsuccessful.

Hence, towards the conclusion of the fourth INC session, a group of forty-three developing countries submitted their own proposal for the entire section on "Commitments."[24] The text refrained from taking a position on the commitments of the developed countries to reduce emissions, pending an agreed pronouncement from the developed countries on this question. However, it called on the developed countries, through assessed contributions, to "provide on a grant basis new, adequate and additional financial resources to meet the full incremental costs of developing country Parties," covering also costs of adaptation and mitigation. It incorporated the South's position on technology transfer on "concessional, preferential and most favorable terms"; adding that protection of Intellectual Property Rights (IPR) should not hinder technology transfer in compliance with the Convention. Developing country Parties were required "in accordance with their national development plans, priorities, objectives and specific country conditions to consider taking feasible measures to address climate change, provided that the full incremental costs involved are met by the provision of new, adequate and additional financial resources from the developed country Parties." Developing countries could also, on a strictly voluntary basis, take additional nationally determined measures.

Thus, it could be said with some qualification that negotiations proceeded on a broadly triangular basis between a group of OECD countries centered around the EC; the United States, which maintained a distinct and separate position; and the developing countries, constituting the Group of 77 and China. The statement would have to be qualified on account of differences within the first and third groups and, in particular, the operation within the Group of 77 of a separate sub-group, AOSIS, which participated in the Group of 77 but also functioned autonomously as a separate coalition.

Dynamics of the Negotiations

We now turn to negotiating mechanics. At the second INC session, the Secretariat produced a useful compilation of elements of the Convention proposed by delegations.[25] These were considered in the two working groups. In Working Group II, dealing with "mechanisms," substantive agreement was possible in several areas such as scientific cooperation, entry into force of the convention, withdrawal from its obligations, etc. Though this did not extend to such questions as the financial mechanism or review of national reports, it was possible for Working Group II to take a decision requesting the co-chair to prepare a single text based on contributions from delegations as well as discussions in the working group sessions. The text would include areas of convergence and, where these did not exist, present alternative versions.[26]

In Working Group I, dealing with the core issues of "objectives," "principles" and "commitments," the extent of disagreement was far greater. Differences related to the most fundamental and crucial aspects of the Convention. This working group, therefore, decided only that a new compilation should be prepared under the authority of the co-chairs. But it also "left to the judgement of its co-chairmen to decide how they might further facilitate discussion" at the next session.[27] Entirely at their discretion, the co-chairs decided to present at the third INC session not only a new compilation but also a so-called "consolidated text" on principles and commitments, with brackets and alternative formulations. This text was "prepared with the aim of simplifying the compilation and to help identify major issues."[28] The debate in this session brought out the wide divergence of views on principles and commitments, even on such basic structural questions as whether the Convention should include sections on "Principles," on "General Commitments," and "Specific Commitments," or national strategies and programs.

At the fourth INC session, the Bureau of Working Group I initiated a process of "streamlining" in order to produce a more manageable text.[29] This entailed exclusion of some of the elements proposed by delegations. Editing of this nature required an implicit determination by the Bureau concerning the relative weight of a country presenting a proposal, the extent of support the proposal enjoyed among delegations and the importance which its supporters attached to the proposal. "Streamlining" thus meant an operational enhancement of the role of the Bureau in the negotiations.

The progress achieved through this procedure was, nevertheless, extremely limited on account of the depth and range of differences on basic issues. Thus, continued "streamlining" in the fourth session resulted in a "consolidated working document" in which the entire sections on "Principles," "Objective," "General Commitments," "Specific Commitments" and "Special Situations" were placed within square brackets. In fact, in these chapters, covering the core of the Convention, only a single word—"Commitments," appearing as a title—remained outside brackets! The Secretariat went on to produce yet another edition of a "revised consolidated text" under the guidance of the INC chairman and the co-chairs of the working groups in the first part of the fifth session. But, as we shall see, the revised document was never considered in the INC, as the dynamics of the negotiations underwent a sea-change in the last part of the final session.

The process outlined above involved drafting and revising the text, rather than bargaining or negotiation. Debate in the working groups was followed by new "streamlined" or "consolidated" texts produced by the Bureau with the assistance of the Secretariat. There were few attempts at bargaining or negotiating to narrow down the wide substantive differences between the North and South. Indeed negotiations did take place within each group aimed at formulating common group positions but there was little effective bargaining between North and South. It appeared that both the EC and the US preferred to continue postponing substantive negotiation with the South, deferring it until they succeeded in arriving at a common OECD position. The limited progress that was achieved resulted partly from the "streamlining" exercise of the Bureau and Secretariat and partly from the fact that the debate helped to sensitize delegations to the concerns of other parties. In a few cases, the process led individual delegations to modify their initial positions.

Thus basic differences remained unresolved even up to the conclusion of the first part of the fifth and final session of the INC. The position at this late stage in the negotiations may be summed up as follows: Differences on "Principles" and "Objective" resulted in these two chapters remaining entirely within square brackets. On "Commitments"—the crucial chapter of the Convention— differences remained very wide and were, in a sense, even accentuated because of the strong sense of divergences among OECD delegations.

There was a fundamental and irresolvable difference between the EC and US positions on stabilization of emissions. The US remained unprepared to accept even a weak formulation which called only for a commitment to take measures "aimed at" stabilization by a specified date. Both the EC and the US continued, however, to press the developing countries to accept commitments which bore no relation to the negligible historic and current contributions of these countries to global warming.

On financial commitments, some progress was made. Differences remained between developed and developing countries on whether financial resources should be provided on a voluntary or assessed basis and whether they were intended to meet the "full" or "agreed" incremental costs of developing country Parties. The EC appeared to have dropped its earlier hesitation, accepting a commitment to provide "new and additional" financial resources. There was even a modest change in the US position. During the fifth session, the US appeared to accept, at least in principle, a commitment to contributing financial resources. Nonetheless, it continued to maintain its reservation on "new and additional" financial resources.

Likewise, the North continued to insist on review of national plans and strategies, including those of developing country Parties. In regard to "mechanisms," progress was more clearly discernible and there was significant cleaning up of texts in several areas. However, basic differences remained unresolved on such issues as dispute settlement, the question whether the GEF or a separate Climate Fund should constitute the financial mechanism, as well as the question whether national plans and programs should be subject to multi-lateral review.

Closing the Gap

It was remarkable that it proved possible to bridge this chasm in ten days of negotiations during the resumed fifth session in New York. How did it prove possible to achieve this in a few days when efforts over the past fifteen months had yielded such limited and disappointing results?

Two new factors enabled the finalization of the Convention apart from the high political priority which delegations placed on completing a Convention before the UNCED meeting in Rio that was scheduled for June. First, at long last, the North arrived at a common formulation of its commitment concerning emissions. Second, connected with this breakthrough and at the initiative of the chairman, a new negotiating procedure was adopted to hasten an agreement.

On the very first day of the final round, Chairman Ripert announced his intention to seek a speedy conclusion of the negotiations on the basis of a Working Paper that he would present to the plenary.[30] The first installment of the draft was distributed immediately following his announcement and the second installment—covering the crucial areas of "commitments" and "mechanisms for transfer of finance and technology"—was made available on the following day, after finalization of a common US–EC formulation on their commitments regarding emissions. The chairman explained that the Working Papers were based on the Revised Text prepared at the conclusion of the previous round and that he had used his best judgment to produce a clean text without brackets. In another major procedural change, it was accepted, at the chairman's suggestion, that negotiations on the basis of this text should initially be confined to an "Enlarged" Bureau. This included about two dozen countries, comprising its original members as well as other countries which, in the chairman's view, were key players.

The chairman explained in the Enlarged Bureau that plenary meetings were unnecessary at this stage since the formulations in his text were not new and had been debated in earlier sessions of the plenary. The Indian delegate observed that, while other elements in the Working Paper had indeed been debated previously, this was not the case with the new formulation concerning emissions for developed countries. On his suggestion, it was agreed, in the interests of transparency, that a plenary session would be held to discuss the new text. This text would require developed country Parties to adopt policies and measures which "will recognize that the return by the end of the decade to earlier levels of anthropogenic emissions of carbon dioxide and other greenhouse gases not covered by the Montreal Protocol would be an appropriate signal by developed countries that long-term emission trends have been modified consistent with the Objective of this Convention." Another sub-paragraph also referred to "the guideline of returning anthropogenic emissions of carbon dioxide and other greenhouse gases not covered by the Montreal Protocol to their 1990 levels." The formulation was thus couched in language which was deliberately vague and was riddled with ambiguities.

The debate emphasized the weak and highly ambiguous nature of the formulation, which did not refer to stabilization or reduction and spoke vaguely only of efforts to "return" to earlier emission levels. The Philippines delegate, on behalf of the Group of 77 and China, raised no fewer than fifteen detailed questions on the meaning of the text. In replying to another request for clarifications from the Indian delegate, the chairman acknowledged that the text was ambiguous. He observed that this ambiguity reflected the lack of agreement among the industrialized countries.[31]

Negotiations within the Enlarged Bureau commenced on the basis of the chairman's Working Paper. The chairman argued that the formulation on emission commitments of developed countries, which had emerged after very difficult and delicate negotiations, should not be reopened. The Indian delegate expressed the view that the paper as a whole was under negotiation, including the new formulation on the emissions commitments of developed countries. The Indian delegate pointed out several deficiencies in the formulation and he presented the chairman with an alternative formulation.[32] In the end, India did not press for action on this proposal, in light of the progress made in resolving other issues in the Convention and bearing in mind also the importance of securing a consensus Convention. The US–EC formulation itself was later revised slightly by its authors. It was finally incorporated into the Convention in this revised form, with only a few very minor changes.

Vigorous negotiations took place in the Enlarged Bureau over the next several days, often lasting until the early hours of the morning. There were also frequent consultations among delegations outside the Bureau. Efforts were made by group representatives to inform and consult countries not represented in the Enlarged Bureau, but these efforts could not keep pace with the rapid development of text within the Bureau. After agreement had been reached within the Enlarged Bureau, texts were circulated to the plenary for wider discussion and approval. But, since some of the crucial elements relating to "Commitments" were finalized only on the final day, a majority of delegations saw the full text of the Convention just hours before its adoption by consensus. The discussion did not significantly alter the text emerging from the Enlarged Bureau.

Thus, late in the evening of the May 9, 1992, the INC adopted the text of the Framework Convention on Climate Change. The Convention reflected "differentiated responsibility" in relevant areas. While it fell short of removing ambiguities concerning the emission commitments of developed countries, it avoided shifting the burden to the developing countries. It recognized that "the extent to which developing country Parties will effectively implement their commitments under the Convention will depend on the effective implementation by developed country parties of their commitments under the Convention related to financial resources and transfer of technology and will take fully into account that economic and social development and poverty eradication are the first and overriding priorities of the developing country Parties." It also provided that the "agreed full incremental costs" of developing country Parties should be provided by the developed countries through provision of "new and additional" financial resources.[33] Review of the national policies of developing countries will not be required. Information supplied by developing countries will be used exclusively to "assess the overall aggregated effect of the steps taken by the Parties in the light of the latest scientific assessment concerning climate change." "Coordination" of policies is restricted to developed country Parties and to countries which voluntarily wish to engage in such measures. A compromise was reached on the financial mechanism under which the Global Environment Facility (GEF) is to function as the interim mechanism under the convention. At the same time, GEF should be restructured in order to meet the requirements of the Convention for transparency and universality of membership. The Conference of the Parties to the Convention will set the policies, program priorities and eligibility criteria. The GEF will have a transparent system of governance, with equitable and balanced representation of all Parties, and "accountability" through regular reporting to the Conference of the Parties.

Unfinished Business

It is important to note that the Convention itself recognizes the unfinished nature of the negotiations on its core Article 4(2), concerning the emissions commitment of developed country parties and "other Parties included in Annex I" (viz., the industrialized East European countries). It specifically charges the Conference of the Parties to complete this task. Thus, Article 4(2), after setting out commitments of the developed and East European industrialized countries in its sub-paragraphs (a) and (b), goes on in its subsequent sub-paragraphs to require that the Conference of the Parties should in its first session:

(1) decide methodologies for calculations of emissions by sources and removal by sinks of greenhouse gases for the purposes of sub-paragraph (b) of the Article. These are crucial questions and the difficulties inherent in coming to an agreement on these complex matters were apparent from the fact that the INC was not able to arrive at a definition even of the basic concept of "net emissions."[34]

(2) take decisions regarding the criteria for "joint implementation" in relation to the commitments under Article 4(2)(a). The concept of "joint implementation" remained undefined during the negotiations and the term was used by developed countries in more than one sense.

(3) Finally and most importantly, the Convention specifically requires under Article 4(2)(b), a review of the adequacy of the commitments undertaken by the developed country parties and the industrialized East European countries under sub-paragraphs (a) and (b). There is an obvious need to remove the ambiguities inherent in the formulation of these commitments. This is recognized in the Convention and a review is specifically required at the very first session of Conference of the Parties. The same sub-paragraph furthermore provides that "based on this review, the Conference of the Parties shall take appropriate action, which may include the adoption of amendments to the commitments in sub-paragraphs (a) and (b)."

Thus, the Climate Change Convention itself indicates areas in which further negotiations are urgently required. These pertain to the core area of the Convention, namely, the specific commitments of developed countries, including the East European industrialized countries. If satisfactory solutions are not found to these unresolved questions, the Convention will fall short of being an effective instrument for combating climate change. Hence calls to initiate negotiations on protocols, etc. are not only premature but would result in a positive setback by delaying progress in clarifying the core provision of the Convention. The international community must now concentrate on completing the task left unfinished in negotiating the Convention—as specifically noted in the Convention itself. This must result in a clear-cut and unambiguous, time-bound program for stabilization and reduction of the emissions of greenhouse gases originating from the developed countries. Without doubt, these countries have appropriated more than their due share of the planet's atmospheric resources and have thereby induced the phenomenon of incremental global warming with all its attendant dangers for humankind.

Notes

1. J. Legett, "The Nature of the Greenhouse Threat," in Jeremy Legett (ed), *Global Warming - The Greenpeace Report* (Oxford: Oxford University Press 1990).

2. WMO, *World Climate Conference Declaration and Supporting Documents* (Geneva: WMO, 1979).

3. UN General Assembly resolution 43/53 of December 6, 1988.

4. UN General Assembly resolution 45/212 of December 21, 1990.

5. INC document A/AC.237/5.

6. These are listed in INC document A/AC.237/6.

7. INC documents A/AC.237/Misc.1/Add.1–9.

8. A/AC.237/Misc.1/Add.3.

9. Statement by the leader of the Indian delegation, June 19, 1991.

10. US submission to INC. Paper No. 11 in A/AC.237/Misc.1/Add.1.

11. *Ibid*. Paper No. 4.

12. *Ibid*.

13. A/AC.237/Misc.1/Add.2. Paper No. 14.

14. A/AC.237/Misc.1/Add.1. Paper No. 3.

15. *Ibid*. Paper No. 11.

16. *Ibid*. Paper No. 10.

17. A/AC.237/Misc.1/Add.6.

18. A/AC.237/Misc.1/Add.7. Paper No. 20.

19. Intervention by the Netherlands delegation on behalf of the European Community and its Member States, June 28, 1991.

20. Intervention by the Netherlands on behalf of the European Community and its Member States, September 1991.

21. Statement by the Portuguese delegation on behalf of the European Community and its Member States, February 28, 1992, cited in *ECO*, February 27, 1992.

22. INC document A/AC.237/WG.1/L.8.

23. Joint Statement of the Group of 77 and China, February 28, 1992.

24. INC document A/AC.237/W.G.1/L.7 dated December 18, 1991.

25. INC document A/AC.237/Misc.1 and Add.1–3.

26. *Report of the INC on the Work of its Second Session.* A/AC.237/9.

27. *Ibid.*

28. INC document A/AC.237/Misc.9.

29. *Report of the INC on the Work of its Fourth Session.* A/AC.237/15.

30. INC documents A/AC.237/CRP.1 and Add.1–8.

31. *ECO*, May 5, 1992.

32. This read as follows:

"Each developed Country Party (for the purpose of this article, the EC shall be considered as a single Contracting Party) as defined in annex — shall adopt national policies and take corresponding measures on the limitation of anthropogenic emissions of greenhouse gases and the protection and enhancement of greenhouse gas sinks and reservoirs. These policies and measures will, as a first step, stabilize its emissions of carbon dioxide at 1990 level, in general by the year 2000, taking into account the differences in their starting points and approaches and the need for equitable contributions of these Parties. Thereafter, each developed country shall progressively reduce emissions consistent with the objective of this Convention."

33. Article 4(3) of the Convention reads as follows:

"The developed country Parties and other developed Parties included in Annex II shall provide new and additional financial resources to meet the agreed full costs incurred by developing country Parties in complying with their obligations under Article 12 paragraph 1. They shall also provide such financial resources, including for the transfer of technology, needed by the developing country Parties to meet the agreed full incremental costs of implementing measures that are covered by paragraph 1 of this Article and that are agreed between a developing country Party and the international entity or entities referred to in Article 11, in accordance with that Article. The implementation of these commitments shall take into account the need for adequacy and predictability in the flow of funds and the importance of appropriate burden sharing among the developing country Parties."

The word "such" in the second sentence is important since it can only refer to the phrase "new and additional" in the preceding sentence. Delegations (e.g., India) have drawn the attention of the Depository to the fact that a discrepancy has crept in between the negotiated English text and its versions in other UN official languages as presented at UNCED, in that the word "such" has been omitted in these versions.

34. The EC proposed to define "net emissions" as the gross emissions originating in a country minus (or plus) the increases (or decreases) in the absorptive capacity of sinks located in its territory. An alternative definition which readily suggests itself would define "net emissions" as gross emissions minus that part of these emissions which can be deemed to have been absorbed by the country's entitled share of natural sinks. The EC definition in effect treats the global sinks as having been appropriated by the developed countries.

The author led the Indian delegation to the INC Sessions. The views expressed in this article are, however, entirely his personal views and they do not necessarily reflect the position of the Government of India.

CHAPTER 7

A Personal Assessment

Bo Kjellen
Ministry of Environment
Stockholm, Sweden

Prologue

An unusually heterogenous group met in the new Westfields Conference Center at Chantilly, VA, outside Washington, DC, in February 1991. There were diplomats and climate experts, people from capitals and permanent UN delegations, old hands at multilateral negotiation as well as newcomers. They came from almost a hundred different countries, with divergent objectives and negotiating goals. And there was considerable uncertainty in the air.

Nonetheless, in the late afternoon of May 9, 1992, just fifteen months later—as Chairman Jean Ripert of France concluded the negotiation and received well-deserved applause—many of the negotiators had become friends and would regret the unavoidable separation. Group dynamics had worked wonders throughout the many negotiating sessions and late-night drafting meetings. It was a special human experience and an example of what multilateral negotiation can achieve, if the conditions are right.

149

One reason for our success was the structure of the negotiations: they took place within the general framework of preparations for the UN Conference on Environment and Development (UNCED), sometimes called the Earth Summit, working under an implied pressure to produce a Convention before the Rio Conference in June 1992. With that sense of urgency came an overriding concern for the global environment and the conviction that important issues, critical to the future of mankind, were at stake. Without pressing the argument too far, even the most experienced or cynical negotiators had to be sensitive to this kind of reasoning. Furthermore, the task was given increased urgency because the Intergovernmental Panel on Climate Change (IPCC) and the Second World Climate Conference had presented unequivocally the scientific case for early action.

Yet all participants realized that no sensational or easy breakthroughs were within reach. International action on the climate issue would have profound economic impacts. and governments would be hesitant to take unpopular action in a time of economic recession and hardship. Thus the participants were faced with a built-in conflict between forces that pressed for early resolution and forces that argued for delay. As the negotiators began to recognize the nature of this conflict and their shared dilemma, a group dynamic evolved that would enable sharp differences of opinion to be rounded off or muted in discussion. The shared understanding of the complexity of the undertaking created a sense of common purpose among the negotiators. All realized that they had to learn from each other, that there are many different skills and types of expertise needed in this extremely complicated and uncharted field, and that no one country or individual possessed the key to success. It was truly a joint undertaking.

I emphasize these "soft" psychological elements, because I firmly believe that they explain quite a lot of the events that occurred during the negotiations. In particular, it seems to me that they explain how it was possible to accelerate the talks at the critical points of impasse when it was most necessary to do so. Perhaps, as a participant, one is tempted to overestimate the importance of some of these elements—but I believe optimism is better than cynicism.

This generally positive evaluation should not conceal the fact that several negotiators certainly had hoped for—and perhaps even expected—a result, i.e., an agreement with binding and specific commitments to future emissions reductions. As all the negotiators were aware, the great majority of non-governmental organizations (NGOs) take a fairly dim view of the outcome of the climate negotiations;

their criticisms have been fairly pointed. As negotiators we certainly have to reflect very carefully on the following point: could not more have been achieved, given the urgency of the threats facing the global climate?

In my view, the political and economic environment in which the negotiation took place did not permit a more far-reaching result. A major objective of the negotiation was to have all the big players sign the convention during the Earth Summit at Rio. We were well aware of the extremely restrictive position of the United States—if we were to reach consensus there could be no commitments on targets and timetables in the final agreed text.

In the last analysis, the Framework Convention on Climate Change does include reasonably firm notions of targets and timetables. This convention is certainly not up to the expectations of several European countries, including my own, but it does include clear obligations to report and evaluate national policies as well as measures to limit the risks of rapid climate change. Furthermore, the Convention contains important mechanisms for periodic review, opening the possibility for sharpening commitments at least twice in the nineties, at the first Conference of the Parties (COP) and then again before the end of 1998.

The real issue, in my view, is of a different nature: will it be possible to carry forward the positive negotiating atmosphere from the first phase of negotiations into the follow-up? Can we maintain the political momentum? This is particularly important since every single nation will probably have to undertake much more decisive and difficult action to reduce the rate of greenhouse gas buildup than has been attempted to date. This will be especially important in the energy and transport sectors. We must not forget that the IPCC has called not only for stabilization in the rates of emission of greenhouse gases, but—beyond that— for significant emissions reductions. These can be achieved only through politically difficult measures and through unprecedented efforts of international cooperation, in particular with regard to developing countries.

Therefore, while expressing satisfaction with what we have achieved during the negotiation, it is necessary to think of our achievement as very modest when one considers the dimensions of the problems and the weaknesses of the multilateral system. This duality was captured by the chairman of the IPCC, Professor Bert Bolin, reacting to the conclusion of the agreement when he observed: "It is fantastic that we have a convention, since nobody could have believed it possible five or six years ago. But of course this convention is not enough."

The Management of a Complex Negotiation

The whole UNCED process, with the need to conclude several difficult and technically complex negotiations over a short period of time, underlined the need for efficient management. The climate negotiations were certainly no exception.

I do not think that it is generally understood to what extent logistics and management influence the substantive outcome in a multilateral negotiation. There are thousands of pages of background documentation to be produced and translated into six languages; there are deadlines to be respected; and there are meeting-rooms and interpreter teams which have to be available. This may sound rather pedestrian, but we all need to respect the importance of these problems: if they are not properly addressed, the whole machinery gets out of order and no meaningful results can be achieved.

In this respect, the decision taken by the 45th UN General Assembly in 1990 to set up the International Negotiating Committee and to provide it with a small independent Secretariat turned out to work in a most satisfactory way. The UN Secretary-General's decision to appoint Michael Zammit Cutajar of Malta to the post of Secretary-General for the INC contributed to the success of the negotiation.

Zammit Cutajar came to the job with considerable UN experience, having served for several years with the Secretariat of the UN Conference on Trade and Development (UNCTAD). The appointment of a Maltese national to the post of Secretary-General was also a recognition of the important role played by Malta in alerting the General Assembly to the importance of the climate issue. Furthermore, Zammit Cutajar brought with him excellent knowledge of the people in the UN system and the necessary grasp of complex UN procedures. Not least important, he was always perceived to be cooperative and generous in his relations with delegations. All of these attributes contributed in a significant way to the positive atmosphere of the talks.

It is, however, the chairman who carries the main management burden in a negotiation like this. He has to set the tone for the negotiation, realize what results can be achieved at what moment, and use his intuition to grasp the initiative when delegations seem to run out of steam. And more generally, he has to maintain momentum towards the desired objectives even though conditions are changing all the time. It is like steering a sailboat to the correct port in conditions of bad visibility and with changing winds.

Chairman Jean Ripert had the experience and the human qualities to manage this negotiation. His long and distinguished career meant that he came to the negotiation with all the necessary contacts. His association with the IPCC meant that he had an immediate grasp of the technical issues involved in the climate debate and, in particular, of the specific problems related to the developing countries. And he commanded great respect.

I believe that style plays an important role for a chairman—and Jean Ripert's style certainly fit the challenges of this negotiation. Most of the time he acted with an almost palpable slowness: but this was on the surface. While explaining technical or legal details pertaining to the negotiation in painstaking (some would say irritating) detail, his mind was searching out solutions and anticipating ways to avoid blocked situations. It became clear that this was his method of work. The overall effect inspired broad confidence in his leadership. In the final stages, when heads of key delegations were invited by the chairman to negotiate the final texts and thrash out the last difficulties, we were all impressed by the sharpness of the picture he laid out before us.

Ripert's leadership style did not exclude the human touch. He was able to remind the negotiators of the real issues beyond the drafting and of their responsibility to the international community without sounding condescending or offensive. And in his concluding words at the end of the negotiation he struck just the right balance between pride and modesty in assessing what had been achieved.

I have dwelt on these issues at some length, because multilateral negotiations today take place in a very different international environment compared with what prevailed during most of the post-Second World War period. In earlier years, the United States provided strong leadership and it had both the political force and economic resources to move the negotiations forward. The situation has changed dramatically. Several scholars and experts have scanned the horizon to see if any other country—or group of countries—could take on the former role of the US This may certainly happen—or the US might move back into a leading position. But for the time being—and this was a characteristic of the whole UNCED process—there was no single national actor capable of providing constant leadership. (This remains true despite decisive steps taken at critical moments by one or another actor, such as the European Community.)

Given this state of affairs, the roles of the chairman and of the Secretary-General become central. Aided by solid Secretariat work, the chairman remains the only person who can give the needed leadership and provide the right sense of direction. The UNCED process was lucky to benefit from chairmen such as Ripert in the Intergovernmental Negotiating Committee (INC) and Tommy Koh in the

Preparatory Committee. These two men demonstrated that very different styles of leadership can lead to the desired results. The climate negotiation underlines the importance of very careful scrutiny of candidates for leading posts in multilateral negotiations in the present period.

It would carry me too far from my main points to go much more into the detailed operations of the Bureau of the INC or the management of the two working groups (with their laboriously negotiated system of co-chairs).[1] However, Working Group II with Liz Dowdeswell of Canada and Robert Van Lierop of Vanuatu as co-chairs displayed a most rational and systematic method of work. No doubt the small Bureau of the INC also played an important role; but I had no personal contact with its work. Yet another example of Jean Ripert's leadership was that he enlarged the Bureau of the INC in the right moment of the negotiations and increasingly moved the most delicate parts of the negotiation to that body. This was a difficult decision, bound to cause some controversy and irritation; but even those who felt excluded from the Expanded Bureau admitted that the chairman had taken the only possible decision in the circumstances.

The Climate Negotiation as Part of the UNCED Process

The decision by the General Assembly to set up the INC came after the UNCED process had already begun. The UNCED Preparatory Committee (PrepCom) had its first session in Nairobi in August 1990, and already at that stage, it was clear that a climate negotiation would be undertaken. The governing bodies of both the UN Environment Programme (UNEP) and the World Meteorological Organization (WMO) had taken decisions to that effect. Still, the PrepCom could not foresee the exact negotiating structure.

I was elected chairman of Working Group I of the PrepCom at its first session in Nairobi. Among other matters allotted to this working group were all issues related to the atmosphere. These issues included climate change, ozone depletion and transboundary air pollution. The discussions in Nairobi were general since there was very little background documentation available. At the same time it was clear that the deliberations on Agenda 21 would be heavily influenced by the evolution of existing or anticipated international instruments.

At the second session of the PrepCom in March 1991, the climate negotiation had already begun, even though the Chantilly meeting in February had been more organizational than substantive. Sitting as the chair of the Atmosphere Working Group, I felt considerable concern among the delegations that the PrepCom not duplicate the work of the INC, since the purposes of the two processes were quite different. Based on proposals by the Nordics and a number of other countries, I therefore suggested that Agenda 21 should focus on individual economic sectors within the "Atmosphere" chapter, particularly in relation to the climate issue. I had the strong support of the UNCED Secretariat in following this approach. In fact, quite a few countries had felt that Agenda 21 should devote more attention to the environmental impact of different economic sectors; but as it turned out it was only in the "Atmosphere" chapter that the method was fully carried out.

This does not mean that the atmosphere chapter goes very far. Real discussion of it started only in Geneva during PrepCom 3 in August 1991. In March, the main attention was directed towards the energy and transport sectors. It was quite evident that these two sectors would be controversial. Immediately, some countries raised objections which would become familiar themes in the climate negotiations. The United States was clearly unhappy with this application of the sectoral approach, as were several of the oil-producing countries—in particular Saudi Arabia. The specific proposed recommendations on energy efficiency and the use of new and renewable sources of energy were considered too far-reaching. As the PrepCom came to an end, the goal of the Secretariat was clearly to present something more general. But this was not the only change of plans: the PrepCom had also added industry and agriculture to the sectors that would be covered in the "Atmosphere" chapter of Agenda 21.

There was no in-depth discussion between Jean Ripert and me on the precise relationship between the atmosphere chapter of Agenda 21 and the Climate Convention. It was clear, of course, that the two documents would have a different legal status: The Convention would be legally binding, whereas Agenda 21 would be a political document, at best an influential program guiding national actions. But the two instruments would appear in the same package in Rio de Janeiro, and therefore must not be contradictory or repetitive, but should be mutually supportive and reinforcing. Most delegations felt that this would not be a major problem, but the issue came to play a significant role in the final phases of the two negotiations.

This close relationship between Agenda 21 and the Climate Convention was underlined by the negotiating calendar of the last months before the Rio Conference. The INC met in plenary February 17–28 and April 30–May 9. The UNCED Preparatory Committee had its fourth and decisive meeting between those dates. All these meetings took place in New York.

This was the period when the INC began its most intense phase of negotiation, particularly among the members of the Organization for Economic Cooperation and Development (i.e., the OECD countries). Underlying all the discussions of commitments on emission control, stabilization by the year 2000, etc., were matters related to energy, transport and industry policies. Thus, it was not surprising that the energy section of Agenda 21 would be scrutinized carefully. In order to move the negotiation forward, it was necessary to use a general disclaimer, stating that nothing in the Agenda 21 chapter would in any way complicate, prejudice or pre-empt the negotiation of the Climate Convention.

The "Atmosphere" chapter turned out to be the most difficult in the whole of Agenda 21. In the final stages of PrepCom 4, Saudi Arabia and some of the other oil-producing countries wanted to introduce far-reaching changes in a text which had originally been presented by the Group of 77 (G-77) and then slightly amended during subsequent negotiating sessions. The Saudi requests created an impasse. With the support of the Arab countries, Saudi Arabia argued that the whole "Atmosphere" chapter should be put in brackets until Rio. One complicating factor was the fact that the negotiation of the Climate Convention was not yet concluded in the INC.

The Saudi and Kuwaiti negotiators to the final session of INC continued along the same line. They made repeated attempts to introduce less stringent and less binding language with regard to the specific commitments to be undertaken by industrialized countries. And they defended their positions with considerable tenacity. In fact, these interventions created a rather serious situation in the very last phase of the INC, temporarily but effectively blocking agreement. In the end, however, the chairman pushed the text through. It remains to be seen whether Saudi Arabia and other Arab states will sign the Convention at a later stage.

There was an epilogue to this episode in Rio de Janeiro. A special working group was set up under my chairmanship to work out an agreement on the atmosphere chapter. Saudi Arabia immediately asked for the floor and surprised all present by asking for the deletion of the chapter in its entirety. They said it would not be needed since the Parties had now agreed to the text of the Climate Convention.

Many objections were raised against this position and after long and difficult negotiations a general agreement was reached, which could be endorsed subsequently by the Conference itself. Saudi

Arabia had originally insisted that the words "safe and" should appear every time "environmentally sound" was used, but this had been unacceptable to a number of other countries. Even though considerable efforts had been made in the process of drafting the text in order to accommodate the Saudi position, the government of Saudi Arabia now felt compelled to lodge a general reservation against the "Atmosphere" chapter of Agenda 21. On some specific points—in particular those related to the use of the phrase "environmentally safe and sound" technologies—they opened the way to an agreed solution only at the very last moments of the negotiation.

There is another sector where Agenda 21 and the Convention were closely related: the transfer of financial resources and technology. This was a key issue in both negotiations: I think it is fair to say that the two sets of debates were mutually supportive, since so much of the discussion focused on financial mechanisms. The convention now contains a conditional endorsement of the Global Environment Facility (GEF): the GEF is designated as the interim financial mechanism for the Convention. The role of the GEF is also confirmed in Agenda 21, but Agenda 21 also refers to other mechanisms and deals with the need for increased transfers of resources in general. No doubt the G-77 will continue to insist on concrete action in these fields during the follow-up to Rio, both with regard to Agenda 21 and the Biodiversity and Climate Conventions.

Initial OECD Attempts at Coordination

The negotiators for the industrialized countries shared a clear understanding that there was a need for regular contacts among OECD countries if a number of difficult issues were to be resolved. However, the negotiations were taking place within the UN General Assembly system, which offers no specific formal mechanism for the OECD countries to meet as a group to coordinate their positions on issues of substance. (Some of the OECD countries cooperate at UN headquarters in the "Western European and others" group, but that cooperation only concerns procedural issues related to elections, etc.) The OECD countries did not want to appear to be "ganging up" on or "bullying" the G-77. Cautiously, OECD countries explored solutions to their dilemma.

In early 1991 Canada volunteered to coordinate a series of OECD meetings that would be limited to exchange of information. Ambassador Barry Mahwinney from the Foreign Ministry in Ottawa chaired these meetings with great skill and tact. These meetings were held during the first three sessions of the INC, and were, in themselves, very useful. Obviously, there were also a number of bilateral contacts. The European Free Trade Association (EFTA) and the members of the European Community (EC) countries had regular consultations, which showed that there was considerable consensus in Europe on the main negotiating issues. The Nordic group also met to prepare for various INC and PrepCom meetings. It is established practice that the Nordic countries try to coordinate their positions. In this negotiation, however, on various occasions there were important nuances of difference among them: during some parts of the negotiation Norway tended to lead on the articulation of specific positions. This was true, in particular, on the notion of cost-effectiveness and the establishment of a "clearing house" mechanism. While the other OECD delegations had general sympathy for the Norwegian ideas—and the Swedish thinking was very close to the Norwegian—opinions were divided on details and tactics. Similarly, Finland was insisting more than the other Nordics on the question of sinks and the corresponding notion of net emissions. However, in the spirit of easy-going cooperation that exists among the Nordic countries, we were mostly able to present a common view.

During the quite disappointing fourth session of the INC, which was held in Geneva in December 1991, the G-77 criticized the OECD for the absence of a coordinated view. In particular, I recall that one of their most brilliant and outspoken representatives, Ambassador Dasgupta of India, clearly indicated that there could be no progress in the negotiation if the OECD countries did not "get their act together." Clearly, the OECD members no longer needed to be concerned about inadvertently intimidating the G-77 countries, and could focus upon coordination of their position.

However, this was easier said than done. There was no real mechanism available for formal coordination, and OECD countries had tended to move in rather disparate ways. The European Community, supported by the EFTA countries, insisted that there should be firm commitments on stabilization of CO_2 emissions at the 1990 level by the year 2000. The United States was still standing very firm in the position that there could be no targets or timetables. Their very constructive chief negotiator, Robert A. Reinstein of the State Department, could not do much more than repeat familiar US positions. In the territory between these positions, the CANZ Group (Canada, Australia, New Zealand) tried to frame a compromise by using the notion of a comprehensive approach as opposed to exclusive concentration on emissions of CO_2. Some of the Nordic countries participated from time to time in these efforts.

As the cold and foggy winter days passed in Geneva, many delegates grew rather pessimistic: no breakthrough seemed to be within reach and the remaining negotiating time before Rio was dramatically short. And yet the group dynamics I have already referred to had started to work; there was a much better understanding of the issues on the table. But the fact remained that if a single negotiating text were to come out of the Geneva meeting, it would have more brackets and open questions than when the December session started.

A Period of Intense OECD Coordination

It was in this gloomy situation, almost at the end of the session, that the EC took an initiative to launch a more effective process of OECD coordination. It was the outspoken and knowledgeable Danish representative of the EC, Jörgen Henningsen, who proposed that a special working group be set up to discuss OECD positions in detail. Henningsen produced a check-list of issues. These included the character of quantitative commitments, the conditions for implementation, issues related to the comprehensive approach, the notion of net emissions, monitoring and reporting, and issues related to financial commitments. The EC initiative opened a new avenue for cooperation among OECD countries. In an important way, it set the stage for the final agreement reached in New York. But the road was to be long and winding, involving many late night sessions and frustrating meetings. The length and frequency of these meetings severely tested the patience of the delegates.

Jorgen Henningsen had not had any previous contacts with me before the setting up of this new group. I was therefore rather surprised to learn that the EC wanted to put my name forward as the chairman. I accepted the task with considerable concern, realizing that no miracles could happen.

The check-list prepared by the EC turned out to be of great help. OECD countries worked on the paper during one and a half days at the end of the Geneva session and decided to continue the work in this ad hoc group before and during the February 1992 session of INC 5 in New York. However—and this is one of the ironies of a negotiation of this kind—it turned out that the detailed discussion of concepts and ideas in the closing days of Geneva did not really continue in the group. Instead, the ad hoc group became the main focal point for drafting the specific text on commitments with respect to sources and sinks that would be submitted to the INC plenary. Thus, the EC initiative and the Henningsen list served as a catalyst to relaunch the negotiation.

In looking towards the future, however, I believe that the checklist and the preliminary discussions that took place in Geneva could well serve as a background to follow-up action for the Convention itself. I shall therefore return to some of these issues in the concluding section of the chapter.

At the INC 5 plenary session in New York in February, the United States introduced a new position and took a more activist role. In the first meeting, Bob Reinstein stated that he could see good possibilities for actually working out a streamlined text on specific commitments, which would replace "with a single paragraph" the unwieldy four-page paper that came out of Geneva. He offered to produce an example and invited other countries to do the same. Some countries felt that the OECD Ministerial Declaration of February 1991 could be a good starting point.

The Central Controversies Emerge

This was the point of departure for a period of intense activity in the OECD Group, which worked in parallel with the main negotiations in the INC plenary. Both the EC and the CANZ group presented new texts. The European Community document would serve as the basis for the drafting. The following points were now central and they continued to be of capital importance for the conclusion of the negotiation:

a)   The US had "opened" this phase of the negotiation by indicating that they were ready to accept language on the adoption of measures to limit emissions of greenhouse gases; the sticking point was to combine this with the EC insistence on the target of stabilization of CO_2 emissions by the year 2000.

b)   How could it be made clear that stabilization of CO_2 emissions remained a central objective while admitting that a comprehensive approach involving other greenhouse gases would also be part of an agreement?

c)   How could the special situation of the European Community with its regional "bubble" approach be accommodated in the Convention?

d)   Some countries, such as Norway and Sweden, had already undertaken very significant reductions of CO_2 emissions. How could the necessary flexibility be introduced in the Convention, so that these countries would not be penalized? In particular, Norway had launched the notion of joint implementation—or commitments in cooperation. This approach would make it possible to include actions outside a country's borders when calculating the figures for the national stabilization target. The Norwegians had made great efforts to elaborate on this cost-effective approach which, some have argued, is a step toward a system with tradeable emission permits. Such a system could also be an additional source of financial resources for the developing countries, and the Norwegian delegation even presented a proposal for a so-called "clearing house" to match such "transferred" commitments and development projects.

e) The problem of accounting for sinks of greenhouse gases had been a difficult issue all through the negotiation. The most ambitious proposals called for the full-fledged implementation of the concept of "net emissions." But the practical difficulty of actually calculating carbon absorption levels in growing forests and other areas makes it difficult to use the notion of net emissions at the present state of scientific knowledge.[2] A minimum first step would be to assume that the burning of biomass releases no net emissions of CO_2 when an equivalent area of new forest is being grown. This concept is actually of great importance for Sweden where the use of biomass is becoming an important element in our energy policy.

f) How would the fulfillment of national commitments be reviewed, and what mechanisms should be used generally to consider the adequacy of the commitments in the Convention? And linked to that question, of course, is the much wider, but unavoidable issue of how the contracting parties would subsequently proceed from stabilization to reduction of emissions.

We had now moved squarely to the central issues—there was no time for general statements of intention any more. Raw nerves were being touched, important national interests were at stake. And at the same time the negotiators were well aware of their global and long-term responsibilities, constantly highlighted by well-informed and militant NGOs.

OECD Negotiations During INC 5—And After

It was therefore not surprising that the ad hoc OECD Group spent long hours trying to reach agreement on a text which could be presented to the INC itself. To chair this group was both a satisfying and a frustrating experience. It was satisfying because the negotiating atmosphere was generally good, and there were many constructive ideas presented. But there was also frustration, not least when the United States at one stage seemed to go back on points which they had previously agreed upon. Despite the immediate cause of frustration, I think we all realized that there were forceful interests at work in Washington—and we were not yet really in the final stage of negotiation.

The ad hoc group met every day, and during a couple of nights, between February 17 and 24. Not surprisingly, at that moment, the G-77 had become rather irritated with the slow progress of OECD countries in negotiating just five paragraphs. Aware of these feelings of irritation when I presented the text to the plenary on February 24, I did not expect unqualified ovations—particularly since there were still a number of brackets in the text. (The final text of the Framework Convention is reproduced in the Appendix of this volume). Reactions, on the whole, were carefully worded. The G-77 representatives realized that more progress was impossible at that stage, and they understood that the OECD Group needed still more time.

I had hoped that this text would provide the basis for concluding the negotiation. This proved true, but the final version of the convention looked rather different from this early draft. Nonetheless the main elements were in place, and the ad hoc group meetings in New York had served as an important stepping-stone on the road to final agreement.

Chairman Ripert called an informal meeting of the Extended Bureau in Paris in the middle of April 1992. The purpose of this meeting was to prepare for the final negotiating session in New York a few weeks later. In cooperation with Barry Mahwinney of Canada, I decided to try to get the delegates from the OECD countries together in the days preceding the Ripert meeting. The discussions at both these sessions proved quite difficult.

The main reason for the difficulty was that the United States indicated problems with any use of the word "stabilizing" in the text, even in the rather vague form it was applied in the paper of February 24. This US position obviously created considerable problems on the European side and the whole atmosphere became rather tense. During informal consultations, Bob Reinstein tried different formulations, based on the replacement of "stabilization" with "reduction of growth." However, there were no prospects for an agreement at this stage, and ultimately the differences were expressed in two alternative texts for the language on specific commitments.

At the same time, the general group chaired by Barry Mahwinney made good progress and contributed greatly to the ultimate success of the negotiation. The final outcome of the April "pre-meeting" was a respectable but not heroic result and it marked the end of the efforts of the ad hoc group.

I presented the results of the ad hoc OECD Group to the informal Extended Bureau meeting of April 15–17. The reactions were mildly positive. India and Pakistan felt that the OECD was moving towards texts that were far too general but other developing countries expressed both understanding and approval. Many appreciated the results, knowing that the central issues were now being clarified. It now remained for the chairman of the INC to manage the final stages of the negotiation.

One reason for the general support of the OECD text was the introduction of a new feature, namely the last paragraph on reporting requirements. This new text opened the way towards practical reporting and review procedures for countries undertaking specific commitments. The importance of introducing the unified OECD text was immediately understood by G-77 representatives on the Extended Bureau. In the final text of the Convention the provisions for reporting and review form an extremely important part, integral to the whole structure of specific commitments. In my opinion, this text constitutes a key element in the future implementation of the convention. Looking back at the various texts, it is interesting to find how the different elements relate to each other and how the final text resulted from a kind of trial and error approach. The negotiators served increasingly as intermediaries in a complicated effort involving many different layers of national interests at work in their respective capitals.

Without doubt, Washington and various EC capitals were the central actors in the final phase of negotiations. My own country, Sweden, took a fairly relaxed attitude at this stage, being able to accept almost any conceivable option, as long as a fair amount of flexibility was assured. This was particularly true with regard to "differences in starting points and approaches": one must bear in mind, for example, that Sweden reduced its CO_2 emissions by around 30 percent between 1973 and 1988.

The Final Round

As the final session of the INC approached, we all realized that high-level political contacts were underway between key capitals. We also understood that the issues would remain very divisive at the EC level. The Commission was simultaneously involved in a difficult effort to reach agreement on a package of energy taxes, including a carbon tax.

But the Paris meeting of the Extended Bureau had given the chairman some clear guidance on how to structure the ultimate negotiating session of the INC. It falls outside the scope of my contribution to comment in detail on this strenuous meeting, but, given the circumstances, the result was wholly satisfactory.

I would like to add a few personal comments before elaborating on the contents of the specific commitments embodied in the final text of the Convention. I offer these observations in order to shed light on the work to be undertaken over the coming years.

- Chairman Ripert maintained exquisite control over the meeting. He had decided that the working groups should not be activated during the session. Instead, "clusters" of issues were negotiated in smaller groups. He allowed a number of informal contacts to develop over the first days, before presenting a consolidated text of his own. The ultimate negotiations then took place among key delegations in the Extended Bureau at the Head of Delegation level. In particular, the two eighteen-hour sessions towards the end finally led to agreement. The use of this method is difficult and problematic. It cannot fail to create problems with delegations which are not participating—but Jean Ripert managed to handle this situation in an efficient and effective way.

- The Paris consultation was particularly useful in facilitating agreement on the controversial article on Principles. The United States had taken a very uncompromising line on this issue from the beginning, trying to avoid having any such section in the final text. Now there was a meeting of minds, which led to the ultimate agreement in New York on the text of a limited article on principles.

- The fundamental question of transfers of financial resources and technology is of paramount importance. For the G-77, this has been a central point not only in the climate negotiation, but in the whole UNCED process. This issue was of particular importance in the context of the Climate Convention, however, since the developing countries could maintain that the OECD Group has been responsible for 75 percent of the atmospheric buildup of greenhouse gases. It would therefore be absolutely unacceptable to force the developing countries to hold back on their national development plans in order to reduce the risks of rapid climate change. To me, this argument is irrefutable.

It is also clearly in the long-term interests of the OECD countries to enable developing countries to choose development paths that will minimize negative effects on the global climate. However, in practical terms the issue is difficult to tackle in an economic climate where few new funds are available, and when the notion of incremental costs is still not sufficiently explored. Against that background, the negotiation has so far centered more on the mechanisms for financial transfers than on the resources to be committed. The issue will not go away, but the solution negotiated so far—that the Global Environment Facility (GEF) will be used as a vehicle for the delivery of new and additional funds on a conditional basis—appears to be rational and workable.

Due to the pressure of time, the institutional arrangements to implement the convention are not yet fully in place. This will put considerable pressure on the work during the period preceding the entry into force of the Climate Convention and on the first session of the COP.

And Now to the Future

The UN General Assembly has set the next phase of discussions on the right track by allowing the INC to continue its work. This phase will emphasize two main lines of action. The first line is focused on the preparations for the first COP. This preparatory process will be intense, with many difficult issues still to resolve before the COP convenes in early 1995. The second line of action will continue to explore the possibilities for constructive work on climate issues at the national or EC level. Such national or regional actions are to be encouraged now and need not wait on any future global agreement.

We also foresee a strong need for continued scientific research. The reorganized IPCC must work closely with the INC. And there is a need for continued inputs from the independent sector. The successful past arrangements for NGO participation in the INC will continue to be appropriate for the foreseeable future.

There may be many diverging views on priorities at this stage of the process. I would like to offer my own analysis, based on a reflection on the contents of the Convention and its relationship to other aspects of UNCED follow-up. My point of departure is Article 4(2) on the specific commitments to be undertaken by OECD countries, since this has been the centerpiece of my own work on the convention. I will, however, also offer a number of comments on other aspects of the Convention as well.

The Nature of the Specific Commitments and the Interim Period

Sub-paragraph 4(2)(a) reflects in an important way the US position in the negotiation, but the language is firmer than the United States originally proposed. In lieu of language on "the reduction of growth of net anthropogenic emissions of the total of all greenhouse gases not covered by the Montreal Protocol," the Convention now calls for "the return by the end of the present decade (i.e. year 2000) to earlier levels of anthropogenic emissions of carbon dioxide and other greenhouse gases not controlled by the Montreal Protocol."

This is certainly not the absolute, firm commitment that was originally sought by the Europeans for stabilization of CO_2 emissions by the year 2000 at 1990 levels. But the paragraph has to be read in conjunction with sub-para 4(2)(b), which commits each of the Parties that have undertaken specific commitments to submit "detailed information on its policies and measures referred to in sub-para 4(2)(a) as well as on its resulting projected anthropogenic emissions by sources and removals by sinks of greenhouses gases not controlled by the Montreal Protocol for the period referred to in sub-para 4(2)(a), with the aim of returning individually or jointly to their 1990 levels of these anthropogenic emissions." These two paragraphs are at the heart of the Convention. How they are implemented in the years to come will determine whether or not the world feels that "developed countries are taking the lead in modifying longer-term trends in anthropogenic emissions consistent with the Objective of the Convention." (See further sub-para 4(2)(a).)

In my view it is essential that the implementation of these paragraphs be discussed and negotiated without delaying the process leading to agreement on the framework of the interim process. One important question immediately comes to mind: should there be binding protocols signed by the states that have undertaken specific commitments, in order to implement the details of national reporting and to address some important subsidiary questions? The OECD countries will have to face that question very soon: in my own view, the balance of the arguments are in favor of a protocol. It would give more stability to the whole process and it would clearly demonstrate that the OECD countries are serious in their approach to the implementation of the Convention.

Whether there is to be an OECD protocol or not, a number of issues will have to be clarified during the interim period. This is necessary to enable the developed country parties to present adequate information within six months of the entry into force of the Convention. There are two different categories of such issues:

- issues directly related to reporting techniques;
- substantive issues related to the character of the commitments.

I do not want to go too deeply into the technical detail of reporting requirements in this paper. There is already some precision contained in Article 12 of the Convention, where sub-paras 2 and 3 deal with the specific reporting demands to be placed on developed countries.

The frequency of reporting will be decided upon by the Conference of the Parties and need not be an immediate concern. On the other hand, it is desirable to begin now to reflect on the procedures for review by the Secretariat of future submissions by the Parties. The reporting requirements are, I believe, the most forceful means of ensuring efficient implementation of the Convention.

Core Issues

Several points related to specific commitments were of capital importance in the negotiation between OECD countries. I believe that the most important ones are:

a) *Joint implementation.* The notion of joint implementation was first presented by the European Community as part of their decision in 1990 to stabilize CO_2 emissions in the year 2000 at 1990 levels. Joint implementation refers to bilateral agreements between countries to share the costs of (and the credits for) emissions reductions. This approach implies a type of "bubble" concept. The EC has its own north–south problem, resulting from the contrast between the development requirements of countries in the south such as Spain, Portugal and Greece and the managed growth objectives of the economic "giants" to the north, e.g. Germany, France, UK, and the Netherlands.

But as the 1990 decision applied to the EC as a whole, it meant that some countries must actually do more than others to reduce emissions. Sub-paragraph 4(2)(a) explicitly allows joint implementation, and sub-para 4(4)(c) states that the first session of the Conference of the Parties shall take decisions regarding criteria for joint implementation. This issue is also of interest to the EFTA countries, which will soon be cooperating with the EC in the framework of the European Economic Area (EEA).

b) Linked to this issue has been the notion of *cost-effective implementation.* Choosing the most cost-effective investments requires countries to cooperate in an organized way. One approach involves using an international "clearing house." Following this approach, country "A" would "credit its emissions account" with reductions achieved in country "B" through the financing of measures — for example, measures to increase energy efficiency in power plants. The clearing house would analyze and rank various project proposals by their cost. Norway elaborated proposals of this type in the course of the negotiation, suggesting a specific mechanism for cooperation with developing countries and the possible establishment of a "clearing house" to match money and projects. Within this particular framework, one comes rather close to the notion of tradeable permits. However, there was no possibility of reaching a complete agreement on a clearing house during the negotiation. The language in sub-para 4(2)(a) opens the possibility for such action and it appears desirable to continue the discussion at an early moment in the next stage of negotiations.

c)   A totally different set of problems is linked to the notion of a *"comprehensive approach."* This concept was the center of attention at INC I in Chantilly. At that time, the EC insisted that since CO_2 is the only truly measurable GHG (given the present state of science), only CO_2 should be part of the developed country commitments in the Climate Convention. At the time, the US and some other countries maintained that *all* greenhouse gases should be counted in the Climate Convention. But the Europeans felt that this could lead to a serious weakening of the commitments, particularly since the US also wanted to include gases covered by the Montreal Protocol. The controversy was softened in the course of the negotiation, and the present language in the convention refers to "carbon dioxide and other greenhouse gases not controlled by the Montreal Protocol." Nonetheless, there remains an important area of uncertainty; the OECD countries must try to tackle this issue as soon as possible.

d)   The comprehensive approach to non-CO_2 greenhouse gases is connected to another issue that loomed large in the negotiation: the *treatment of sinks and reservoirs*. Countries with important forest coverage, such as Canada, Finland, Sweden, and Russia, have repeatedly raised the question of "net emissions." There are very important scientific uncertainties to cope with in this area at the present time, and the language of the Convention does not now state any particular relationship between sources and sinks. The text merely indicates that the reports should cover both sources and sinks. A specific problem was raised by Sweden with regard to the burning of biomass, an important element in that country's energy policy. We tried to limit the complexity of the problem somewhat by introducing the notion of emissions of "greenhouse gases based on fossil fuels." No agreement could be reached on this issue before Rio '92.

e)   With regard both to comprehensiveness and to all the questions related to sinks, there is need for *more scientific research*. This is acknowledged in sub-paragraph 4(2)(c) which also states that the Conference of the Parties shall consider and agree on methodologies for assembling national inventories of greenhouse gases at its first session. This means that the sink issue has to be tackled early in the interim process. The IPCC continues to work on these and other important matters related to the implementation of the Convention.

f) A different issue is the degree of *flexibility* that should be allowed to countries undertaking specific commitments as required by sub-para 4(2)(a). The somewhat excessive list of disclaimers in this paragraph could lead to a serious weakening of the convention if not properly managed and circumscribed. As an optimist, I believe, however, that it will be possible for the countries concerned to define the terms of flexibility in such a way that the effectiveness of the Convention is safeguarded. The Precautionary Principle[3] requires that the Convention be implemented in good faith and with a common determination to do as much as possible, not as little as possible. In any case, it is necessary to start analyzing which countries deserve postponements or exemptions based on the elements listed in sub-para 4(2)(a) following the words "taking into account . . ." The signatories must soon agree on a restrictive implementation of this list in order not to dilute the effect of the commitments in this Article.

g) One final issue linked to sub-paragraphs 4(2)(a)-(c) should be noted: *the list of countries in Annex I* of the convention. This list contains OECD countries and a number of former socialist countries undergoing the transition to market economies. Paragraph 4(6) allows these countries a certain degree of flexibility in the implementation of their commitments. The OECD countries will have to open early discussions with the representatives of economies in transition. These discussions are necessary in order to define the degree of flexibility that these countries should be allowed in fulfilling their commitments under Article 4.

This seems like quite an extensive list of issues to tackle in the interim period, even though I have only dealt with problems flowing directly from the specific commitments on sources and sinks. I have already indicated that it would be possible to consider a separate protocol for countries undertaking these specific commitments. In order to facilitate such an initiative I believe that OECD and agencies belonging to the OECD family, such as the International Energy Agency (IEA) and the European Conference of of Ministers of Transport could lend valuable support. All these organs are already involved in the climate issue: I would hope that the member countries could work out a coherent way to use these intellectual and analytical resources to facilitate the negotiations.

North–South: The Development Imperative

So far, I have dealt almost exclusively with the problems related to the specific commitments on sources and sinks. Yet there is no doubt that the relationship between North and South will continue to be a principal focus of attention during the interim period. The real impact of the general commitments undertaken by developing countries, as well as the reporting obligations which they accepted, remains to be seen. In my view, these impacts will emerge slowly owing to the general lack of new funds in a depressed world economy.

The leading spokesmen of the G-77 have repeatedly pointed out that the fulfillment of their obligations will be dependent on the amount of financial and technical resources that their countries receive to cover the "agreed full, incremental costs of implementing measures that are covered by Paragraph 4(1)" (i.e., the so-called general commitments). If only small amounts of funds are forthcoming, then the commitments could not be interpreted very broadly. But, no precise definition is given in the Convention for incremental costs. The GEF is now addressing the issue of defining incremental cost, but it is clearly more complicated in the climate context than it is under the Montreal Protocol.

Many important questions affecting North–South relations lie, to a large extent, outside the climate issue *per se*. These include: what will happen to the International Development Agency (IDA) replenishment? Will the GEF become operational at a sufficient level within a reasonable time-limit? Are there any possibilities for other sources of financing to be conceived? One important question will be the relationship between the GEF and the Convention. We must avoid the extremes: the GEF must not be allowed to set the broad guidelines for financing of projects under the Convention. But, on the other hand, the Conference of the Parties must avoid being drawn into the micromanagement of specific projects. Many of these questions remain highly controversial. They were the subject of the final negotiation in the INC in May 1992, and they will not appear in a more favorable light as we move into the interim period. But they cannot be brushed under the rug. They have to be addressed. And some practical arrangement between North and South must be reached if the objectives of the Convention are to be achieved.

Financing problems will loom large in the follow-up to UNCED. The serious economic slowdown in the OECD will discourage bold, new initiatives. Yet it is quite clear that new financial resources must be provided to the South if developing countries are expected to contribute in full measure to solving the global climate problem.

Rapprochement between North and South must be achieved if the developing countries are to be full partners in the Convention today and to undertake specific commitments in the future. The problem is how to make implementation of the Convention support national development strategies. Although this is a problem that most people would prefer to avoid, I think that we have to be both realistic and optimistic: we can achieve both environment and development objectives within the framework of the Convention. Doing so can promote a more general understanding of the need for a new partnership between North and South. We should begin to frame such new partnerships immediately. This, hopefully, will involve the transfer of new technology and the development of innovative forms of financing, including automatic transfers, e.g., through some forms of international taxation. Furthermore, I won't give up the idea of a real "peace dividend": there must be a way of avoiding a continued stalemate. We all share the same small planet and the climate issue has more than symbolic value.

In this context, the Parties to the Convention must come to grips with an issue not adequately covered during the negotiation: namely, differentiation of responsibilities. The differentiation of national responsibilities was first addressed in a Swedish initiative at the second session of INC. The proposal was based on the idea that one should establish a number of categories of countries with different specific commitments. At INC 2, France supplemented the Swedish proposal by calling for the use of objective, quantitative criteria— per capita income, GNP, emission levels, etc.—to define categories and govern passage from one category to another. Like many other ideas that were proposed during the negotiation, it would not fly. What we now have in Paragraphs 4(8) and 4(9) simply refers to the need to give special consideration to countries that are particularly vulnerable (Paragraph 4(8)) as well as to the least developed countries (Paragraph 4(9)). To be complete, I also mention Paragraph 4(10), which deals with the particular situation of countries that are highly dependent on income generated from the production and consumption of fossil fuels. It is not clear what the special consideration implied in such paragraphs can—or should— be. But if I may risk a guess, I expect that the issue of differentiated commitments will come to the surface again before the end of the decade.

One group of countries requires a special comment at this stage: the small island states. Many of them are not only small but also low-lying. For these countries, the prospect of sea-level rise becomes a question of national survival. The delegates from the Alliance of Small Island States (AOSIS) gave a special urgency to the climate negotiation. Since we are working with long lead times, it may well be that even if fundamental trends can be reversed, the question of assisting countries to adapt to sea-level rise must still be addressed. The rights of these particularly vulnerable countries have been clearly acknowledged in Paragraph 4(7) of the convention.

The Climate Convention and Agenda 21 Follow-up

I have not dealt with institutional issues in this personal assessment of a fascinating negotiation. But they no doubt will play a very important part in the negotiations during the interim period and in the first Conferences of Parties. Satisfactory solutions to many of the issues raised here will be a prerequisite for the smooth operation of the Convention. I did not participate very much in the discussions of institutional issues during the negotiation, so here I can only express the hope that the negotiators will be capable of devising practical solutions without losing too much time on details.

One issue is particularly important: the operation of the Subsidiary Body for Scientific and Technological Advice (SBSTA), and its relationship to the IPCC. Sound scientific backstopping is a precondition for the application of the precautionary principle—and the Climate Convention is highly dependent on the precautionary approach. The IPCC has been instrumental so far, and it is essential that its role be maintained intact until we can be sure that any other structure can deliver better service to the negotiations.

The need for a solid scientific background is underlined by several of the provisions in the Convention. In particular, sub-paragraph 4(2)(d) stipulates that a review of the adequacy of commitments should be undertaken at the first meeting of the contracting parties and then "once more not later than the end of 1998." Thereafter, such reviews shall occur at regular intervals determined by the Conference of the Parties, until the objective of the Convention is met." This last phrase in fact refers to the ultimate need—as clearly stated in IPCC's reports—to go beyond stabilization into a stage of significant reductions of greenhouse gas emissions.

We cannot foresee today what measures must be undertaken to carry out such reductions: no doubt the effects on energy and transport systems, in particular, will be significant. However, since the long-term future is at risk, these problems have to be faced, and our societies have to be ready to consider far-reaching changes in life-styles and consumption patterns. This can only take place only if there is a general understanding of the need for such changes. They cannot be imposed on people in democratic societies. Such changes will have to be spread over a long period of time, say twenty to twenty-five years—but the effort has to begin today.

This is the point where Agenda 21 can, and will, support the climate convention, since several chapters address precisely these problems. There are the general recommendations in the chapters on economic policies or consumption patterns, but the key is the more specific language on transport, energy and industry sectors in the Atmosphere chapter. To take one example: in the transport section the possibility of regional conferences on transport and the environment is mentioned. Sweden is right now trying to get agreement on such a conference for Europe, Japan, and the US. If a Conference of this kind can be held, it would obviously assist countries in taking measures which would help them in meeting their commitments in the climate convention. Similarly, if the statement of principles on forests and the corresponding chapter in Agenda 21 leads to an increase, rather than a reduction, in the Earth's forest cover, there would be a favorable impact on the efforts to combat global warming. This outcome would strengthen the Convention.

Epilogue

This assessment reflects my personal perceptions of the negotiation and of the prospects for the coming years. Others will certainly look at these issues from other angles, and some of my statements and conclusions can no doubt be challenged. But I do believe that all of us who had the privilege to participate in the negotiation felt that it was a very special experience. Over all the years that I have dealt with multilateral cooperation, I have never before felt the same need to balance the interests of my country with those of the planet as a whole. Nor had I been faced with the need to balance the interests of today with those of tomorrow. Perhaps that is the real fascination with the climate issue: it forces us to reflect on and to take responsibility for things that will happen far beyond our own brief lifespan.

Notes

1. In the early part of the negotiation, controversy on the Chairmanships in the working groups became a very difficult issue. It was not until the end of the second session, in Geneva in June 1991, that a solution could be found.

2. IPCC is working on this issue.

3. The Precautionary Principle encourages nations and individuals to take early actions to prevent environmental damage even when these actions are not economically attractive.

CHAPTER 8

The Road to Rio

José Goldemberg
Professor
University of Saõ Paulo
Saõ Paulo, Brazil

Prologue

Brazil's participation in the United Nations Conference on Environment and Development (UNCED, sometimes called the Earth Summit or Rio '92) began in 1988. Then-President Sarney responded to the outcry in the international press on the "burning of the Amazonia forest" by offering to host the Earth Summit. Until then, global environmental concerns had been absent from discussions in Brazil. Typically, the reports appearing in the international press of events occurring in the Amazonia were most often regarded in Brazil as attempts to interfere with national sovereignty.

Moreover, the environmental movement was not very strong in Brazil at this time. Urgent, more immediate problems—such as the very serious water and air pollution problems in the cities—were consuming the attention of the fledgling environmental movement. Their focus was on such cities as Cubatao, near the Port of Santos, which became popularly known as the "valley of death."

It was the Montreal Protocol on Substances that Deplete the Ozone Layer that first raised awareness of global environmental problems in Brazil. Political leaders and private citizens began to recognize that serious global problems existed and that events taking place inside national borders could influence the climate in other countries.

The work of a few Brazilian scientists helped to raise public consciousness about national environmental problems and their connection to global issues. These scientists argued convincingly that the indiscriminate felling of the Amazonia forest would not only contribute appreciably to global CO_2 emissions but that it could also significantly harm the regional climate in the northern and center-west areas of the country. At the same time, it became apparent that the process of deforestation in the Amazonia was not as extensive or as catastrophic as was originally reported by international observers, including Myers, Setzer and others. This realization gave confidence to the Brazilian government that it could face the external criticism that was sure to accompany a major international conference on the environment to be held in Brazil.

Thus, the initial Brazilian position denying any wrongdoing in Amazonia gave way to the more realistic national position—that deforestation was a major problem but measures could, and would, be taken to reduce the extent of the damage. In so doing, Brazil became, in effect, a responsible actor in the field of international environmental problems. With the new policies in place, the government was able to counterattack in the media, pointing out that the great majority of contributions to global greenhouse gas emissions occurred in industrialized countries.

In 1990, President Collor gave strength and visibility to Brazil's new environmental policy. At the international level, this policy included requesting "new and additional resources" and the "transfer of clean technologies" from the industrialized countries in order to limit additional contributions to the greenhouse effect. The political objectives of President Collor were (1) to obtain some national advantages from international cooperation and (2) to obtain a clear personal victory by transforming Rio '92 into a success. The dynamics of events leading to Rio '92 were such that the second objective proved easier to achieve than the first.

Dynamics of the Climate Negotiations

To make Rio '92 a success, Brazil chose a conciliatory approach in the international discussions on the Climate Convention. The dynamics of the debate emerged from the political maneuvering that occurred in two principal fora: the Intergovernmental Panel on Climate Change (IPCC) and the Intergovernmental Negotiating Committee for a Framework Convention on Climate Change (INC).

The IPCC was an extraordinary exercise in international cooperation that involved hundreds of scientists, economists, diplomats and public servants. The IPCC Scientific Assessment raised the awareness of politicians and citizens worldwide to the seriousness of the "greenhouse problem"—just as the discovery of the Antarctic Ozone Hole had stimulated a general recognition of the dangers associated with the depletion of the ozone layer. As a consequence of the IPCC process, many people became educated to the intricacies of a complex scientific phenomenon, its consequences, and the possible policy response strategies that could be employed to cope with it.

The INC held a series of negotiating sessions to clarify the policy consequences of the IPCC Assessment. The INC process was influenced by politics right from the start. Developing countries, working together under the umbrella of the Group of 77 (G-77), realized that the existence of a "greenhouse effect" gave new leverage to their claims for a better distribution of wealth among nations. The G-77 tried to insist that industrialized countries raise the target level of Official Development Assistance (ODA) to 0.7 percent of GNP. The argument frequently used was that poverty was the principal cause of environmental degradation and therefore national economic development goals had to be achieved first —for only then could developing countries contribute to the solutions of global environmental problems.

The principal tension during the UNCED preparatory process concerned its emphasis: would the Earth Summit emphasize development and poverty or environmental protection and sustainability? In my view, the idea that the Rio Conference could become a conference on development and not environment was a "midsummer night's dream"—nothing more than a naive fantasy. Global environmental degradation, and, in particular, the greenhouse problem, is a consequence of affluence, principally the burning of oil and coal. Local environmental degradation, on the other hand, is intimately linked with poverty. The industrialized countries were not particularly interested in addressing the root causes of poverty, which had been the focus of the North–South confrontation for the last thirty years. Limiting the outcome at UNCED to a much less ambitious target—reducing dangerous emissions of greenhouse gases at moderate cost—seemed to be the unspoken goal of most of the leading industrialized countries in Rio.

From the point of view of the developing countries, most global environmental problems result from over-consumption and excessive energy use in the North. Nonetheless, throughout the climate negotiations, the US argued for a companion convention on the protection of forests. This tactic was seen by the developing countries as an artful diversion and an attempt to spread the "blame" for the greenhouse problem.

Most of the industrialized countries recognized that the vast majority of greenhouse gas emissions result today from commercial energy use in the developed countries. By the year 2020, emissions from commercial energy use in developing countries will have grown dramatically and will probably surpass the emissions from the industrialized countries. However, today, the main contribution to global climate change from developing countries is the result of deforestation and changes in land-use patterns. Thus, the emphasis on deforestation by the US was seen by the Brazilian government as a direct threat to national sovereignty, because it could easily lead to interference in the internal affairs of Brazil. From the Brazilian point of view, such a situation could be acceptable only if it were coupled tightly to binding limits on the emissions of greenhouse gases from industrialized countries.

Midway through the negotiations, the Organization of Petroleum Exporting Countries (OPEC) introduced a new measure of merit, the concept of "net emissions." "Net emissions" refers to the difference between the annual rate of uptake of CO_2 or other greenhouse gases by the biota and the rate of anthropogenic emissions of the same gases. Since it is virtually impossible to measure biotic rates of uptake today (and will remain so for the foreseeable future), one could conclude that this approach was introduced by the OPEC countries in order to downplay the importance of fossil fuel-derived emissions of CO_2. While the measurement of biotic uptake is impossible today, I think it is fair to consider all the other greenhouse gases which contribute to the risks of rapid climate change. However, to put these on a common footing with CO_2, a suite of suitable conversion factors is necessary—such as the Global Warming Potential factors (GWPs) developed by the IPCC. Continuing uncertainty concerning the direct and indirect warming effects of particular gases has precluded the emergence of any clear consensus in the scientific community as to what precise values should be assigned to the GWPs of various greenhouse gases.

Fortunately, the members of the Organization for Economic Cooperation and Development (OECD) were split on a number of issues. The European Community (EC), as well as some individual

European countries, favored the inclusion in the text of the Convention of strict, explicit targets and unambiguous timetables for the reduction of greenhouse gas emissions. The Netherlands, Germany, and the Nordic countries favored this approach. Others, such as the United States, and to a lesser extent Japan and Great Britain, were skeptical about including specific targets and timetables. These governments argued that the enthusiasts for greenhouse gas emissions reductions had no real idea of the costs of such measures in terms of money and jobs. As the negotiations moved forward, the position of the United States hardened to the point that many questioned whether President Bush would even attend the Rio '92 Conference.

In the latter stages of the first phase of the INC negotiations, Norway proposed the creation of an international clearing house for emission reductions and damage abatement strategies. This concept, referred to as "joint implementation," would allow an industrial country to fund emissions reductions in a developing country and to share the credit for the reductions thus achieved between the two countries participating in the project. I think it is a good idea to accept some version of "emissions trading," although some developing countries resist the idea because—in their opinion—it legitimizes the high emissions of industrialized countries. Legitimizing these disproportionately high emissions rates is considered wrong on ethical grounds. This is a very naive view of the world; the same principle could apply to income but does not.

The recession in the United States (and other industrialized countries) dampened some of the sense of urgency about the climate problem. At INC 5, in February 1992, members of OPEC began to filibuster the negotiations. This led to an increase in the rhetorical noise of the G-77, replete with the usual arguments on "guilt," "historical responsibility," "compensation for past deeds" and the "right" of the poor to 0.7 percent of the GNP of the rich in the form of ODA.

One development that, in my view, worsened matters in the climate negotiations was the introduction of some over-simplified economic analysis during the preparations for Agenda 21. The UNCED Secretariat estimated that the "new and additional" resources needed to fulfill the promise of Agenda 21 would cost approximately US$125 billion per year. Not surprisingly, this announcement frightened many industrial country participants in both UNCED and the INC.

The true measure of what the industrialized countries were willing to do to protect the global environment can be gauged by the low level of funding of the Global Environmental Facility (GEF). This new financial vehicle operated jointly by the World Bank, UN Environment Programme, and UN Development Programme was funded in the pilot phase at a level of approximately US$1 billion over three years. The commitment of the United States can be measured by its minimal offers of money to the GEF and INC processes—US$50 million to GEF plus US$25 million for bilateral agreements to assist in the preparations of national inventories of greenhouse gas emissions.

The GEF will probably be the only vehicle for delivering "new and additional resources" until the first meeting of the the Conference of the Parties (COP). The COP will meet within six months after ratification of the Convention by fifty countries. This is unlikely to happen before 1995. The key element of the GEF program with respect to global climate change is the commitment by the developed countries to pay—through the GEF—"the agreed full incremental cost" of measures taken by the developing countries to advance the overall objectives of the Climate Convention. I accept the definition given by the GEF of the meaning of "full incremental costs": "incremental costs are the extra costs incurred in redesigning an activity vis-à-vis a baseline plan—which focuses on national benefits—in order to address global environmental concerns."

There are some potential problems with such a definition. For example, it could be understood to exclude actions by the private sector and actions taken to increase local capacity. Such a narrow definition would be short-sighted and sub-optimal. If applied by the GEF to decisions about project selection, it could quickly damp out the enthusiasm of developing countries for participation in the climate negotiations.

Two Revealing Episodes

Two events that took place a few months before the Rio '92 Conference led me to conclude that the high expectations held by many observers for commitments at Rio '92 were probably groundless. First, at their annual economic summit held in July, 1991, in Houston, Texas, the Group of Seven (G-7) industrialized countries offered US$1.5 billion in grants or highly concessional funds to save the Amazonia forest, reportedly at the insistence of Chancellor Kohl of Germany. It is thought by many that Kohl was worried about the upcoming parliamentary elections in Germany and that he hoped by engineering this declaration to capture the sympathy and neutrality of the German "Greens." In response to this offer of new funds, the Brazilian government proposed a number of projects but all were turned down. The reason given for this unequivocal rejection was that the first phase of the $1.5 billion offer was to be limited to only $250 million. This was quite a surprise to the Brazilian government because the Houston communiqué does not mention a "first phase" at all. Subsequently, we discovered that further conditions would also be applied: approximately $50 million of the $250 million in the so-called "first phase" funding were allocated to a "trust fund" for grants, and the remaining $200 million were to be negotiated bilaterally between Brazil and individual members of the G-7. The rest of the $1.5 billion was to be discussed at some unidentified point in the future.

The second episode occurred at the beginning of April 1992. I visited Washington at this time and had a revealing interview with a high officer in the Bush administration. He told me that the US would insist on the following preconditions in concluding any agreement on the text of a Climate Convention:

(1) There should be no "targets and timetables" for *any* greenhouse gases (especially not CO_2). A "pledge and review" process should be implemented to monitor voluntary, non-mandatory commitments by signatory countries.

(2) The GEF should be the sole channel for "new and additional resources." The governance of GEF should be made more representative but final authority for disbursements of funds should be left to the Executive Directors of the World Bank.

(3) There must be no rhetoric about "guilt," "crimes," etc. in the Earth Charter. Rather, the charter should be a concise, objective and positive text.

(4) A "statement on principles" on forest preservation should be adopted in Rio. This statement should include commitments leading the way to a future convention on forests.

Reviewing and analyzing the dynamics of the early climate negotiations, we became convinced that the preparatory process for Rio '92 was heading towards a potential disaster. We feared that the deadlock that stymied the Convention on Biological Biodiversity (which ultimately was not signed by the United States) would be repeated in the INC. It seemed imperative that efforts be made to reconcile the divergent national positions.

Brazil Moves Toward Conciliation

Following my meetings with the high officer in the Bush administration in Washington, the government of Brazil decided to send a small delegation of high-level officials to Japan, China and India.

The purpose of these visits was to obtain support for a Convention which would try to reconcile the widely opposing views of the G-77 and the United States.

In addition, Brazil and Switzerland began preparations for an informal meeting of the key actors in the climate negotiations. This meeting would be held in parallel to the second "resumed" session of INC 5 and would deliberately be kept small so as to avoid the heavy rhetoric always present in the large meetings of the INC.

We believe that this combination of initiatives quickened the pace of negotiations in New York. We also believed that they led directly to the agreements embodied in the Convention, although the April 1992 meeting of the G-77, which took place in Kuala Lumpur just before the final session of the INC in New York, attempted to torpedo these "parallel" negotiations. In Kuala Lumpur, the G-77 reaffirmed the usual rhetoric of "guilt," "crimes," and "historical responsibility." Ultimately, India, China and Brazil—the largest of the developing countries—succeeded in toning down the rhetoric of Kuala Lumpur. Thus, the Convention signed in Rio preserved a reasonable compromise between the hard-line positions of the US and the views of the 153 other signatories.

Specific Achievements of the Climate Convention

In my view the following are the most important elements of the final compromises embodied in the Climate Convention:

a)   lack of full scientific certainly should not be used as a reason for postponement (Article 3, Para 3)
b)   the "precautionary approach" should take precedence over adaptive strategies to global warming (Article 3, Para 3)
c)   serious, systematic efforts should be made to "return by the end of the present decade to earlier levels of emissions" (Article 4, Para 2(a)).

Lessons for the Future

A number of factors contributed to the success of UNCED and led to the near-unanimous agreement on the Climate and Biodiversity Conventions. (The US was the only important exception to this consensus—refusing at the last minute to sign the Biodiversity Convention.) In my view the following are the most important factors contributing to our success in these negotiations:

(1) Adequate preparatory work is essential. In the case of the climate issue, important preparatory work was done by the Brundtland Commission. These efforts raised the visibility of global environmental problems and highlighted the need for actions to support the process of "sustainable development."

(2) Public expectations of cooperation increase the likelihood of reaching agreement. Great expectations were raised by the lengthy process of preparation for the Earth Summit. These preparations received wide media coverage. The fact that there was a deadline to reach agreements (June 1992) helped convince many countries to pursue the negotiating process actively. Ultimately, the momentum of the negotiations propelled many governments to go to Rio and to sign the Conventions. In the next phase, the international discussions should proceed directly to inter-governmental negotiations. Workshops should be convened only when deemed necessary to clarify technically complex issues that arise in the negotiations. Otherwise, the scheduling of workshops will delay implementation of the Convention and might even be used deliberately for that purpose. Setting target dates for completing negotiations is as essential for the success of the next "protocol" phase as it was in concluding agreement on the Convention in time for UNCED '92. By early 1994, just before the COP meets, negotiations on the first set of protocols should have been completed.

(3) In some circumstances, domestic political events in a few key countries can shape the outcome of international negotiations. The presidential campaign in the United States encouraged President Bush to take positions that were chosen, arguably, to improve his standing with the US public in the November 1992 elections.

(4) Non-governmental organizations play a critical role: NGOs focus the spotlight of public attention on key issues. The "Global Forum"—a gathering of NGOs at Rio's Flamingo Park that was held simultaneously with the official conference—influenced attending politicians from many countries. Since the press coverage of the Global Forum was sometimes greater than the coverage of the formal meetings, wise political leaders paid attention to the concerns that were raised in Flamingo Park. To maintain a high level of public awareness and honesty in the process, NGOs and academics should continue to participate in the INC process and its follow-up activities. A "pledge and review" process is very weak and its success depends on the

presence of great determination on the part of the Secretariat to the Conference of the Parties. On the other hand, binding targets and timetables set at realistic levels give shape and substance to the international process. They allow a measure of accountability to be maintained as the negotiations proceed. My view is that without "targets and timetables" the implementation of the Convention will depend on the good will of OECD countries.

In addition to that, in the absence of "targets and timetables," the private sector will not feel inclined to do anything until forced to do so by governments. In the first phase of the negotiations, the Brazilian government worried that accepting "targets and timetables" might eventually result in limitations to the development of our country. Nonetheless, as the process of negotiations moves toward protocols, commitments to specific emission reductions by the developed countries will help to give credibility to the overall process. For advice on these matters, the INC should not set up a separate advisory body and could perfectly well rely on the Scientific and Technical Advisory Panel (STAP) of the GEF on the Scientific Working Group of the IPCC. The unnecessary multiplication of advisory bodies will lead only to delays.

Epilogue

History is littered with international agreements that took many years to negotiate but were never implemented. The Law of the Sea is a clear example of what we must avoid if the Climate Convention is to be kept out of this category. But without a large measure of attention and publicity in the future, the next phase of the climate negotiations may lack some of the urgency and drama that so dominated the months of discussion preceding Rio '92. We will need this sense of urgency—with its attendant energy and commitment—to maintain momentum in the negotiations and to reach agreement on practical protocols to the current Framework Convention.

The Framework Convention must be swiftly implemented in order to prevent rapid global warming. Fortunately, most of the necessary measures are of mutual benefit to industrialized and developing countries alike. Technology transfer and even direct grants to cover the costs of activities that will reduce emissions in developing countries can only help developed countries in their efforts to open new markets and to control global environmental problems.

In my view, the interests of Northern industrialized countries in maintaining an environmentally healthy planet can, and must, be mobilized to help developing countries; if these efforts can be addressed simultaneously to meeting the national development objectives of the Southern countries, the process of international cooperation for environmental protection can only be strengthened. The developing countries' goal of eliminating poverty will be achieved only if the motivation for international cooperation is based on a shared sense of enlightened self-interest, rather than on some vague appeal to moral virtue and humanitarian relief.

CHAPTER 9

A Failure of Presidential Leadership

William A. Nitze
President
Alliance to Save Energy
Washington, DC

Prologue

The framework convention on climate change was signed by 154 countries at the Earth Summit in Rio de Janeiro in June 1992. It represents—simultaneously—a success for United States diplomacy, a long-standing tension in American foreign policy, and a failure of presidential leadership.

There is little question that the climate change issue presents a difficult foreign policy challenge to the United States. The US is in the unenviable position of being at the same time the world's largest emitter of carbon dioxide and other greenhouse gases, the recognized leader of the international system, and the home of the world's most influential environmental organizations. The US therefore has no place to hide from the climate change issue at home or abroad and is universally perceived as both the largest part of the problem and the largest part of the solution. As much as President Bush and his closest associates may have wished for this issue simply to go away, it has refused to do so. In these circumstances, it is perhaps surprising that the final agreement signed in Rio is as well-designed and as forward-looking as it is.

A Success for US Diplomacy

At one level the US achieved its major negotiating goals in the final agreed text of the convention signed in Rio. It avoided a binding commitment to hold its CO_2 emissions below a specified level by a date certain, won agreement to the so-called "comprehensive approach" (whereby mitigation strategies would encompass all greenhouse gases rather than just CO_2), and persuaded all of the world's major countries to participate in an ongoing process for addressing the problem that is consistent with the initial US proposals. The US did agree to a far-reaching long-term objective: "stabilization of greenhouse-gas concentrations in the atmosphere at a level that would prevent dangerous anthropogenic interference with the climate system . . . within a timeframe sufficient to allow ecosystems to adapt naturally," but this wording is sufficiently flexible to give all parties considerable room to maneuver in the short term.

The US success in achieving its major negotiating goals resulted from the unwillingness of either the other OECD countries or the major developing countries to sign an agreement without the participation of the US. These countries determined for themselves that an otherwise well-structured convention with non-binding language on short-term targets that could be signed by the US was preferable to a similar convention with binding language that was not signed by the US Many delegates argued that this approach was probably justified by the need for active US participation in shaping a successful international strategy for addressing climate change. The US success in achieving its goals, however, still begs the question of whether it set the right goals in the first place. I will return to this question later.

Background to the US Position

The US Administration advanced a number of arguments to support its position that the convention should focus on a process for developing cost-effective mitigation strategies rather than on short-term targets and timetables for emissions reductions. Although there is broad scientific consensus that we will experience a significant warming of the Earth's surface due to the continuing atmospheric build-up of greenhouse gases, serious uncertainties remain about the extent, timing and distribution of that warming and the associated changes in climate. With the possible exception of emissions from the burning of fossil fuels in the industrialized countries, we do not yet have reliable inventories of greenhouse gas sources and sinks. Data concerning developing countries or economies in transition are sketchy or unreliable; there is no credible baseline against which to measure compliance with agreed targets and timetables. Furthermore, the Administration argued that it did not have sufficient assurance that it could comply with the proposed target of stabilizing US CO_2 emissions at 1990 levels by 2000 without serious harm to the domestic economy.

None of these arguments justifies the US refusal to agree to a binding short-term stabilization target. Scientific uncertainty about the extent, timing and distribution of global warming in no way reduces the risk of adverse impacts or the need for action to reduce those risks. More reliable information about anthropogenic sources and sinks of different greenhouse gases would indeed be helpful in designing a comprehensive international strategy to address greenhouse warming, but we already know that fossil fuel burning releases about 6 billion tons of carbon into the atmosphere every year and that the US accounts for almost one-quarter of that total. Finally, there are many who believe that the policy changes, if any, required to hold US CO_2 emissions at 1990 levels in 2000 would not harm the US economy given the admittedly large potential for cost-effective energy efficiency improvements.

Unfortunately, the Bush Administration's climate change policy was not based on a rational assessment of the national interest in light of these facts. It was based instead on a volatile mixture of ideology and politics. As discussed at greater length below, the internal policy debate was driven by the ideological preoccupations of a small circle of presidential advisors led by former Chief of Staff John Sununu. These men believed that the climate change issue was being used by environmentalists to impose their "anti-growth agenda" on the US economy. This perception made it difficult for them to see that the Administration could have linked increased investment in energy efficiency, renewable energy and other technologies to reduce greenhouse gas emissions with a market-oriented, pro-growth agenda. It was left to Governor Clinton and Senator Gore to make that connection in their successful presidential campaign.

Beneath these ideological preoccupations lay a core of hard politics. Any serious US policy to address climate change will require a gradual reduction in the production and use of coal and oil and a gradual increase in our reliance on energy efficiency, natural gas and renewable energy. But the states in which George Bush won a plurality on November 3, 1992 are on the whole more dependent on the production and use of coal and oil than the states in which Bill Clinton won a plurality. Unable or unwilling to present the transition to a new mix of energy sources in a politically positive manner, the Bush Administration felt the need to accommodate the wishes of industrial interests dependent on the production and use of coal and oil by opposing any targets and timetables for CO_2 reduction for as long as possible.

The Climate Change Issue as a Symbol of a Long-Standing Tension in US Foreign Policy

The Bush Administration's political response to the climate change issue in turn reflects a deeper aspect of the long-standing tension between Wilsonian idealism and realpolitik in US foreign policy. Americans have always thought of their country as being special and in some ways exempt from the constraints to which other nations are subject. As a result, we Americans place a high value on individual autonomy and freedom from state intervention in their lives. The sense of being exempt grows out of the vastness of the North American continent and the breadth of the Atlantic and Pacific Oceans. From Thomas Jefferson to Woodrow Wilson the central goal of US foreign policy was to avoid entanglements in the affairs of other nations while we went about the business of conquering and exploiting the great lands that history had bestowed upon us.

Starting with our entry into the First World War during the Wilson administration, however, the United States has increasingly been drawn into world affairs as a bulwark against the spread of authoritarian and totalitarian regimes and as a promoter of democratic values throughout the world. Despite a number of instances where American ideals clashed with short-term American political and commercial interests, this new "manifest destiny" abroad appeared to be in happy harmony with our "manifest destiny" at home. To paraphrase "Engine Charlie" Wilson, the former chairman of General Motors, what was good for America was good for the world.

Paradoxically, our very success in achieving our "manifest destiny" abroad has undermined this happy harmony. Our two great (formerly) authoritarian opponents, Germany and Japan, have become successful democracies and formidable economic competitors —in significant part due to US protection and assistance since 1945. Furthermore, the collapse of our great totalitarian opponent, the Soviet Union, has removed the primary justification for our leadership role in the world and has made a large portion of our military-industrial complex obsolete. As former Secretary of State Dean Acheson once said of Britain, we have lost our defining purpose but we have not yet found a new mission.

Our complementary success in achieving our "manifest destiny" at home has also undermined the harmony between our domestic and foreign policy goals. The United States has constructed a unique society powered by the consumption of large amounts of cheap energy, most of it derived from fossil fuels. High levels of consumption of both gasoline and electricity have been the primary means of satisfying the demands of citizens for the personal autonomy and freedom from government interference referred to above. Combined with a legal and political tradition hostile to land use restrictions, increased use of personal vehicles and electrification have produced the vast urban sprawl and "edge cities" that now dominate the American landscape. Americans have grown accustomed to driving to things they like, driving away from things they do not like, and using electricity to build autonomous little environments for themselves.

For better or for worse, much of the rest of the world would like to follow the United States down this ultimately unsustainable path. Both personal vehicle use and electrification have risen rapidly in Europe, Japan and the other OECD countries, where they now approach historic US levels. More significantly, people in Russia, China, India and Brazil aspire to the same patterns of energy use. Their success threatens to increase greenhouse gas emissions to dangerous levels and potentially to harm the global environment in other ways.

It is on this point that the climate change issue becomes the symbol for an underlying tension in US foreign policy. Sparing future generations from irrevocable degradation of the global environment has become a moral imperative for the world community. If the United States is to be true to the Wilsonian strain in its foreign policy, it must lead the world in finding ways to satisfy our common aspirations for economic development without risking such degradation.

However, since the poorer countries will be reluctant to cooperate in this endeavor unless the US changes its own patterns of energy consumption, such leadership requires fundamental changes at home. These changes imply both a degree of interdependence with other countries and a reduction in individual autonomy that will appear to some as contrary to American values and traditions. If Americans are to accept these changes and the limitations on national and personal autonomy that they may entail, they must be persuaded that the costs and benefits will be distributed fairly and that the country as a whole will benefit.

Fortunately, a strong case can be made that these changes—particularly increased reliance on energy efficiency, natural gas and renewable energy sources—will not only reduce greenhouse gas emissions but will also result in greater economic growth, a net increase in jobs, reduced oil imports and increased US competitiveness. Governor Clinton and Senator Gore made this case convincingly in their successful presidential campaign. If President Bush had understood it and possessed the capability to communicate it to the American people, the US might already have adopted a more progressive position on the climate change issue, engaged broad international support and been able to use the climate negotiations to make greater progress towards a sustainable energy future for the whole world.

A Failure of Presidential Leadership

This brings us to the basic explanation for the Bush Administration's negative posture during the climate change negotiations—a failure of presidential leadership. President Bush's failure to come to grips with the climate issue is in large part the result of personal ignorance. In contrast to Margaret Thatcher and other world leaders, the President never took the time to be thoroughly briefed about the science, impacts or policy implications of climate change. Instead he chose to rely on the knowledge, opinions and judgment of a few key advisors, particularly his former Chief of Staff, Mr. Sununu.

There was an easily available and rewarding alternative. The President could have arranged for a comprehensive briefing on the science of climate change by his Science Advisor and a group of carefully chosen experts. He could have reached out to a broader variety of experts in considering possible policy responses and their respective costs and benefits. If he had, he would probably have adopted a position closer to that taken by Governor Clinton and Senator Gore. Ironically, the Administration initially positioned itself well on the issue with Secretary of State Baker's endorsement of the "no-regrets" approach to climate change in his first official speech on January 30, 1989. But after it became apparent that Governor Sununu was strongly opposed to the spirit and much of the substance of that speech, Baker reversed himself on the issue in early 1990 and withdrew from the field.

With both the President and the Secretary of State disengaged, Sununu was free to shape US policy to his liking. In theory, the White House Chief of Staff does not himself make policy but rather makes sure that the bureaucracy presents the President with a range of options on key issues that are consistent with the President's basic agenda. In practice Chiefs of Staff have considerable leeway to introduce their own views into the policy process. Believing that he knew more about the issue than anybody else in government and that he was protecting the President from an environmental cabal that was committed to stopping further economic growth in the US, Sununu used this leeway to the hilt. He opposed any initiatives that might require the US to take steps to reduce its greenhouse gas emissions beyond those it was already committed to for other reasons.

Starting in 1989 Sununu faced resistance to this approach from William Reilly, the Administrator of the Environmental Protection Agency (EPA), and his colleagues and from elements in other agencies, notably the State Department's Bureau of Oceans and International Environmental and Scientific Affairs (OES). Sununu was able to neutralize this resistance in late 1989 by engineering the resignation of two key officials in OES and transferring responsibility for the climate change issue away from the State Department to a White House working group under his direct control. The White House retained control over the climate change issue through the Earth Summit and beyond.

The US head of delegation during most of the climate convention negotiations, Robert Reinstein, in theory reported to his superiors at the State Department. But in practice, he reported directly to the White House, and ultimately to Sununu. EPA Administrator Reilly played a critical role in getting Sununu and Bush to commit to the negotiations on the climate convention in May 1989, but was subsequently relegated to a marginal role on the issue. To their credit, Reinstein and his colleagues at the Department of State succeeded in negotiating provisions in the convention that created an international agreement for periodic review of the Climate Convention based on national reports and regular scientific assessments. This process will set the stage for farther reaching commitments on emission reductions later on.

With Sununu's resignation as Chief of Staff in early 1992, the climate change negotiations came under the control of Samuel Skinner, his successor, and Clayton Yeutter, the short-lived domestic policy coordinator, neither of whom shared Sununu's strong opinions on the issue. Initial hopes for a change in US policy were frustrated, however, by the continuing influence of OMB Director Richard Darman and Economic Advisor Michael Boskin, who had been Sununu's allies on this and other issues. It was only when it became apparent that George Bush would be totally isolated at the Earth Summit if the US did not compromise on climate change that the US Administration showed some flexibility—and signed on to the compromise language on emissions stabilization negotiated by British Environment Minister Michael Howard, Baker's deputy Robert Zoellick, and Robert Reinstein.

Although it received the most public attention, the concept of targets and timetables for CO_2 emission reductions was not the only issue on which the US found itself at odds with other countries during the convention negotiations. There were at least two other important issues as well: the level of financial assistance that would be provided to developing countries to offset the "full incremental costs" of any steps they might take to reduce emissions, and the degree of access to US and other advanced-country technologies on non-commercial terms. In contrast to the targets and timetables issue, however, the US had considerable support from other OECD countries on the issues of financial transfers and intellectual property rights concerning new technologies.

The US positions on these issues have remained constant since the start of the Bush Administration in 1989. With respect to financial assistance, the US has argued that "new and additional" development assistance for projects related to climate change cannot be justified until the specific need for such assistance had been better identified. With respect to non-commercial access to advanced technologies, the US has argued that the private sector controls most of the technologies concerned and that protection of intellectual property rights is in the long-term interest of all countries. Rather than commit large amounts of government money to a "climate fund," the US proposed that the OECD countries make available more modest amounts to assist developing countries in preparing inventories of their sources and sinks of greenhouse gases, in developing national plans for reducing their emissions, and in implementing specific agreed projects. The US and the other OECD countries prevailed on these issues and the approaches described above are reflected in the provisions of the convention dealing with financial assistance and technology cooperation.

What Could Have Been: US Leadership Misses the Boat

The bottom line appears to be that the US got the kind of process-oriented convention that it wanted and that, with the exception of binding targets and timetables for CO_2 reductions in the OECD, the Climate Convention is, on the whole, a reasonably well-constructed document. George Bush's leadership failure has, however, had a real cost in terms of missed opportunities.

Those opportunities do not relate principally to the lack of a binding stabilization commitment in the convention. First, the convention contains the strong objective of stabilizing greenhouse gas concentrations in the atmosphere at a level that would prevent dangerous anthropogenic interference with the climate system. Second, none of the other OECD countries is likely to go back on its prior stabilization commitments. Third, the special obligations language of the convention, although not legally binding, is sufficiently strong to constitute a political commitment to stabilization of emissions, even for the US.

Where the US refusal to agree on a binding stabilization target imposed a real cost on the process was in making it almost impossible to engage in constructive discussions on specific short-term steps for reducing emissions. US stonewalling on targets and timetables gave other countries a perfect excuse to avoid discussing specific steps that they could easily take to reduce emissions. China, India, Brazil and other developing countries were able to justify deferring any serious obligation to develop climate mitigation strategies on the basis of the US refusal to make a serious commitment to reduce its own emissions. The European Community, Japan and the other OECD countries were likewise able to use the US position as a screen for not getting too specific on how they were going to fulfill their own political commitments.

Thus, by failing to agree in early 1991 to join other OECD countries in stabilizing CO_2 emissions at 1990 levels by 2000, the US missed the opportunity to gain more specific agreements on the details of:

- an infrastructure for inventorying and monitoring greenhouse gas sources and sinks, particularly in developing countries;
- the process of forming national strategies, including internal policy reforms;
- a mechanism for financing projects to implement national strategies that leverage the concessionary financing of the Global Environment Facility (GEF) and other public funds with the much larger

amounts of private investment available in the next decade; and
- a process for setting longer-term greenhouse gas reduction goals and allocating responsibility for achieving those goals, in light of new information about both the science and impacts of greenhouse warming and progress in implementing national strategies.

It would not have been easy for the US to obtain consensus on such provisions given the positions taken by China, India, Brazil and other key developing countries. These countries distrusted the US and other rich industrialized countries on at least three grounds. First, they perceived rich country efforts to get them to reduce their greenhouse gas emissions as a device to gain control over their development and a threat to their sovereignty—a form of eco-colonialism. Second, they saw the climate change issue as a pretext for the OECD countries to divert development assistance funds from something that was more important to them—economic development—to something that was more important to the rich countries—preventing long-term climate change. Third, they believed that rich country demands on developing countries to restrain their emissions were hypocritical. The rich countries had created this problem through their own polluting growth patterns and their continuing over-consumption. Therefore, argued developing country delegates, they should clean up their own act before asking the developing countries to do anything.

These suspicions manifested themselves in the debates on two specific issues. The first was whether the rich countries should pay the "full incremental costs" of poor countries installing and using less-polluting technologies. The US and other OECD countries finally agreed in principle to paying "agreed full incremental" costs without committing to specific amounts or modalities of payment as a condition to getting the developing countries to sign the convention. The developing countries insisted on this provision only in part because they believed the less-polluting technologies were inherently more expensive. In larger part they pressed for it as leverage to ensure continuing foreign assistance flows. The end of the Cold War and the preoccupation of the US and Europe with rebuilding Eastern Europe were perceived as diverting OECD resources and attention from the developing countries. Rich countries' concern over the global environment was one of the few levers the developing countries had left to obtain assistance from the North.

A second ground for developing country mistrust was the role of the World Bank in managing the Global Environment Facility (GEF), which became the interim mechanism for providing financial assistance to the developing countries under the convention. Although the GEF had been established under the joint management of the Bank, UNEP and UNDP, it is perceived by many developing countries as an arm of the Bank. The Bank, in turn, was seen as an institution controlled by the US and its allies that dictated the development policies of the developing countries themselves in the interests of the rich countries. The developing countries argued strongly for a separate earmarked fund that gave them greater certainty on the overall level of funding available and the manner in which that funding would be spent. Once it became clear that the OECD countries would not accept a separate climate fund because of their concern about creating a potentially open-ended obligation, however, the developing countries had no choice but to accept the GEF as the interim mechanism.

Even those OECD countries pressing for a binding CO_2 stabilization commitment may not have been entirely unhappy at the US refusal to accept a binding emissions reduction target. There was little question that Norway and Finland faced considerable difficulties in implementing their emissions reduction commitments. The Benelux countries, France and Italy had not yet identified adequate policy responses to projected growth in CO_2 emissions from their transportation sectors. Europeans' appetite for the convenience, mobility and status derived from private automobiles remains strong even in the face of high gasoline taxes and excellent public transport systems. Even Germany did not have a fully convincing plan for meeting its ambitious CO_2 emissions reduction targets (25–30 percent reduction by 2005). The US refusal to accept a binding target therefore put the more activist European countries in the happy position of being able to claim political credit for supporting a binding target without being under any obligation actually to meet that target.

Despite these attitudes on the part of other countries, US acceptance of a binding CO_2 stabilization target would probably have enabled it to obtain more specific agreement on the mechanisms referred to above. As a first step, the US could have formed a common front with the other OECD countries and gained their support for its proposals. As a second step, the OECD countries together could have probably persuaded the developing countries to agree to more specific mechanisms by pointing out their potential for increasing investment flows. These agreements will now have to be sought in subsequent negotiations.

This is the consequence of the US going into the negotiations with the wrong set of goals. Instead of playing "defense" against imposition of commitments to reduce its CO_2 emissions, the US should have taken the initiative in doing what was necessary to get other countries to reduce their emissions. The economic cost to the US of stabilizing its CO_2 emissions at 1990 levels by 2000 is probably small. The environmental cost to the US if China and India do not start now to reduce their emissions could, by comparison, be large indeed. In this sense the bargain struck by the US is actually contrary to its long-term interests. The US succeeded in making its own actions to reduce emissions the center of world attention, while missing an opportunity to bargain for commitments from other countries that would reduce the overall environmental risk to the US.

If the US had agreed to a CO_2 stabilization target at the start of the INC negotiations in February 1991, it could have made common cause with the other OECD countries in negotiating with the developing countries on the specific mechanisms indicated above. These mechanisms might in turn have enabled the developing countries to make minimal, perhaps conditional, commitments to start reducing their own emissions. Instead the US chose to try to justify its stonewalling on targets and timetables by accusing the West European countries of bad faith in making political commitments that they had no specific plans to honor. This tactic not only made cooperation among the OECD countries difficult, but relieved the Europeans of US pressure to come up with specific plans in cases where they had not already done so.

Fortunately, the Clinton–Gore Administration has begun to reverse this situation and to put the US in the position it should have been in almost two years ago through President Clinton's commitment in his Earth Day address on April 21, 1993 to reduce US greenhouse gas emissions in 2000 to their 1990 levels. The Administration will develop a revised National Action Plan on Climate Change to fulfill this commitment. These steps should in turn clear the way for negotiating more specific agreements on the mechanisms referred to above.

Next Steps

Even before formal diplomatic negotiations reconvene, the highest priority should now be given to actually producing the inventories and national strategies contemplated by the convention. The US must be prepared to lead the OECD countries and the multilateral development banks in providing whatever assistance is necessary to get Russia and the major developing countries to produce these inventories and the associated strategies. The US offer to provide $25 million over two years to help developing countries and countries with economies in transition to prepare studies on how to address climate change is an important start, but much more needs to be done.

Work on financial mechanisms, structural reforms and technology transfer should proceed in parallel with this effort, but negotiation of protocols that set ambitious targets and timetables for greenhouse gas emission reductions should probably be deferred until the current convention is fully implemented. The US's own national strategy will be of particular importance because it can serve as a model for other major countries and regions.

Lessons to be Learned

There are at least four lessons to be learned from the above story. The first is that the US must recognize its own self-interest in getting other countries to alter their environmentally damaging behaviors. Throughout most of the negotiations, the US delegation focused on the "defensive" goal of warding off pressure from other countries to commit to a change in US behavior that would necessarily mean reducing CO_2 emissions. As has been suggested above, this was the wrong goal. The right goal would have been for the US to seek the greatest possible commitments by other countries to reduce their emissions without risking undue sacrifices itself. The latter goal would have been achievable if the US had been willing to make a commitment to CO_2 reductions at the start of the INC negotiations in February 1991. The new Administration has finally made a commitment to reduce US greenhouse gas emissions, but unfortunately, valuable time has already been lost.

The second lesson is that without better integration of its foreign and domestic policies, the US can no longer exercise the type of leadership that it has exercised since 1945 and that is still desperately needed. During most of the Cold War period, the US was able to pursue its foreign policy objectives without worrying about domestic impacts and vice versa. The growing interdependence between the two first became evident in the trade area. The combination of allowing the US to be other countries' market of last resort and ignoring the erosion in US industrial competitiveness began to impose real domestic costs in terms of lost jobs and declining living standards. The interdependence of foreign and domestic policy has now become even more apparent with respect to global environmental issues, particularly climate change. The US will only be able to achieve its critical national security objective of curbing pollution and natural resource depletion abroad if it takes the lead in curbing them at home.

A third lesson is the importance of governments working closely with non-governmental organizations (NGOs), particularly industry and environmental groups. It is ironic that the US decisions to sign on to a stabilization "goal" in the convention was made possible by anticipated energy savings from voluntary energy efficiency initiatives in the private sector. Imagine what could happen if the new Administration aggressively reaches out to industry and citizens' groups to craft new partnerships for the promotion of green technologies. Because of the unique status of NGOs in the US, these partnerships could serve as a model for creating the type of participatory decision-making processes needed to achieve lower emissions growth in other countries.

A final lesson is the need for a more "bottom-up" process for developing national and international mitigation strategies. One of the problems with the INC negotiations was that developing countries were largely represented by professional diplomats accustomed to taking stereotyped positions in North–South debates instead of by people who really understood the potential for cost-effective policy, organizational and technological improvements in all countries. The relative speed and success of the ozone negotiations was, by contrast, related to the active participation in the negotiating process of governmental and non-governmental experts familiar with CFC-based technologies and CFC substitutes. Fortunately, the Convention review process will give the parties to the climate convention a greater opportunity to bring such expertise into play.

Conclusion

US intransigence on targets and timetables has caused much time to be lost, but with the right US leadership, we can catch up rapidly during the period in which national action plans and national reports are being prepared. Once the parties begin to inventory their emissions and identify cost-effective options for reducing them, we will be able to consider negotiating emissions reduction targets that go far beyond what we can imagine today.

CHAPTER 10

Looking Back to See Forward

Tariq Osman Hyder
Director General
International Economics and Environment
Ministry of Foreign Affairs
Islamabad, Pakistan

and presently

Ambassador of Pakistan
to Ashgabat, Turkmenistan

Prologue: Why Revisit the Negotiations?

In any negotiation, the parties often begin with fundamentally different perspectives and with very different objectives. The final agreement represents an acceptable middle point between the "best case" expectations that each party held at the starting point of the talks. It is somewhat useful to revisit the negotiating procedure after an agreement has been concluded because without doing so one may miss some lessons that suggest how the subsequent process should move forward. Such an exercise, however, carries with it the risk of dwelling too much on what the various sides have gained or lost at a time when it is the common future vision which must be stressed. In this context there are three specific reasons for reviewing the history of the recently completed negotiations of the Climate Convention. First, it is useful to understand the final text of the agreement and the process that led to it. Second, it is important to learn lessons for future international negotiations. And third, it is possible to gain insights on how the agreed instrument should be implemented in order to achieve its objectives.

While the first reason would merit a review by itself, in my experience the final text of international negotiations include such minutiae that only law professors and candidates for legal doctorates take the trouble to interpret them. Yet, such an exercise is of great value when the frontiers of international law are being extended, taking into account parallel developments in a broad field.

International legal instruments reflect the common position and the political will of the international community at a given point in time. But, in order to reach consensus on some important issues, the final act or instrument emerging from a negotiation process may be deliberately ambiguous. This allows the agreement or treaty to be successfully concluded while all sides maintain that their national interests have been safeguarded and that their national positions are reflected in the text. But in the implementation phase, an ambiguous agreement must give way to operational clarity or no practical results can be achieved. For this to happen, an understanding of what took place is important, but not as important as goodwill, international cooperation and a flexible process that aims to fulfill the over-arching objective of the agreement.

In contrast to the two years required to negotiate the Climate Convention, the Law of the Sea negotiation took over a decade and still did not lead to a signed agreement. Several aspects of the work of the Intergovernmental Negotiating Committee for the Framework Convention on Climate Change (INC) contributed to the success of the climate negotiations.

One important factor was the relationships that developed among the delegates. The intensity of the negotiations led to excellent formal and informal contacts among the various negotiators. These contacts occurred both within traditional groupings—such as the Organization for Economic Cooperation and Development (OECD) and the Group of 77 developing countries and China (G-77)—and also among some new groups and coalitions. The human dimensions of the negotiating process are important, but one should not make too much of this factor alone.

During the two years of the INC process, public concern about the global environment raised expectations worldwide for unprecedented cooperation at the international level. All countries taking part in the INC process were under pressure to complete a Convention in time for signature at the UN Conference on Environment and Development (UNCED), also called the Earth Summit, held in Rio in June 1992. No country or group of countries wanted to be in the position of appearing to ignore or disregard this concern for the environment.

The structure of the INC process also helped to achieve such a substantial success in so short a time. It was, for example, extremely important that United Nations General Assembly Resolution 45/212—the resolution that set up the INC—provided for universal participation in the negotiations. To assure the participation of developing countries, this resolution established a trust fund to defray the costs incurred in their attendance at the negotiating sessions. While participation by so many states tends to mean that the process of negotiating an agreement requires more effort before a consensus is reached, the final agreement is likely to find a far broader acceptance than would have been possible otherwise.

For developing countries, it is important that the precedent for universal participation be sustained as the climate negotiations move ahead. It is for this reason that, after the draft text of the Convention was accepted by the INC on May 9, 1992, the G-77 argued for the inclusion of language in the supporting resolution—Resolution INC/1992/1 on Interim Arrangements—assuring that all INC participants should be invited to the meetings held to organize the first Conference of the Parties. The G-77 urged that these meetings not be limited only to countries that had signed the Convention. The developed countries suggested an interpretation that would have limited future participation in the INC process to signatories. But, in the end, all parties accepted the logic of the G-77 position.

This precedent with universal participation in international environmental negotiations has a significance that goes beyond the climate issue. It illustrates the advantages and disadvantages of the UN process in its entirety. Many observers complain that the United Nations is too much of a "talk shop" where little of practical importance gets done. Some critics argue that the developing countries act as a Greek chorus lamenting the injustices and inequities of the world's socio-economic system while the "real" decisions take place "behind the curtain" of G-7 deliberations or within the Bretton Woods institutions associated with the World Bank and the International Monetary Fund (IMF). There may be some truth in such critiques, but in the UN process as a whole, the medium is often more important than the message. The UN system permits all sides to express their opinions from a position of sovereign equality and, therefore, to maintain self-respect. Countries acknowledged to have dominant economic, political and military power are forced to take into account the contrasting views of many other countries, however weak those other countries may be. This balance promotes a more equitable dialogue. As a result, all sides come away with a better understanding of the differences among their positions. They are therefore in a better position to adjust their own policies or to reconsider their options.

Universal participation and equitable discourse are essential to the success of the UN process as a whole. The same approach may be appropriate to assure the success of the Global Environment Facility (GEF). Following its designation as the financial mechanism for implementing the Climate Convention, the GEF has to be restructured so that it can fulfill its new mandate.

The Objectives of the Convention: Protecting the Atmosphere while Sustaining the People

One of the most important achievements of the Convention process was the agreement reached on the objective of the convention: to stabilize greenhouse gas concentrations in the atmosphere at a level that would prevent dangerous anthropogenic interference with the climate system. (For the complete text of this article and the remainder of the Framework Convention, see Appendix, eds.) Thus in agreeing to the Framework Convention on Climate Change, 154 countries have taken a public commitment to control a potentially dangerous international environment problem, even before the expected damages have been experienced or observed. But the Climate Convention goes beyond narrow protection of the environment, to link the goals of environmental protection to the requirements for sustainable development.

The negotiations of the Framework Convention on Climate Change and, in particular, the Objective of the Convention must be understood in its historical context. The last United Nations exercise on a similar scale was the UN Conference on the Human Environment, held in Stockholm, Sweden, twenty years ago. The principal outcomes of that meeting were the Stockholm Declaration on the Human Environment, the establishment of the United Nations Environment Programme (UNEP), and the initial endowment of the Environment Fund. Although the Stockholm Declaration advanced the frontier of international environmental law, the expectations raised in Stockholm remain largely unfulfilled.

Hence, in 1992, the international community was determined to broaden the scope of the debate from a singular focus on environmental protection. The relevant question had become one of how to sustain the quality of the global environment while meeting the need

for balanced economic development. Furthermore, it seemed necessary to ensure that specific agreements resulted from these negotiations, measures which were broadly acceptable, were generally
effective and which provided concrete results that could be easily
observed and clearly weighed by both citizens and governments, in
the balance of their expectations.

Linking Environment and Development: the North–South Dimension

The entire UNCED process can be seen as a struggle between the
developing and developed countries to define sustainable development in a way that fits their own agendas. The developed countries,
most of which are relatively rich, put environment first. By contrast,
the developing countries, most of which are poor and still struggling to meet basic human needs, put development first.

The developed countries divide environmental problems between
"global" and "local" impacts. By global they mean environmental
impacts which already have effects worldwide or are likely to have
such effects in the foreseeable future. These effects are generally
caused by excessive consumption (particularly in the developed
countries) and include stratospheric ozone depletion and rapid climate change from global warming due to the atmospheric buildup
of carbon dioxide and other greenhouse gases. By local impacts they
refer to the local environmental issues that are of primary concern to
the developing countries and result principally from the burdens of
poverty. The effects of poverty on the environment include water
and soil pollution, desertification, aridity, and air pollution. Historically, these problems have been left for developing countries to solve
as best they can with their own limited resources, supplemented by
the limited inflows of Official Development Assistance (ODA).

But the link between environment and development was recognized at the beginning of the UNCED process. The starting point
was UN General Assembly Resolution 44/28 of December 22, 1989.
In its preamble to this resolution the General Assembly, noted inter
alia, that it was:

> Gravely concerned that the major cause of the continuing deterio-
> ration of the global environment is the unsustainable pattern of
> production and consumption, particularly in industrialized
> countries.

> Stressing that poverty and environmental degradation are closely
> inter-related and that environmental protection in developing

countries must, in this context, be viewed as an integral part of the development process and cannot be considered in isolation from it.

In the operative part of that resolution, the General Assembly, inter alia:

> Affirms further that the promotion of economic growth in developing countries is essential to address problems of environmental degradation.

> Affirms the importance of a supportive international economic climate conducive to sustained economic growth and development in all countries for the protection and sound management of the environment.

> Notes that the largest part of the current emission of pollutants into the environment, including toxic and hazardous wastes, originates in developed countries, and therefore recognizes that those countries have the main responsibility for combating such pollution.

The UN resolution also indicated that the Earth Summit should include among its objectives:

> Ways and means of providing new and additional financial resources, particularly to developing countries, for environmentally sound development programmes and projects in accordance with national development objectives, priorities and plans and to consider ways of effectively monitoring the provision of such new and additional financial resources.

The developing countries felt that the resolution and the process it had set in motion adequately reflected the responsibilities of the present situation with respect to global environmental degradation. The responsibilities for developed countries thus included not only curbing their own profligate patterns of consumption and production but also providing the new and additional financial and technological resources that developing countries would need to combat a situation not of their own making.

As the global environment negotiations unfolded, however, it became obvious that the developed countries, having acquiesced to Resolution 44/228, no longer felt bound by its language or its intent. In every negotiating forum, they sought to give primacy to environmental protection at the cost of the universal right to development. The words "new and additional financial resources" proved unacceptable to some developed countries: few developed countries were ready to enter into an immediate and specific commitment in this regard. There were exceptions to this general trend: the Nordic countries and Japan proposed specific timetables for increased development assistance flows and offered to commit more funding for the environment.

Yet, despite their acceptance of the language of Resolution 44/228, the developed countries would not accept responsibility for their historical contribution to the buildup of carbon dioxide in the atmosphere or agree to divide the global responsibility in proportion to past emissions. They proclaimed that reiterating such issues would amount to "finger pointing" and such unnecessary accusations would make it difficult for them to sell any resulting convention or declaration to their citizens and to their treasuries. The developed countries accused the developing countries of using the environment as a "club" with which to beat additional financial resources from the recession-strapped economies of the North.

Endgame: An Irresolvable Paradox

The repetition of these narrow positions by the developed countries led to a somber, pessimistic, and introspective mood in many developing country delegations attending the end of the first part of the fifth INC session in New York in February 1992, and the early phase of the fourth meeting of the UNCED Preparatory Committee in March 1992, the final meeting held before the Earth Summit. It appeared to many that the developed countries had accepted Resolution 44/228 only to draw the developing countries into the negotiating process. Once engaged in the process, the developing countries would have to continue until a compromise was reached. It seemed that the developed countries assumed that their superior economic strength would win out over the core interests of the developing countries.

The developed countries saw the global warming scenario essentially as a zero sum game in which their own high per capita emission levels and consumption and quality of life would suffer if the developing countries were allowed unlimited economic growth. To forestall this eventuality, the developed countries' position throughout the climate negotiations was that the developing countries must also acknowledge their responsibility to combat global pollution and adverse global environmental change. As a step in this process, the developing countries would be required to prepare reports on their national sources and sinks of greenhouse gases. These reports would be monitored and reviewed and the developed countries would use these reports as a criterion not only for giving additional assistance, but presumably also for assessing and reprogramming the current levels of economic assistance. This linkage seemed particularly disturbing to developing countries which, in general, believe that existing assistance flows are already inadequate. They note that, due to unfair trade practices, there is currently a net negative flow of resources from developing to developed countries.

Hence, many developing country delegations worried about whether the whole process would have gained them anything of value, or if, on balance, they would be net losers. In January 1992, during an informal meeting called one evening to break the impasse on financial resources in the climate negotiations, I had a conversation with an OECD colleague who had attended the same university that I did. This conversation made a deep impression on me as we went into the final phase of negotiations. We were discussing our different viewpoints on the Global Environment Facility (GEF) as a possible funding mechanism under the Convention and I ventured to suggest some compromises. He told me that while these proposals seemed logical to him, he doubted that his colleagues who had responsibility for that particular aspect of the negotiation would agree. He cautioned me that I would find them to be, in the idiom of his country, "hard men."

I never allowed myself to forget that remark. In the negotiations that followed, as Pakistan chaired the G-77 (representing over 120 countries at that time), I was determined to learn as much as possible and to make our delegation a part of the most hard-working and capable teams of G-77 negotiators. I would seek to achieve this while transparently involving the whole group in every decision-making process so that the strength of a common position on all important issues would weigh against our organizational, institutional, and other weaknesses. Many observers in developing and developed countries had predicted that on important issues, particularly those affecting the potential for energy use and economic growth, the G-77 would fall apart, dividing along regional or economic groupings with each group cutting the best deal it could for itself. In view of these expectations, we were all very proud when not only the two Conventions but also the process leading to the Rio Declaration demonstrated the opposite to be true. Although the developing countries could not always reach their primary objectives, they maintained a common front on the major issues. This decisive advantage helped us to accomplish what we did achieve.

The Dynamics of the Climate Negotiations

Broadly speaking, there were three parties to the negotiation. The United States of America, the other developed countries which are members of the OECD, and the developing countries. The US had certain important differences with its OECD partners, particularly the European Community (EC). For example, the US opposed any specific timetables for stabilization or reduction of greenhouse gas emissions. But on many other main issues, especially those affecting the developing countries, the developed countries held negotiating positions very much in common.

The early phases of the negotiations were preoccupied with procedural questions, such as which countries would constitute the guiding Bureau and who would be the chairman of the INC itself. It was eventually decided that two co-chairs would lead each of the two working groups. Working Group I focused on national commitments, on the preamble, principles and the overall objective of the Convention. Working Group II was given the responsibility for institutional arrangements, reporting requirements, and legal provisions including dispute resolution.

The INC initiated serious work on these issues during its second session, held in June 1991 in Geneva. At this session, a number of delegations came up with draft texts for the Convention, reflecting both the developed and developing countries' points of view. These draft texts were further refined during the third session of the INC in Nairobi in September 1991. India and China, each heavily dependent on their large domestic coal reserves for energy production, had been among the first to carry out substantial work outlining how the Convention could be structured in an equitable manner. The small island states, the countries most at risk from sea-level rise due to global warming, were also quite quick to suggest their own strategic approach on how to proceed.

It was during the fourth INC session held in Geneva in December 1991 that the G-77 managed largely to harmonize its position on the issue of principles. The key to this early success was the leadership of Ghana, the chair of the G-77. As the incoming chair of the G-77, Pakistan was asked at this time to also chair the open-ended drafting group.

The G-77 presented concrete formulations for the section of the Convention devoted to principles at the fourth session of the INC held in Geneva in December 1991. These suggestions highlighted sovereignty, right to development, sustainable development, equity, common but differentiated responsibility, special circumstances, the precautionary principle and international cooperation. Many OECD countries refused to agree that a section on principles was needed. These delegations were prepared to concede, at most, that some reference to these principles (albeit in a watered-down form) should be included in the preamble to the Convention.

Despite reaching agreement among the G-77 on principles, there was not sufficient time during the fourth INC session to harmonize a common position of the G-77 on the core sections of the Convention that dealt with general and specific commitments. But forty-four developing countries used the work of the G-77 drafting group to present their own proposal on commitments. This proposal divided the section on "Commitments" into three parts: general commitments, developed country commitments, and developing country commitments. This developing countries' proposal stressed that the fulfillment of commitments by developing countries must be in accordance with their national development plans and priorities. Any effort made toward fulfilling these commitments will necessarily be dependent upon the provision by the developed countries of new, adequate and additional financial resources required to meet the full incremental costs involved.

During the fifth INC in New York in February 1992, the G-77 and China achieved a common position on the commitment that they would require of the developed countries. During the fifth INC, Pakistan, speaking on behalf of the G-77 and China, stressed that a section on guiding principles would provide a foundation for what would follow in the Convention. Therefore, the suggestions of the G-77 required a substantive and positive response from the developed countries.

The Paris Meeting

During INC negotiations, a range of formal and informal contact groups evolved both within the meetings and outside of them. The chairman of the INC, in addition to the Bureau meetings, developed a pattern of convening an informal meeting toward the end of each session of those that he considered to be the main actors in the negotiations. The agenda for this informal meeting was to decide how to proceed until the beginning of the next formal session.

By the end of the first part of the fifth INC session in February 1992 in New York it was clear that little progress had been made. But, the pace appeared too slow to conclude a Convention in time for the Rio meeting in June. The fifth session of the INC was to resume in New York for only nine days, from April 30 to May 8, 1992. To advance the agenda before May, the chairman decided to call a special informal meeting in Paris from April 15 to 17, 1992, following the conclusion of the fourth meeting of the Preparatory Committee for UNCED (PrepCom), so that the Convention could be salvaged.

Before this meeting in Paris, the developed countries had been unwilling to make any firm commitments to transfer advanced technologies or new and additional financial resources to developing countries. Similarly, these countries had failed to agree on emissions reductions. In fact, the draft proposal on future emissions reductions by the developed countries had been bracketed by some of the OECD countries themselves!

As the G-77 coordinator, I was authorized to attend the informal Paris meeting, upon the invitation of the chairman, in order to safeguard the interests of the Group during the meeting. After consultation with the other G-77 delegates, I stressed the following themes at the Paris meeting:

a) The Convention must be viewed as an integrated whole, not as a collection of bits and pieces that simultaneously address various issues.

b) The US proposal of a process-oriented approach must be viewed in the context of balanced commitments being made under the Convention by developing and developed countries. These commitments must take into account the common but differentiated responsibilities of these two groups of countries and these very different levels of capability to respond to climate change, as recognized explicitly in UN Resolution 44/228.

c) As the principal sources of the greenhouse gas emissions that contribute most to increasing the risks of rapid climate change, the developed countries, as the main polluters, must make a firm and explicit commitment to stabilize and eventually reduce their future emissions levels.

d) The developing countries could not accept a reporting and review system which would subject all their development plans and projects to outside review and approval. Instead, adequate funds should be provided as an incentive for "good" projects and programs—projects and programs that move the implementing country toward the overall objective of the Convention, i.e., limiting future concentrations of the offending gases to levels that do not threaten human societies or natural ecosystems. Virtually all members of the G-77 agreed that the preferable approach, in terms of both acceptability and effectiveness, would emphasize the carrot rather than the stick.

e) Unfortunately, in the negotiations prior to the Paris meeting, the developed countries had given no clear indication of their willingness to come up with even part of the financial resources and technology which developing countries would require in order to achieve the objectives of the Convention.

f)   Finally, to the developing countries, the section of the text on "Principles," which emphasized the developmental context in which the climate issue must be placed, was essential both as a guideline and as a safeguard. This text was the minimum necessary to vouchsafe the long-term interests of the developing countries, who, through a Climate Change Convention, would be accepting international legal obligations with incalculable implications for fiscal policy and national development strategies.

There was a free and frank exchange of views at the informal meeting in Paris. Many negotiators from developed countries revised their original formulations on the reporting and review procedures. Because I had just coordinated the successful negotiations of the Rio Declaration on behalf of the G-77 and China during the fourth UNCED PrepCom in New York, the chairman asked me for some thoughts on how to proceed on the question of the text on "Principles."

I responded that during the negotiations on the Rio Declaration the developed countries had, for the first time, accepted the principle of a universal right to development, which should be reinforced in the Climate Convention. Despite certain reservations, we had negotiated a formulation on the different responsibilities for causing and combating global pollution. The Rio Declaration resolved the issue of a common responsibility for our global heritage by recognizing that the responsibilities of various countries and, indeed of groups of countries, are different. In the negotiations of the Rio Declaration, there had been a fear on the part of some of the developed countries concerning the inherent dangers of conceding the right to development in an international declaration. As one of their negotiators put it, if 70 percent of the world's people begin using up the world's resources at the per capita average of developed countries, the finite carrying capacity of the globe would be quickly overwhelmed. Although I understood his point, this has to be considered an unlikely outcome. Furthermore, from an equity perspective, there is no moral basis for introducing such an argument into the international debate on climate change. Despite their acceptance of this principle in the Rio Declaration, the developed countries tried to weaken the final text of the Convention concerning the right to development.

Throughout the INC negotiations, the US had opposed the incorporation of any section on "Principles" in the text of the Convention. At the Paris meeting I suggested to Chairman Ripert and to the US representative at the Paris meeting, that we should work toward adapting the Convention's section on "Principles" in light of the considerable work that had produced consensus position in the Rio Declaration. The US representative observed that the Rio Declaration, unlike the Climate Convention, constituted soft law and was not as yet legally binding. Significantly, however, the US delegate agreed that his country would show flexibility on this issue. Ultimately, the "Principles" section of the Rio Declaration was used (with a few modifications) for the Convention's section on "Principles." Moreover, the formulation in the Rio Declaration on the responsibility of the developed countries to take the lead in reducing emissions was not only used in Article 3(1), "Principles," in the Convention but also partially repeated in Article 4(2)(a), "Commitments."

Two main procedural issues dominated the debates at the Paris meeting. The first concerned the pressure of time on our ability to generate a clean, complete text of the Convention. It was clear to everyone that the number of alternatives and brackets remaining in the proposed draft text would not permit a Convention to be agreed before the deadline of Rio 1992. Should the chairman, therefore, be authorized to come up with a single, integrated text which would be used in the final phase of the negotiations? A second question related to the modality of the negotiations. The Secretariat had divided the text into eight cluster groups. So far, the delegates had been meeting in two main working groups supplemented by a variety of informal contact groups. How could the remaining issues be divided effectively among the working groups?

In all negotiations the chairman tended to be cautious about coming up with his own text. This is always the last card in his hand, available if necessary to save the negotiations from an irresolvable deadlock. If the chairman plays this card too soon, or without a prior consensus, he loses the pivotal authority on which the success of the negotiation often rests. Aware of this, Chairman Ripert had resisted making any dramatic intervention until various delegations beseeched him to come up with a chairman's text. It was now clear, at least to G-77 delegates then present, that there was no alternative to relying on a chairman's final negotiating text.

As the G-77 coordinator, I informed the chairman during the closing of the Paris meeting that it was the request of the G-77 that he prepare an integrated draft text of the full Convention. Since the developing countries wanted a meaningful Convention, the members of the G-77 and China expected that this text would rationalize the existing, bracket-ridden document, streamline it, and propose compromise formulations, particularly relating to the commitments to be undertaken by developed countries. The very possibility of such action by the chair would surely exert moral and political force on the OECD countries, encouraging them to come up with meaningful commitments before the end of the resumed fifth plenary session. The delegates from developed countries agreed in general with this proposal by the G-77. The chairman accepted this responsibility.

Decisions on the modalities and limits of the final negotiating sessions were left for the last day of the Paris meeting. In periods of intense negotiations, dividing the subject among many small groups dissipates the focus of the delegations. The G-77 delegations, unlike those of the developed countries, usually have a limited number of members and it is difficult for them to split themselves up to service a large number of small, simultaneous sub-meetings.

A Fateful Dinner

As we ended the consultations late on the second day of the Paris meeting, I walked out with Michael Cutajar, the Executive Secretary of the INC Secretariat and with Ambassador Bo Kjellen of Sweden. Cutajar is a consummate professional whose capabilities we had all come to admire. Kjellen and I had served together in our respective Embassies in Hanoi during the war years. The three of us decided to walk to a nearby café and to have dinner together. As we were discussing how to proceed in the negotiations, an idea came to me. I drew a chart on the paper tablecloth atop our table, in order to help me visualize the concept.

Why not, I suggested, divide the final stage of negotiation among three negotiating groups, in order to focus on inter-related issues without the artificial division we had been employing until now. The existing system had led to some confusing anomalies. For example, the financial mechanism came under Group II while the discussion of financial resources took place under Group I.

Under my new proposal, the first set of issues would include the preamble, definitions, objective, and principles. The second set would incorporate the most controversial of the core issues: commitments, the mechanism for financial resources and for technology transfer, and the reporting and review procedure. The third set would include whatever issues were left over, among them the institutional and final clauses. Michael Cutajar tore the edge off the tablecloth and took the strip of paper to his hotel to think about it overnight.

The next day, I discussed this approach with colleagues from both developing and developed countries. Following these consultations, I made a similar proposal in the meeting. With the concurrence of the other delegates, the chairman accepted my proposal.

In essence, my proposal was adopted as the negotiating procedure for the resumed fifth plenary session of the INC. Of course, in addition, the chairman also convened a number of informal meetings of the Expanded Bureau to resolve some of the remaining issues. In such cases, success depends upon limited but representative participation in the informal, semi-private consultations initiated by the chair. The closed form of this process always leads to complaints about a lack of transparency from delegations which are not part of the informal consultations. Unfortunately, given the pressure of the upcoming deadline of the Earth Summit, there seemed to be few alternatives in April 1992.

While all delegations and interested groups had specific observations regarding the chairman's text, the G-77 and China formulated an integrated and concise response to it. This position represented a constructive revision of the chairman's text. The position of the G-77 was circulated to other delegations and given to the chairman. The G-77 would judge the extent to which the final text reflected the core objectives of the G-77 and China in light of this formulation.

The statement presented to the chairman by the Group of 77 and China considered the following elements as essential to the Climate Convention:

a) The principles of:
   - sovereignty of states,
   - right to development,
   - common but differentiated responsibility,
   - differentiated environmental standards.
b) A balance between general and specific commitments, with commitments by developing countries reflecting their national developmental goals, policies and poverty eradication priorities.
c) Access to adequate, new and additional financial resources for developing countries and transfer of technology to developing countries on preferential and/or concessional terms.
d) Revision of a text on the funding mechanism that is broadly acceptable because it is a balanced and neutral text, leaving detailed arrangements to the Conference of Parties.

Toward the end of the resumed fifth session, the Chairman circulated his final revised texts. The G-77 and China were satisfied that the new text took into account the main concerns of the developing countries even though not all their concerns had been incorporated. The developed countries seemed to reach a similar conclusion. While many delegations wanted to make amendments to the revised text, as happens in such negotiations, this could not be allowed. To do so would have resulted in the unraveling of a delicate compromise that carefully balanced the basic concerns of all groups. The plenary decided that the text must be viewed as a total package and accepted or rejected en bloc. Hence despite some strongly held and expressed objections and reservations, on May 9, 1992, the Framework Convention on Climate Change was adopted.

Assessing the Outcome

The end result did not completely satisfy the objectives of the different parties. For example, there was no explicit reference to the "right to development" in the section on "Principles." There was no specific commitment on the part of the developed countries to control and reduce greenhouse gas emissions within a specific timeframe. Perhaps most important, the Convention did not fully take into account the vital developmental interests and concerns of the developing countries. Its provision on the financial mechanism fell short of their expectations. However, the Convention does provide a sufficient framework to evolve further procedures for subsequent action at the global level.

These clear and obvious weaknesses notwithstanding, the developing countries did manage to gain significant concessions from the developed country partners. Reference in the Convention to the Global Environment Facility (GEF) which was originally in Article 11 on "Financial Mechanism" were shifted to Article 21 on "Interim Arrangements." The G-77 had entered the negotiations with a clear preference for an independent climate fund that would avoid the inherent inequities of the weighted voting system associated with the World Bank and related Bretton Woods institutions. As the negotiations progressed, however, the G-77 shifted the debate from a focus on the name of the financial mechanism to its desired characteristics. What did it matter what it was called as long as it was universal in scope, transparent, had an equitable and balanced governance, and was responsive to the needs and priorities of the recipients?

Furthermore, the G-77 and China had also insisted that the financial mechanism must be under the guidance of and be accountable to the Conference of Parties (COP) of the Climate Change Convention. It was for the COP to decide on the policies, program priorities and eligibility criteria of the financial mechanism in relation to the Convention. It was this constant pressure from the G-77 and China which motivated the GEF and its then limited Participants to think about restructuring while the negotiations progressed in order to make the GEF more acceptable. No doubt the conditionalities of the

Convention relating to GEF will lead to further changes in the right direction.

The developed country Parties committed themselves in Article 4(3) to provide new and additional financial resources to meet the agreed full costs incurred by developing country Parties in their reporting requirements under Article 12(1).

Of more significance in this the most crucial article of the Convention (Article 4(3)), the developed countries committed themselves to provide such financial resources, including for the transfer of technology, as would be needed by the developing country Parties to meet the "agreed full incremental costs" of any implementing measures they take to fulfill their obligations under the Convention. This paragraph also specifies that the implementation of these commitments shall take into account the need for adequacy and predictability in the flow of funds.

By contrast, the extent to which developing country Parties were bound to implement their commitments under the Convention was specifically conditioned in Article 4(7) on two very important provisos. First, the developing countries will be required to implement their commitments under the Convention only to the extent that the developed country Parties effectively implement and fulfill their own commitments related to financial resources and transfer of technology. Second, the developing countries will condition the implementation of their commitments on the notion that economic and social development and poverty eradication are their first and overriding priorities. Hence it is clear that progress towards the objective of the Convention depends on the provision of adequate funds and technology to the developing countries and on the recognition that their developmental objectives cannot be subordinated to purely environmental goals.

These commitments are the foundation of the Convention. After the careful groundwork that had been laid in the negotiations, it was not surprising that 154 states and one regional economic integration organization signed the agreement in Rio during UNCED.

Looking Ahead: Working Out a New Structure for the GEF

The Convention will enter into force ninety days after the deposit of the fiftieth instrument of ratification, acceptance, approval or accession. At the time of the Secretary General's report to the UN in November of 1992, two states had ratified the Convention. By mid-1993, eleven states had ratified the Convention. In November the Secretary-General prudently estimated that the Convention might enter into force by September 1994. Thus, the first Conference of the Parties could take place as early as the period March–September 1995. In the period before the official entry into force of the Convention, substantial preparatory work will be required at the national, regional and international level.

One set of unresolved issues concerns the emerging role of the GEF. Many parties to the Convention have expressed a need to continue the work of defining the mandate and guidelines in the GEF, prior to the first Conference of the Parties. Most parties agree that the GEF has to be restructured so that its membership is made universal and representation in decision-making is equitably balanced, and the system of governance is transparent. As we move toward a new modality somewhere between the UN and the Bretton Woods systems, the Parties have a number of models to draw upon. Personally I do not think that working out an acceptable system for the governance of the GEF should present a major problem, given a certain flexibility and vision on the part of the major donors. The guiding principles are already clearly laid down in the Climate Change Convention and in Chapter 33 of Agenda 21.

The current GEF restructuring exercise is occurring in response to the specific provision of the Climate and Biodiversity Conventions. As the interim financial mechanism under the Climate Convention, the GEF must be appropriately restructured and its membership made universal. In this regard, Section 4, ("Means of Implementation"), in Chapter 33 ("Financial Resources and Mechanisms"), of Agenda 21, specifies the following guidelines and conditions for GEF restructuring:

Financial Resources and Mechanisms: Para 33.14(A)

(iii)   *The Global Environment Facility*, managed jointly by the World Bank, UNDP and UNEP, whose additional grant and concessional funding is designed to achieve global environmental benefits, should cover the agreed incremental costs of relevant activities under Agenda 21, in particular for developing countries. Therefore, it should be restructured so as to, inter alia:

- Encourage universal participation;
- Have sufficient flexibility to expand its scope and coverage to relevant programme areas of Agenda 21, with global environmental benefits, as agreed;
- Ensure a governance that is transparent and democratic in nature, including in terms of decision-making and operations,

by guaranteeing a balanced and equitable representation of the interests of developing countries and giving due weight to the funding efforts of donor countries;
* Ensure new and additional financial resources on grant and concessional terms, in particular to developing countries;
* Ensure predictability in the flow of funds by contributions from developed countries, taking into account the importance of equitable burden-sharing;
* Ensure access to and disbursement of the funds under mutually agreed criteria without introducing new forms of conditionality.

There are two main issues related to the governance and framework of the GEF. The initial negotiations on decision-making tried to reconcile the apprehensions of the donors and recipients. The donors do not want to put their money in a mechanism in which their numerical inferiority will permit them to be outvoted in a one-country, one-vote ballot devised along traditional UN lines. The donors prefer the weighted voting system of the Bretton Woods agreement used by the World Bank and the IMF. Reflecting a similar fear of being marginalized, the developing countries do not want a Bretton Woods-type decision-making process utilized by the GEF.

Quite early in the climate negotiations Pakistan had suggested that one possible model for the governance of the GEF might be that of the Interim Multilateral Fund operated by the Executive Committee of the Montreal Protocol. In this mechanism, the developing and developed countries effectively each control 50 percent of the voting power for decision-making and the positions of both sides are reasonably safeguarded. As negotiations continue, it would appear that this suggestion has found some favor in terms of its conceptual framework, if not in terms of an exact replication. It is also vital that the INC send a clear signal to the GEF as to what it would consider to be "acceptable" in a restructured GEF in terms of governance.

To avoid duplication and unnecessary expense, the operational side of GEF's activities will remain mainly within the World Bank and to some extent with UNDP. However, the degree of independence of GEF from the World Bank will have to be monitored very closely. At present the legal foundation of GEF, upon which its tripartite organizational structure and very existence rests, stems from the enabling Resolution No. 91-5 of the Executive Directors of the World Bank. This foundation must be broadened in order to meet the needs of the Climate Convention.

The relative weight and influence of the World Bank, UNDP and UNEP within the GEF structure will merit the attention of both the Participants Assembly of the GEF and to a lesser degree the COP of the Convention. Currently, the Participants Assembly has the competence to approve free standing projects up to a set financial limit and the World Bank Executive Directors approve projects above that limit. The GEF Secretariat has suggested that the restructured Facility be established by another Resolution of the Executive Directors of the World Bank. They argue that this Resolution could specify that the Participants Assembly would make decisions on various issues within its competence without review by the World Bank's Board of Governors or the Executive Directors. The GEF Secretariat has suggested further that this Resolution could also clearly specify that no changes would be made to this Resolution without prior consultation of the Participants.

There are other important issues besides governance that pertain to the GEF and link it to the Convention. In the Implementation Committee of GEF, countries with projects under consideration are not part of the presentation, discussion, and the decision process for these projects. A way must be found to engage these most interested parties in the discussions.

At the recent Participants meeting of the GEF held in Abidjan, from 3 to 5 December 1992, a most constructive discussion occurred. GEF was given until the end of 1993 to complete its restructuring. A consensus emerged that, in order to promote universal membership, the present membership fee of around $5.2 million would no longer be required (although administrative expenses could be recovered on a reasonable basis from member countries).

One of the important aspects of these GEF meetings has been the two-way educative process for both the Participants and the GEF Secretariat. The developed country Participants are coming to terms with a new situation removed somewhat from the Bretton Woods framework. The developing world Participants have been actively working towards this goal. At the same time the GEF Secretariat (which comes mainly from the World Bank) is getting used to the rough and tumble of a process more like that of the UN system.

The World Bank staff—with its constituency system and superb professionalism—is used to going into meetings in which its officials preside and its objectives are achieved with the end result preplanned. This is still the case in the GEF where the chairman comes from the World Bank, and the meetings follow a World Bank agenda.

All three sides are adjusting well to the new situation. The developing countries want to increase membership within GEF before giving their final approval to its restructuring. The donor countries want to assure themselves that their money will be well utilized. The GEF wants to retain its administrative independence, avoiding any debilitating political gridlock. Undoubtedly, there will be vigorous debate on issues of governance and procedure in relation to the chairmanship and the possibility of its rotation, the role of the three agencies, the composition of the Scientific and Technical Advisory Panel (STAP) of the GEF, the role of NGOs and on other issues as well.

Another key issue for the future of the GEF is the definition of "incremental costs." The climate negotiations left the issue of what constitutes "incremental costs" rather vaguely defined. Nonetheless, the resolution of this issue is fundamental to any operational agreement both for individual project financing and also for the allocation of limited funds. While an acceptable definition of the technical and economic parameters for this analysis will take some time to reach, it is nonetheless important to set eligibility criteria quickly in order to get some crucial projects off the ground. In this debate, the developing countries will argue that incremental costs include all those costs which are incurred as a result of the transition from current development plans, programs and projects to the mode of sustainable development.

Looking Ahead: Other Actors

During the interim phase before the first Conference of Parties, the INC must ensure that the Secretariat of the Convention is fashioned in such a way that it provides an intellectual and analytical foundation which would help to resolve this and other issues by working together with the GEF and other international bodies. Of these other bodies, the Commission on Sustainable Development must also be given a key role in the study of the definition of incremental costs towards sustainable development.

The IPCC has already made an effort to strengthen the rationale for its continued existence by modifying its structure and by adding a new committee to look at the economic implications of climate change. One could well take the view that eventually the Convention's Subsidiary Body for Scientific and Technological Advice will completely replace the IPCC. An argument for its continuation is the fact that while all states are members of the IPCC, membership of the Subsidiary Body will be confined to countries who are parties to the Convention. Most probably the IPCC will remain to collect scientific information and to forward its analysis to the COP. The Subsidiary Body would then concentrate on formulating recommendations for policies that should be implemented. Hence, a complementary process would ensue.

How the Commission on Sustainable Development and the IPCC will relate to the interim INC is not yet clear. Nor is it yet clear how the INC/COP will tie in to the restructuring process that the UN is now undergoing. This question is especially important in respect to the reporting procedure under para 33.13 of Agenda 21. In this paragraph of Agenda 21, the Sustainable Development Commission was charged with responsibility to monitor the flow of development assistance towards the 0.7 percent GNP target for developed countries. How all these related roles will be divided so that a mutually beneficial set of relationships become firmly institutionalized remains to be worked out.

At its last meeting, INC 6, the Committee, keeping in mind the draft resolution at that time before the UN, confirmed the need for close working relationships with the IPCC and the GEF. The Committee must take fully into account the work relating to climate change being carried out within the United Nations system, with a view to promoting a coherent and coordinated program of activities aimed at supporting the entry into force of the Convention and its effective implementation. The Secretariat was charged to provide information on these activities regularly and to coordinate the work of the INC with the work that should take place under the "Atmosphere" chapter of Agenda 21. Thus, the Executive Secretary of the INC was requested by the Committee to organize a "clearing house" in consultation with other concerned entities for the exchange of information and experience on relevant technical and financial cooperation activities, bilateral and multilateral, including GHG inventories and country studies, and to report to the Committee. The Executive Secretary was also charged to pursue relevant information and training programs.

Finally, there was a general agreement that the Bureau of the INC should remain as it is, with replacements for those members who had to leave, including the chairman whose voluntary departure will be missed by all INC members.

Subsequent Developments

The 7th INC took place in New York from 15 to 19 March 1993. The Secretariat has prepared a good working paper highlighting the issues which have to be resolved between the INC (and subsequently, the Conference of Parties) and the financial mechanism. These issues concern the governance of a restructured GEF, how GEF will function under the guidance of the COP, and how technical and financial support to developing country Parties will be ensured.

There is also an annotated provisional agenda for the organization of work. If there is any weakness in the schemata of the Secretariat, it is related to the scope of the assistance which will be required by the developing countries.

The implicit parameters for providing technical and financial support to developing country Parties mainly related to the assistance which the developed country Parties are required to give to the developing country Parties in complying with their obligations under Article 12, Paragraph 1. These obligations relate principally to the preparation and presentation of data constituting national inventories of sources and sinks and the steps taken to implement the Convention.

Of course this reporting process constitutes the primary operational objective of the developed countries. However, in Article 4(3), the developed country Parties also agreed to provide such financial resources, including for the transfer of technology, needed by the developing country Parties to meet the agreed full incremental costs of implementing measures that are covered by Paragraph 1 of Article 4 and that are agreed between developing country Parties and the international entity or entities referred to in Article 11 (the Financial Mechanism).

The scope of measures that can be implemented and undertaken under Paragraph 1 of Article 4 is potentially extremely broad, when read in the context of Article 2, the objective of the Convention. This Article specifies the overall objective as the stabilization of greenhouse gas concentrations in the atmosphere at a level that would "prevent dangerous anthropogenic interference with the climate system within a time frame sufficient to allow ecosystems to adapt naturally to climate change, to ensure that food production is not threatened" and to enable economic development to proceed in a sustainable manner. Therefore, for the Convention to encourage sufficient ratifications and in order for it to fulfill global expectations, we will have to raise our fund-raising sights beyond the $2.5–$3 billion target now being discussed as a realistic figure for the first replenishment of the GEF.

The GEF participants held a constructive meeting in Rome from 3 to 5 March 1993. Three main results emerged from this meeting. First of all, the participants agreed on the composition and terms of reference of an Evaluation Committee, which would review the experience of the pilot phase so that the appropriate lessons could be drawn for the next phase.

Secondly, the procedure for replenishment was agreed upon. The developed countries wanted to restrict the replenishment meetings to donors who would put up at least $5.4 million for the next cycle. The developing countries did not want to be excluded from these meetings for a number of reasons. Their experience in other international financial institutions had been that these limited donor bodies, as in the case of the IDA deputies, moved beyond the role of resource mobilization to key policy decisions. In the IDA replenishment process the IDA deputies decided upon eligibility criteria, geographical distribution, etc. Since the primary GEF replenishment meeting would discuss funding needs, it seemed essential to the developing countries that they would be represented. In the compromise that was reached, it was agreed that those countries with new contributions over the $5.4 million threshold would participate, with all other countries being permitted to send non-participating observers. Furthermore, the original ten developing countries which are also donor Participants would participate in the replenishment meetings without having to come up with fresh contributions. Thirdly, the main topic of the meeting concerned decision-making and governance issues for a restructured GEF, a GEF modified to meet the expectations of the Convention and the Conference of Parties.

The GEF Secretariat had come up with three basic voting models for those rare occasions when consensus would give way to voting. The first model was based on the Montreal Protocol with its straight 50:50 voting structure divided between developing and developed countries. Its simplicity and elegance appealed to the majority of the developing countries. The second model allocated each group not only an equal number of PA (Participants Assembly) members but also an equal number of votes amongst its members, leaving it to each group to ensure an equitable internal allotment. This alternative appealed to most developing countries, although each stressed that during this meeting no final position was being taken; it also received support from some developed countries.

The third model allocated to each Participant a number of votes derived from a combination of basic votes and contribution-related votes. This alternative set a prescribed balance between donor and non-donor countries, while allowing Participants to choose their PA representation through a system of constituencies which could, if desired, be mixed. Indeed, some developing and developed Participants could form mixed constituencies. As in other models, the suggested size of the PA was around thirty, which would mean that there would be thirty constituencies and no General Assembly, in which all participants would be present. This option was favored by most of the developed countries.

The discussion on decision-making was extremely detailed and candid. It became obvious that there was a considerable difference of perspective between the developing countries and the developed countries. The latter were used to dominating the governance of the international financial institutions. In international negotiations the norm is for the Parties themselves to come up with negotiating proposals. The GEF Secretariat had worked hard but the presentation lacked empathy. For instance, under some variations of the third, weighed voting model, it would have required either ninety developing countries or four developed countries to block a particular proposal!

It became apparent that no consensus on governance could emerge at the Rome meeting and the discussion then began on how the chairman would sum up the debate both as guidelines for the Secretariat to work on before the next meeting and as a foundation of the next meeting. It was too late for either of the two main Parties to convert the other side at this stage. At the Rome meeting there were many more developing countries represented than before. Nonetheless, both the Secretariat of the GEF and the G-77 will have to take a more activist role in ensuring the universalization of the GEF.

On Pakistan's behalf, I proposed to the chairman that in his summing up he should conclude that in a constructive discussion the Participants had divided the guiding principles for the governance of a restructured GEF into three baskets. The first was a set of operational principles relating to simplicity (and therefore easy to understand and explain), efficiency and automaticity, on which there was a consensus.

The second basket contained proposed principles which were strongly favored by a number of developed countries but which did not enjoy universal acceptance at this stage. These included: incentives to donors to contribute more, mixed constituencies and measures to reduce the likelihood of polarization. Some countries felt that these did not constitute guiding principles but rather objectives to work towards.

The third basket consisted of the over-arching principles on which there was complete consensus. Being guided by the language in the Convention in Articles 11 and 21 and in Chapter 33 of Agenda 21, the chairman should reflect them in the following way. The GEF in its system of governance should be:

- Transparent and democratic in nature, including in terms of decision-making and operations, by guaranteeing a balanced and equitable representation of the interests of developing countries and giving due weight to the funding efforts of donor countries. I suggested to the chairman that if the carefully negotiated language of Agenda 21 were used including the phrase, "and giving due weight to the funding efforts of donor countries," the developed countries should be satisfied and might not wish to press for a principle through which incentives for donor participation were necessary. The chairman and a number of developed and developing countries felt that such a summing up would reflect the existing state of play.

It is now, of course, important for the INC to send a clear signal to the GEF as to its expectations.

The Imperative of Moving Forward

The imperative of moving ahead requires that we now regard the Framework Convention on Climate Change as a new model of international cooperation. It was entered into in good faith by all Parties and represents a dynamic, action-oriented process. All Parties and institutions must now live up to their commitments to each other in good faith. The developing countries have welcomed the Convention and many of them, within the limits of their strained resources, are already taking the preliminary steps required of them. It is now up to the developed countries to live up to the spirit of the Convention in regard to their obligations. These countries must reduce the burden they continue to put on the atmosphere by their emissions. Equally important, they must fulfill their commitment to share and to provide the "new and additional resources" and technology which their developing country partners will require. The success of the Convention depends on this.

Expectations for the Clinton Administration

In terms of recent developments, it may be noted that the developing countries have high hopes for the policy program and stance of the new Democratic Administration of US President Clinton. Therefore, they look to the Clinton Administration not only for leadership in terms of environmental policy within America, but also to provide the impetus for leadership and change within the United Nations and Bretton Woods institutions, change linked to both development and the environment.

PART III

The Outside Edges In

CHAPTER 11

Some Comments on the INC Process

Hugh Faulkner
Executive Director
Business Council for Sustainable Development
Geneva, Switzerland

Prologue

The negotiations of the Framework Convention on Climate Change dealt with very complex issues. The potentially profound worldwide impacts of any measure to reduce greenhouse gas emissions made these negotiations qualitatively more difficult than the negotiations that led to the 1987 Montreal Protocol on Substances that Deplete the Ozone Layer.

Business participation in the climate negotiations, and in international negotiations within the UN system in general, is a very new phenomenon. Traditionally, business has not taken part in such negotiations, nor has it particularly pressed to be included. To an "outsider" from the business community, the whole UN system has appeared to be a rather ponderous political machine that inches slowly along and accomplishes relatively little. Consequently, many business leaders have concluded in the past that an attitude of "benign neglect" for the whole process is most appropriate.

This attitude appears to be changing, albeit slowly. The UN Conference on Environment and Development (UNCED, also called The Earth Summit or Rio '92) and the preparatory process leading up to it provided new opportunities for business leaders to participate in a critically important international debate. Maurice Strong, Under-Secretary General of the UN and chairman of the Earth Summit, and others worked hard to open up the process, engaging as many non-governmental organizations (NGOs) and others as possible. Strong made a special effort to include representatives of business and industry. These organizations are considered NGOs by the UN. As a result of such efforts to broaden participation, UNCED was a success—engaging hundreds of people and organizations outside of governments.

The question of how business representatives participated in and affected these international negotiations is important. For fairly obvious reasons, the roles of business representatives varied among national delegations to the climate negotiations.

In comparison to other parts of the preparations for the Earth Summit, the negotiations of the Framework Convention on Climate Change involved extensive participation by NGOs. Jean Ripert, chairman of the Intergovernmental Negotiating Committee (INC), decided to open all formal negotiating sessions to observers from NGOs, including representatives of the business community. NGOs were also allowed to speak at all of the negotiating sessions. While actually getting the floor was difficult, as presentations by government delegates often filled all the available time, nonetheless with patience, an NGO representative could eventually find an opportunity to speak. The important point is that there were no closed plenary sessions in the INC process. This contrasted with, for example, the Preparatory Committee meetings for UNCED and with the parallel negotiations on biodiversity, which often reverted to closed negotiating sessions, followed by briefings. In my view, the structure of the negotiating process contributes to its ultimate success. The complete openness of the INC process was critical to the success achieved in these negotiations.

The role of the business community

The business community welcomed the openness of the INC and participated in the negotiating process in a variety of ways. Industry representatives were included as full members in several national delegations. This represented a new approach at the country level:

engaging identifiable interest groups in the development of national positions. In addition, broad groups of business interests were also represented in the INC by trade associations such as the Chemical Manufacturers Association and the American Forestry Association, who were recognized as official observers in the negotiating sessions. Groups of companies with a special or shared interest joined together to form new organizations, such as the Global Climate Coalition. Umbrella groups like the International Chamber of Commerce were represented as well. Thus, there were numerous channels through which business could air its views on the climate issue. Nonetheless, many business representatives felt that, at best, governments only tolerated—rather than encouraged or actively sought—their views. Despite the variety of vehicles available for participation in the negotiating sessions, NGOs are confined to the margins, rather than invited to join at the heart of the discussion.

Ultimately, the INC negotiations were conducted among governments, which retained the responsibility for the fulfillment of national commitments. This meant that the best opportunity for the business community to shape the results was to contribute to debates within their respective national governments.

As was to be expected, differences and tensions developed during the negotiations among members of national delegations. This also occurred within global groupings of business leaders. On reflection we should expect this—and probably encourage it. Diversity of opinion is a source of strength. Business should not be expected to speak with one voice. There needs, however, to be a group of business leaders who can rise above the narrow constraints of national, sectoral, or corporate interests to address the broader issues, and recommend a compromise direction. This facilitating role was played by an organization called the Business Council for Sustainable Development (the BCSD) in the Rio process.

The tension between a national or sectoral business perspective and the broader concerns that shape the conclusions of a global business group is quite healthy. Governments must pay close attention to those differences, as well as to the issues where the views of national and multinational businesses converge. If they do so, they will more readily find the path to successful compromise. We need to maintain the integrity of the negotiating process if the negotiations are going to succeed.

How did the role and opinions of business representatives in the INC evolve over the period of the first phase of the negotiations? The biggest change may have been the growing sense of comfort with the process that developed among the players. Furthermore, as a result of their participation, business leaders were persuaded that the climate and energy issues under discussion were serious. These issues could no longer simply be wished away. This realization occurred despite continuing skepticism among business leaders concerning the magnitude of the projected climate change and the ability of scientists to forecast the regional impacts with confidence.

Some Comments on the INC Process

The issue of business participation in international negotiations raises many questions, but also offers some constructive opportunities. The most important issue today is how to structure and manage the participation of the business community in future negotiations so as to maximize the likelihood of achieving equitable and practical outcomes.

Openness

Based on the experience of the climate negotiations, a case can be made for openness in the diplomatic process. Where there are divergent interests, it is important that each side knows what the other side is saying. Such openness is likely to contribute to confidence in the integrity of the process, as well as to expand and refine the debate. Whereas tendentious and selective use of evidence will merely elicit rebuttal, repudiation and distrust, openness encourages representatives to be careful and more thorough in their contributions. Open participation, as it evolved in the climate negotiations, also leads to the negotiators hearing evidence, analysis, and impact assessments to which they otherwise might not have access.

For all their resources, governments still tend to look at an issue from a particular perspective, often with incomplete information. Indeed, those who will actually be affected by the policies often have a deeper understanding of their effects than do the government agencies that promote them. Hearing evidence from a variety of sources allows the negotiators to develop a broader view of the problem and to identify possible commonalities among the participants. The more open the process, the more likely it is that facts will be tested, evidence debated, and assumptions exposed. Thus, it is clear that having good and complete information is critical to successful negotiations and decision-making.

Dialogue

Let me focus for a moment on dialogue as a process. We all know that few of us are very good listeners. And most of us tend to gravitate towards people holding positions with which we are comfortable. For example, in diplomatic circles, it is not uncommon to find the Scandinavians huddled together, or for delegates from Canada, New Zealand, and Australia to be seen talking to the representatives of the United Kingdom. Any subset of these countries might try to form a coalition with the United States on a position of mutual interest. Of course, the purpose of these informal discussions is to advance the negotiations—but who are they listening to? Mostly to themselves or their echoes. In a world of global diplomacy, dialogue has to be more universal and "listening" has to extend to those who do not necessarily share the same point of view. If we do not search for new forms of dialogue, we may remain prisoners of past habits and obsolete solutions.

Looking Ahead to a Two-step Process

Ultimately, the climate negotiations are seeking a global strategy to reduce greenhouse gases. In practice, that means choosing a global strategy to change national patterns of energy production, transmission, and use. Reaching an international consensus on how to implement such changes will require at least a two-step process: first, preliminary discussion of the issues and second, renewed negotiations on protocols to implement emissions reductions or other necessary measures revealed through the first consensus-building process.

To encourage efficient negotiations, it is necessary to have a schedule and a deadline for completion of the talks. We have seen how important this has been in other international negotiations. Target dates have been important, for example, in all rounds of negotiations on the General Agreement on Tariffs and Trade (GATT), even when the target has been passed. Business leaders, including the members of the BCSD would therefore support the introduction of target dates for the next stage of INC work.

Phase I: Consultation Prior to Negotiations

One way to strengthen the consensus-building process would be to organize an extended period of consultations and workshops prior to resuming formal negotiations. This pre-negotiations phase could involve the business community along with other expert NGOs and leading members of the scientific community more directly and sub-stantively than a "purely" diplomatic process. Direct diplomatic negotiations ultimately occur between nation-states, and we cannot realistically expect changes in that format beyond procedural improvements.

Among the leaders of the international business community are individuals who are committed to the changes required, have expe-rience in managing global operations, and have a practical sense of what will work. Some of this practical experience should be fac-tored into the pre-negotiation phase to help shape and sharpen the options that are presented to the actual negotiators. The business community can also help to address some of the questions that currently divide North and South. All this was evident from the 1992 report of the Business Council for Sustainable Development (BCSD) titled *Changing Course*.

The BCSD could facilitate participation by business leaders at the highest level, identifying responsible corporate executives and soliciting their participation. The Business Council could share this role with an ad hoc committee of other NGOs that might include, for example, the International Union for the Conservation of Nature (IUCN), the Tata Energy Research Institute (TERI), and the Stockholm Environment Institute (SEI). This committee could work with UNDP and with the support of the Secretariat of the INC to organize the workshops, seminars and other discussions that would constitute the consultative pre-negotiations phase of future talks.

The format of the pre-negotiation phase should be informal and limited to individuals and institutions committed to the goals of the Climate Convention. Each participant should bring a special exper-tise to the issue. In this way, leadership, continuity, and competence can be developed around particular discussion points and options.

Ultimately, international negotiations are about compromise. Such compromises are often built around the lowest common denomina-tor, but something more ambitious is also possible. I would certainly aim at a general agreement, or a priority list of recommendations, as the desired outcome from the next stage of INC talks. Consensus may be too much to expect, but it is very important at least to make specific recommendations concerning alternative response strategies. Let me sum up my view of the elements of an optimal approach:

(1) The impact of alternative policy options must be rigorously explored and understood. This will require a more proactive

role at the analytic level involving industry, expert NGOs and, particularly, the scientific community. It will mean devoting more time to the pre-negotiation phase. A series of structured meetings and workshops would be useful to sort out the facts, weigh the evidence, and analyze the cost-effectiveness of various policy alternatives or technological approaches.

(2) When dealing with complex planetary problems, the national interest must yield to the global interest. This is easy to say but difficult to accomplish. The leadership for achieving this must come from the developed world. Many studies have shown that, in the next stage of negotiations, there are some "win-win'" possibilities. But even "win-win" measures will have costs. The question is how do we set the balance between emissions reductions and adaptive responses, between changes in energy technology and improvements in forestry or agriculture. These and related questions can be answered in pragmatic terms outside the formal negotiations and fed back in.

(3) The BCSD supports the designation of the Global Environment Facility (GEF) as the principal financial mechanism of the Conventions. We see considerable advantage to concentrating the necessary catalytic skills and experience in one place, particularly given that the GEF's role will be to finance the incremental cost of measures taken to advance the objectives of the Convention. This means the GEF will be leveraging other sources of funds, most notably from the private sector. Our view is that the role and capacity of the private sector has been underestimated as a source of investment capacity and as a facilitator of technology cooperation. The GEF will address this question jointly with BCSD in the near future.

The critical question in this context is the definition of "incremental cost." That question deserves the urgent attention which the GEF is devoting to the task. Again, a practical businesslike approach will be useful. Business does not need to be protected from normal commercial risk. But, in some projects there could be identifiable, additional, or incremental costs which do make the project uneconomical. If these costs are covered, the projects will advance. The BCSD also strongly urges that the GEF be given a high level of operational autonomy. The donors and recipient countries should set budget policy objectives, but leave the GEF and its consortium of partners to act. The goal is to stimulate performance and positive results quickly.

(4) We must find a non-disruptive way to get from today's energy system to tomorrow's. Today, commercial energy systems around

the world rely on depletable supplies of fossil fuels. Alternative systems based on renewable resources are economically competitive in only limited applications. Meanwhile, Third World countries have made economic development their primary objective. They hope not only to improve the standard of living for their people but also to limit population growth. In this context, all participants in the INC process understand how radical and global the impact of measures to reduce greenhouse gas emissions could be. So we must search together for smooth and successful transition strategies, ultimately identifying a process of change which will allow us to reach a safe level of greenhouse gas concentration but in a way that minimizes economic costs and social disruption.

But the members of the BCSD are very skeptical of an institutionalized approach to technology transfer. If that means that governments, or agencies of governments, will be engaged in a formal approval process, we worry about the bureaucratic burden of inefficiency that will be added to an already difficult task. The BCSD emphasizes technological cooperation, and we know that this is most effectively achieved on a business-to-business basis involving long-term partnerships. Where there will be a need for help, in my view, is in the domain of financial intermediaries who can facilitate the process and provide, where necessary, a share of the project financing. These intermediaries could work with the GEF, with the International Finance Corporation, with bilateral aid programs, and with others to identify projects and to evaluate appropriate technologies. Timely performance and results must be the aim.

The key to success in the next round of the INC negotiations will lie in our ability to devise transition strategies that can be easily explained to the general public and supported by the body politic. Here again there is a crucial role for the business community to play.

Phase 2: Tough Bargaining Ahead

For the next phase of negotiations, business should be able to provide to the diplomatic negotiators both insight and assistance on practical proposals covering a broad front. Managing future emissions will involve choosing among energy and tax policies. These policy options include carbon taxes, tradeable emission permits, and emission caps. The question is which, if any, of these approaches are more likely to facilitate market developments that move us towards the new goals of economic development and environmental sustainability. The business community can help assess these options.

We have learned one lesson from the international negotiations on measures to protect the stratospheric ozone layer—that setting political targets without real understanding of how these targets are going to be achieved leads to conflict, acrimony, and economic inefficiency. Mostafa Tolba, the Executive Director of the UN Environment Programme, was quick to recognize this and bridged the gap successfully.

The next phase of the climate negotiations will be vastly more complicated. Therefore, I would suggest that the international community use 1994 as a consultative, pre-negotiation phase of structured meetings and workshops on technical, environmental, and economic issues. This period should be used to study the options and to design implementation strategies. It is during these activities that leaders of the business community can be most helpful.

This analytic exercise must be rigorous and focused on the costs and benefits of implementing the convention. It cannot get sidetracked into the politics of targets and timetables. Not everyone in industry will be excited by their role, nor will everyone agree with the conclusions. It will be necessary, therefore, to choose participants who are able to exercise solid business judgment about the evidence. The formula is novel, but workable. We in the BCSD have seen it demonstrated. The BCSD recommends that INC use the Scientific and Technical Advisory Panel of the GEF (the STAP) as a source of technical advice and consultative support in order to build competence and accumulated experience. If the STAP is too slow or not up to the task, the INC should then move quickly and publicly to set up its own STAP equivalent. A little competition will not hurt.

The members of the BCSD stand ready and available to mobilize around specific issues. A Task Force on Joint Implementation has already been formed. Business leaders have very little time for regular meetings of an advisory board. The Business Council can be more responsive to precise tasks, defined around clear issues.

If this type of process were to emerge in the INC, I believe that business generally would take a more active and intelligent interest in these negotiations. Business leaders would become more involved and willing to provide their judgment and advice. Sectoral associations would become engaged in the process. The odds are high that we would be able to mobilize private resources in support of the effort.

In the end, business leaders might not be able to agree on everything. But with their input into the consultative process, the possible impacts of competing options would be better understood and impractical and naive options would be culled out. Most importantly, in this way, I believe business leaders could address the new linkage between political targets and implementation strategies. Moreover, business leaders would confront the global implications of the climate problem and would encourage new and constructive partnerships between private enterprises and national governments. Building these new partnerships will be difficult, but, given the risks of rapid climate change, we must try.

CHAPTER 12

A View from the Ground Up

Atiq Rahman
Director
Bangladesh Centre for Advanced Studies
Dhaka, Bangladesh and
Coordinator, Climate Action Network South Asia (CANSA)

and

Annie Roncerel
Coordinator
Climate Network Europe
Brussels, Belgium

Introduction

The Framework Convention on Climate Change has been signed by over 150 states and, as of December 31, 1993, has been ratified by more than fifty countries. It will enter into force in 1994. Environmental non-governmental organizations (NGOs) from the North and South contributed in a substantial manner to these intergovernmental negotiations. They continue to engage in the climate negotiations, despite the limits of the existing text. The NGOs have a common aim: that the process should produce additional commitments leading to real emissions reductions. Through the implementation of such commitments NGOs hope that the objectives of the Convention itself—i.e., stabilization of greenhouse gas concentrations in the atmosphere at levels that do not threaten dangerous anthropogenic modification of the global climate system—will be achieved.

Representatives of numerous NGOs were officially present as observers to the plenary sessions of the Intergovernmental Negotiating

Committee for a Framework Convention on Climate Change (INC). Only a few participated in official government delegations. Nonetheless, it was widely accepted by all the parties involved that environmental NGOs played a key role in the negotiations, both formally and informally.

From both a historical and a current perspective, developed countries are responsible for the bulk of greenhouse gas (GHG) emissions. NGOs have criticized the Framework Convention on Climate Change (referred to herein as the Climate Convention) for the absence of any serious commitment to emissions reductions by these countries. NGOs also have great reservations about the lack of specific targets and timetables in the Convention for achieving changes in consumer behavior in the wealthy industrial nations. Nonetheless, from a pragmatic standpoint, NGOs recognized from early on in the negotiations that the Climate Convention is only the beginning of a long international process and offers a useful basis on which future negotiations can take place.

All major international NGOs were represented at the INC meetings; however the participation of other NGOs coming from various national organizations allowed a more diversified geographical representation. Their capacity to exert influence on the climate negotiations has been enhanced by cooperation within the Climate Action Network (CAN), a network organized and run by regional groups of NGOs who are committed to tackling global climate change, environment, development and poverty issues. Participation in the climate negotiations has proven to be a valuable experience in which NGOs from North and South could work together to establish common and cooperative mechanisms—despite all the complexities of their own positions and the different interests of their home constituencies. Thus, the negotiations have been a major learning process for the NGOs and provided an effective vehicle for interaction between NGOs and government delegates.

This chapter surveys the various dimensions of NGO interaction within the climate negotiations. It observes the historical relationship of NGOs to climate change issues, examines aspects of NGO activity throughout the plenary sessions, and discusses options facing NGOs from the North and South in the future rounds of climate negotiations.

NGO Contributions to the Climate Debate

Some of the most notable advances arising from the INC negotiating process have occurred owing to the quality of NGO participation.

This can be witnessed in the evolving relationships between NGOs from the industrialized North and the developing South. Though always problematic, the expanding and improving discourse led towards a strengthened institutional capacity among NGOs and an effective contribution to the climate negotiations. NGO expertise in the science and politics of climate change made notable contributions to the early rounds of negotiations. At times, NGOs enhanced the proceedings with experience drawn from their own information bases that were not available to governmental negotiators.

Although the contributions of NGOs did not always materialize as effective documentation and influential advocacy, there are clear examples where NGO pressure and authoritative contributions can be linked to shifts in delegates' views and policies. Therefore, the first portion of this chapter traces the seemingly inchoate NGO involvement in the negotiations and analyzes the impact of NGO participation and movement toward greater coordination of activity.

Emergence of Climate Change Issues and the Role of NGOs

Scientific studies in the early 1980's by a few NGOs in the North and then the South concluded that issues of climate change required urgent attention from policy-makers. The impact of temperature increase from rising GHG concentrations drew everyone's attention. The threat of inundation to small island states and major low-lying deltas stirred the concern of many NGOs, as well as scientists and policy planners.

Gradually concerns about the risks of rapid climate change have been accepted and recognized by a large number of NGOs throughout the South. This recognition reflects a widely perceived need to integrate environment and development into a global agenda. As the process progressed, the issue of global climate change became one of the main focal points for negotiations worldwide. Active research by a large number of Northern organizations such as World Resources Institute, the Woods Hole Research Center, Environmental Defense Fund, the Stockholm Environment Institute and the International Institute for Environment and Development reflected this high degree of concern.

Concern for rising temperature due to GHG buildup expanded to include concerns for its indirect impacts, particularly sea-level rise and intensification of natural hazards such as cyclones. These concerns became heightened as scientific evidence presented further gloomy prognoses. The threat of multiple disaster scenarios has

elicited grave concern from the countries considered most susceptible to the effects of climate change, including island states such as
the Maldives and the low-lying deltaic regions of countries such as
Bangladesh and Egypt. Southern NGOs began to discuss these
major issues within the areas of their influence and interacted with
their Northern counterparts. Thus NGOs worldwide emerged as a
major force meeting a whole range of issues. Their responses
included:

• undertaking scientific and policy research, and participating in
  the scientific debate;
• developing the database on historical emissions and improving
  understanding of greenhouse gases; and
• underscoring the urgent need to address the issues of ecological
  limits to sustainable development and the effects of rapid climate
  change.

NGOs and the Climate Convention

Understanding of climate change progressed rapidly from the scientific world to the policy arena over a relatively short period of
time (in the 1980s). Many Northern environmental research groups,
and therefore advocacy groups, were well prepared to respond to
the rapid diffusion of information. Soon, Southern NGOs followed
with their own research agenda, and started linking up with activities in the North. This complementarity culminated in global
networking activities.

Environmental NGOs in the Second World Climate Conference and INC

The Second World Climate Conference (SWCC) held in November
1990, raised global consciousness of the climate issue and culminated for the first time in a statement on the risks of rapid climate
change from the ministers representing over 150 countries. This ministerial statement stressed the urgent need for action regarding the
risks of rapid climate change. The statement prompted a response
in the UN General Assembly which established the Intergovernmental Negotiating Committee for a Framework Convention on Climate
Change (INC/FCCC). But before the Second World Climate Conference, many Northern and Southern NGO groups had worked for
years to pave the way for a climate convention.

  The SWCC demonstrated clearly that NGOs from the North and
South could work together more effectively than that of either group

acting alone. Cooperation in the climate change negotiations was a natural follow-up to the activities of NGOs working in the Second World Climate Conference. Procedures for rapid information collation, analysis and dissemination allowed consensus-building on strategy to occur very quickly. As a result, NGOs were quickly responsive to the speedily changing textures and foci of the international debate.

A New World Order and an Enhanced Role for NGOs in the Post Cold-War Era

NGOs played other critical roles as well. Governments, with their own traditional negotiating processes, are often tied up by jargon-laden UN procedures. The division of nations into regional groups— Group of Seven (G-7), Group of 77 and China (G-77), the European Community, etc.—has led to different sub-group practices for decision making. The various regional groups have developed their own historical priorities. But in the post-Cold War era, the previous groupings based on bipolar relations between the United States and the Soviet Union no longer seemed so important to many governments. Likewise, the degree to which various countries would be affected by global climate change and its impacts did not divide easily along the traditional North–South lines. Some small island states were threatened with total annihilation because of sea-level rise. Their concern was one of survival. Thus, for the first time, in the SWCC, the small island states were looking for genuine partners to voice their concerns.

The NGOs have traditionally voiced equity concerns which would otherwise remain unheard, with the importance deserved, in the international political processes. Potential victim states discovered that the NGOs participating in the climate negotiations could be constructive partners and conduits of their concerns to the greater community of nations throughout the UN processes. The NGOs on the other hand did not have a clear and simple constituency limited by geographical or political boundaries. Most of them were committed to the Intergovernmental Panel on Climate Change (IPCC) objective of limiting greenhouse gases, to establishing ecological limits and to promoting the joint survival of the human species and natural ecosystems. Thus NGOs were vocal about the concerns of victim states, affected regions, and ecosystems. These concerns were subsequently accepted by all the governments and finally spelled out in the preamble to the Framework Convention on Climate Change.

North–South Relationships amongst NGOs

A welcome aspect of the developing relationship between Northern and Southern NGOs was its open character. This encouraged a pattern of mutual respect and appreciation of each other's positions.

Once the information flow of scientific materials on global climate change began to emerge from a multitude of sources—including government bodies, NGOs, research institutes, and the voluntary sectors—government leaders began to take note. For example, the President of the Maldives, Maumoon Abdul Gayoom, addressed the United Nations General Assembly and called on world leaders to save his low-lying island country from extinction. The government of the Netherlands started the process of responding to sea-level rise. After NGOs such as the Bangladesh Centre for Advanced Studies produced preliminary analyses which showed that the future of 11 million Bangladeshis would be seriously jeopardized by uncontrolled sea-level rise, the government of Bangladesh recognized that the country's coastal region could be destroyed by the impacts of climate change. The governments of the North started to sharpen their position towards future negotiations, while Southern governments (including India, among others) identified their own national interests in the debate, taking their cue from research work at NGO institutions such as the Centre for Science and the Environment and the Tata Energy Research Institute.

Networking amongst NGOs in the Climate Negotiations

The NGOs participating in the climate convention benefited from the coordination of a unique organization. In the first INC session in Washington, DC, great doubts surrounded the future of the INC and many were discouraged about the possibility of agreeing on a potential climate convention. Even in those early days, the NGOs had some degree of cohesion and effective communications. There was a clear recognition amongst NGOs of their complementary roles in climate questions. The NGOs from the North enjoyed a comparative advantage resulting from their traditional environmental activities and existing intra-group contacts. Southern NGOs, on the other hand, had long traditions of involvement in national and regional development issues, but little experience on the international front.

Special areas of expertise eventually emerged. Scientific knowledge in the area of global climate change was gained by studying the impact of GHGs and sea-level rise on national populations.

Diplomatic skills were gained through experience in negotiations on the Montreal Protocol and related legal issues. Information was then expediently exchanged locally. NGOs in Bangladesh and India worked together developing detailed scientific bases on a regional level. Through the larger consultative process these views became representative of the Southern regional position. Such embryonic coalitions were not just confined to climate change but extended also to biodiversity issues and the evolution of the Global Environment Facility. Much of this was stimulated by the meetings preparatory to the UN Conference on Environment and Development (UNCED).

International NGOs in the Climate Debate

The leading international environmental organizations—Friends of the Earth (FOE), Greenpeace, and World Wide Fund for Nature (WWF)—played a very important role in advocacy and policy research. They have been an integral part of the NGOs involved in the climate issues, in the SWCC, IPCC, INC, and UNCED process. These three international organizations have a very large membership and an important political presence in several countries. They have each worked to develop their comparative advantages and play useful and complementary roles. For example, Greenpeace had produced one of the early syntheses on the science of climate change, published in the volume *Global Warming*, edited by Jeremy Leggett. Greenpeace further developed and maintained a special relationship with several countries, particularly amongst the Association of Small Island States (AOSIS) and worked with them on understanding scientific, policy, and strategic issues. Greenpeace also recognized global climate change as an area of increasing importance within their own activities and has succeeded in attracting several key NGO individuals to their core staff.

Friends of the Earth has a strong presence in several Southern countries and succeeded in encouraging their member groups in both the North and the South to work on global climate change issues. One of their more interesting innovations has been to involve local groups in public awareness and information dissemination. This has helped to develop workable solutions to global climate change from grassroots sources. Another such initiative is the Alliance for Climate and Cities which established linkages and partnership between cities of the North and South. The Alliance works at local levels to develop concrete, practical responses to the risks of rapid climate change.

WWF has focused more attention on biological conservation and policy issues related to climate change. Their mapping of ecological impacts of climate change has increased awareness and stimulated research activities in many areas. WWF national groups in the US and elsewhere, working in tandem with WWF International, played a leading role in the Climate Network. WWF International contributed significantly to the analysis of financial mechanisms under the Climate Convention, particularly the role of the GEF. Their publications on global climate change have enhanced the NGOs' understanding of the basic background information on climate policy issues.

Climate Action Network

Climate Action Network (CAN) was formed by a group of environmental NGOs who agreed to work together in the area of global climate change. Their joint activities began with the Second World Climate Conference. But prior to the start of the intergovernmental negotiations, the groups gathered together to form focal points to circulate information and to discuss a common platform. CAN groups were established in Europe (Climate Network Europe and CAN UK); CAN US was established for North America in the United States.

Southern NGOs participating in the climate negotiations also recognized that it would be to their advantage to join this growing force of organized NGOs. Southern CAN regional groups were established in Dhaka, Djakarta, Nairobi, and Santiago. In addition, the presence of international NGOs in CAN has been mutually beneficial. It has increased the capacity of the international NGOs to cooperate with one another and has also given CAN more opportunity to affect the IPCC and INC process. As the national NGOs of the North and South and the large international NGOs aligned together, they slowly grew to appreciate each others' priorities.

Northern NGOs Learn the Concerns of Southern NGOs

While the development-oriented Southern NGOs showed an increasing willingness to work on climate issues, there was initially a major lack of understanding among the Northern NGOs about the developmental issues of particular concern to the South. Southern NGOs emphasized that the debate must go beyond just the discussions of climate change as a scientific issue, to include questions of energy use and greenhouse gases. Specifically, the Southern NGOs insisted that the debate had to include energy concerns as well as other related issues, such as lifestyle and consumption. The debates had also to consider the survival of communities. Protecting ecosystems had to be connected both to saving the planet and to setting levels of consumption that could be maintained within ecological limits. The social and moral dimension of global climate change—including its equity implications, and the rights to development and the associated use of energy to meet development needs—were all major areas of consensus for the Southern NGOs.

Over the stages of the INC process, Northern NGOs increasingly grew to appreciate these concerns and priorities. Northern NGOs began to understand that while Southern NGOs were committed to the concept of ecological limits, they were also bound by the development imperatives of their own countries. The implication was that the developing countries wanted to establish the principle that economic development would necessarily lead them to higher emissions of GHGs in the short term. Southern NGOs also supported the Group of 77 position that demanded new and additional funding for the environment.

Environmental NGOs from the South stressed that the total GHG emissions from a country is a function of both population and per capita emissions. With pressure from the Southern NGOs, the debate surrounding per capita entitlement to the atmospheric sink capacity for greenhouse gases became central to the question of the population/consumption dilemma. Northern NGOs began by addressing the South's population problem, but soon the agenda of the Southern NGOs (per capita GHG entitlement, financial transfer, etc.) began to permeate Northern NGO thinking. Where before divergence had been expected, some convergence of views became possible.

Multiple Roles of the NGOs in the Climate Negotiation

Some NGO roles were unique and others were parallel and complementary to those of the official government negotiators in the INC. The following is an analysis of the different roles NGOs played in the INC.

Environmental NGOs as Scientists and Experts

Working in the areas of science and policy response in the IPCC process, research NGOs developed independent scientific analyses. Participants amongst the NGOs included high-level professionals from natural science, social science, law and other related disciplines. Through their continuing access to outside experts, NGOs were able to get immediate second opinions on some of the more complex issues, making it much easier for NGOs to resolve some issues. Nonetheless, on several scientific concerns the NGOs could not take a simple unified stand or quickly develop a consensus view. NGOs from different parts of the world will always hold different views and divergent predilections for action based on their own individual and national interests. However, these differences were harnessed to enrich rather than depreciate the INC process.

Environmental NGOs as National Policy Groups

Environmental NGOs had their greatest influence on the climate debate in the national policy arena. In the United States, the World Resource Institutes, Natural Resource Defense Council, Environmental Defense Fund, Union of Concerned Scientists, Woods Hole Research Center and Audubon Society worked with UN agencies and US policy-makers. From the South, the Tata Energy Research Institute, and the Centre for Science and Environment worked on research, analysis, and development of policy options for India. The Bangladesh Centre for Advanced Studies initiated early work on the impact of sea-level rise in coastal areas and published scientific papers which helped the Bangladeshi government to initiate sound policy responses. In Africa, ENDA Senegal had a pioneering role in GHG emissions inventory. These are just a few examples.

The NGOs took part in writing many scientific papers, exchanging them with their counterparts in different regions and also with government agencies. They thus became facilitators and communicators with many UN missions. By linking up at various levels, NGOs extended their influence and expertise.

Environmental NGOs as Active Participants in the Negotiations

In the climate negotiations, NGOs of the Climate Action Network maintained continuous contacts with their respective country delegations, and with other delegations which were interacting closely during the negotiating process. Many environmental NGOs assisted their respective countries in developing national policies and negotiating positions.

In the earlier stages, southern NGOs had access to the leaders of the Group of 77 and in some cases they were even allowed to participate in high-level discussions as observers. Southern NGOs were then able to raise Southern views to their Northern counterparts, who could in turn relay a measure of how far Northern governments were willing to yield on key issues, as for example, on the question of financial commitments or on the idea of "prompt start" as a mechanism for early follow up to the convention.

Publishing ECO as a Medium of Communication and its Special Role during the INC Negotiations

The NGOs participating in the INCs published regularly and circulated widely the journal *ECO, the Daily News*. This periodical, first published at the Second World Climate Conference, became an almost daily newspaper with feature stories, articles, editorials, and a biting political commentary. *ECO* reflected the views of all the major actors in the climate negotiations—often with a sharply pointed sense of humor. It synthesized complex issues and long-winded debates, presenting them in simple journalistic terms. *ECO* arrived at the breakfast tables of the negotiators almost every morning before the beginning of the negotiations. It summarized updates, visions, scoops, new thinking, corridor discussions and potential directions and alternatives. Subsequently, *ECO* became an informal forum for debate and an arena for the airing of ideas. At the same time, electronic mail carried the issues of *ECO* beyond the corridors of negotiations and reached communities outside. Use of E-mail encouraged some feedback from the member countries and from other NGOs.

ECO also became a very effective means of facilitating the process of negotiations during all the INC sessions. The editorial debates in *ECO* were a microcosm of the general debate proceeding in the INC forum, as, for example, in its coverage of the differences on joint implementation. *ECO* also added humor to some very tense moments in the formal negotiation process. Because of the nature of the editorial process, representatives of North and South were both present for the production of each edition. As the negotiations at INC plenary sessions progressed, the number of Southern representatives increased in numbers and their input to ECO expanded.

Environmental NGOs on the North–South Equilibrium

For each debate that went on in the official forum between the governments, the NGOs held parallel debates. The NGO participants were often able to arrive at a consensus position rather rapidly. Sometimes this was owing to the informal relationship that existed amongst NGOs, their genuine commitment to ecological limits and the recognized need to disagree on some issues which would require extended debate to resolve. The UN General Assembly process is a one-country one-vote process where North and South each plays its own role usually following traditional voting patterns. For the South, these patterns are set by the established processes of the Group of 77; and for the OECD, by the Group of Seven. In the CAN discussion process, the Southern NGOs were accepted as active partners in a very difficult negotiation process. Outside of CAN, other fora where such ease of rapid understanding has been developed between the NGOs from the countries of the North and the South are rare.

The openness of this process saved substantial time. As a result, NGO debates could make rapid progress, coming to firm negotiated statements which could be quickly circulated both amongst the NGOs and the representatives of governments participating in the climate negotiations.

A Model for North–South NGO Collaboration

The climate negotiations offered a unique opportunity for NGOs of both North and South to develop a better appreciation of the broader issues of environment and development. NGO debates had broadened the agenda of the climate debate from one focused just on energy use to one that addressed the whole set of issues involving sources and sinks, complexities of science and databases. An example of this came in the form of a paper by the Centre for Science and Environment (CSE), New Delhi, titled "Global Warming in an Unequal World." This paper provided a very sharp critique of the index of GHG emissions developed by the World Resources Institute (WRI). This criticism and the draft report of the CSE were circulated at INC 1 by Southern NGOs. Both the substantive conclusions and the rationale behind the report were initially doubted by many— including some of the representatives of Northern NGOs and several governments, both from the North and the South. But very soon it became a major subject of debate and succeeded in raising the issues of equity and per capita entitlement to greenhouse gas emissions. These issues later became central to the debate of global climate change during the sessions of INC. This is not to say that these issues would not have come into the negotiations at a later time or in some other way without the NGO debate. However, the CSE paper highlighted the issues of equity and the legitimate rights of the poor. As a result the concept of per capita entitlement was debated time and again. Furthermore, in the final Convention the issue has now been addressed in the preamble and in Article 4(7).

The Acceptance of NGOs by the Participants

INC, in the spirit of UNCED, developed a unique and healthy working relationship between NGOs and government negotiators. The INC process is, in principle, an intergovernmental negotiation. But because of the variety and complexity of issues involved, the nature of the debate and the diversity of the opinion, the delegates themselves found it quite useful to discuss and test out some of their initial ideas with NGOs.

Some delegates found discussions with NGOs to be time-saving: the most controversial issues were not even brought to the main forum of negotiations after consultations between the delegates and the NGOs. For example, there were detailed discussions between many NGOs and several government negotiators on controversial issues such as the pledge and review process and per capita GHG entitlement. There were also discussions involving NGOs pertaining to differences within the Northern (Group of Seven) governments, such as the US and the EC, and within the Southern government positions (Group of 77), such as the Association of Small Island States (AOSIS).

NGO Issues and Statements in INC as a Vehicle for Setting Targets

The Formal Statements of NGOs during INC Meetings

Thanks to the negotiators and the INC Bureau, NGOs as a group were invited to make one formal statement at each INC session, usually at the end. Formal statements were presented in front of the full plenary. Their contents were intensely debated amongst NGOs prior to presentation and were based on consensus reached by all the NGOs present. In this formal statement, NGOs often set the scene, specified targets and outlined their own strategy for the next phase of the debate.

NGO statements varied widely in their contents and style—reflecting the personalities of the NGO representatives who presented them. The NGOs attempted to maintain regional and gender balance in choosing their representatives. However, when it came to making important and strategic statements, the NGOs made sure that well-known personalities, such as Nobel Laureate Henry Kendall, were available to make them persuasively and effectively.

Issues Raised During the Climate Negotiations

Summarized below are some of the NGO statements presented during each INC session. These statements illustrate the dynamics of the debate and the emphasis placed on key issues by the NGOs in each of the sessions. These vignettes illustrate the NGO perception of the debate at different stages and how the focus shifted as the negotiation progressed.

First INC Session

The first NGO statement in INC 1 was presented by Atiq Rahman (Bangladesh Centre for Advanced Studies). He passionately personalized his experience and brought in the whole North–South debate by recounting the personal statements of a Bangladeshi islander who is likely to lose his livelihood owing to sea level rise. The statement can be summarized as follows:

> When I explained to him (the threatened islander) that the impact of global warming, sea-level rise and increased cyclones would be that this island would be inundated, he asked me "Why me? What have I done to cause it?" He was surprised to hear that many communities of the world use 200 times more energy per head than his people do. He again asked me "Why? Why are their wants more important than our needs to survive?"

In this first statement, the NGO representatives raised the critical issue of ecological limits as an objective of INC. The aim was to stabilize the atmospheric concentration of greenhouse gases—based on the conclusions of the IPCC on global ecological limits. Atiq Rahman also pointed out that the industrialized countries must reduce their CO_2 emissions immediately and identified the important role of developing countries. Rahman also spoke to a number of critical issues, including the role of biomass, methane and farming mechanisms, institutional structure, and rules of procedures for the negotiations.

Second INC Session

In INC 2, held in Geneva, the NGO spokesperson was Chris Rose from Media Natura and the UK. He set the scene by comparing the INC negotiating process to a climbing exercise full of difficulties to be encountered and overcome. He emphasized the responsibility of politicians to the future generations. Rose highlighted the urgency for reaching an agreement on a framework convention and meeting the expectations of the world as follows:

> Seven million people who are already under the threat of sea-level rise, desertification and cyclones will be mightily relieved. Your mandate is not to treat the likelihood of climate change as a theory, but as a fact. The target must be to stabilize greenhouse gas concentration at a level which will prevent dangerous anthropogenic interference with climate.

Third INC Session

In INC 3, Gilbert Arum of Kenya and Climate Network Africa observed that the financial issues were complex, but nonetheless could not be used as excuses for not acting fast enough or for the lack of financial commitments. He emphasized the following:

> Notwithstanding the fact that the economics of climate change are considerably less well developed than climate science, some sectors of government and industry persist in citing allegedly unacceptable costs of reducing GHG emissions, and ignoring the many studies which indicate that this might be achieved at relatively less cost. On technical and financial cooperation/assistance, think of what forms and amounts are needed to begin to confront this problem. The report is set against the goal of achieving a climate convention by June 1992 which protects the planet . . . from dangerous anthropogenically induced climate change.

Fourth INC Session

Michael Oppenheimer, from the Environmental Defense Fund of the US, was the NGO speaker in INC 4. He emphasized the role and uncertainties of science while addressing the negotiators:

> It was science, not politics, which brought you here in the first place. Do each of you remember what scientists have been saying about global warming? Simply this: Despite large uncertainties in the

rate and degree of future warming, despite uncertainties in the size of future sea-level rise, and despite our limited understanding of regional effects, we are sure that large and potentially disastrous changes are afoot.

Fifth INC Session

In INC 5 the statement on behalf of NGOs was presented by Henry Kendall, Nobel Laureate from MIT, and the chairman of the Union of Concerned Scientists. He categorically emphasized the urgency of completing the task facing the INC. He was straightforward in putting the blame on the United States for delaying international progress on the issue.

> I am saddened that the main responsibility for this state of affairs belongs to my own country. The refusal of the United States to make a commitment to reduce, or even to stabilize, its huge emissions of carbon dioxide and other greenhouse gases has been the principal roadblock to progress. In early 1990, I transmitted to President Bush an appeal for action to prevent global warming on behalf of some 55 American Nobel Laureates and more than 700 members of the National Academy of Sciences. It said, in part, that uncertainty is no excuse for complacency. Only by taking action now can we assure that future generations will not be put at risk.

Sixth INC Session

Annie Roncerel of Climate Network Europe presented the NGO statement on behalf of CAN at INC 6. At a moment of intense activities within the GEF, she stressed the importance of the role to be played by the Conference of the Parties on financing mechanisms and made the following points:

- INC should play a decisive role in defining . . . 'agreed full incremental costs' . . .
- The concept of incremental costs should not be used as an excuse for externalizing environmental and social costs from projects . . .
- NGOs should be given observer status in participants meetings . . .
- It is curious that NGOs can sit in INC meetings . . . discussing finance and GEF, but not—as yet—in GEF participants' meetings!

The positions expressed by NGOs did not immediately become the focus of the debate but they succeeded in setting the tone and maintained the pressure towards a meaningful and universally acceptable convention.

CAN Newsletters from the South Affect Global, Regional, and National Policies

Regional CAN groups from the developing world were formed at various stages during the Climate Negotiations (CAN Africa, CAN South Asia, CAN South East Asia and CAN Latin America). CAN Africa and CAN South Asia (CANSA) started bringing out newsletters called *Impact* and *Clime Asia* respectively. These highlighted many of the issues in the negotiations. These initiatives were subsequently joined by another newsletter *SEA News* from CAN South East Asia (CANSEA). At the last stage, CANLA (Climate Action Network–Latin America) brought out its own newsletter in Spanish. These newsletters were widely distributed during each of the negotiating sessions. Each regional group highlighted its own interests, at the same time keeping the focus on the global issues and the specific concerns raised during each of the negotiating sessions.

A Case Study: Clime Asia

As an example, let us take the case of *Clime Asia*, the newsletter of Climate Action Network South Asia (CANSA). CANSA was formed just after the first negotiating session. In its first issue, *Clime Asia* focused on several key concepts, including the controversy concerning the concept of Pledge and Review. This first issue, prepared to coincide with INC 2, highlighted the notion that global climate change is a source of further destabilization in an unequal world. In this post-Cold War world, equity issues, including access to and optimum use of scientific data, become major concerns for many countries. In its second issue, *Clime Asia* highlighted the question of excessive consumption in the North as the key activities determining the level of global environmental stress and enhanced risk of global climate change. It also analyzed why the GEF would not be the best financial mechanism for the climate convention. The issue concluded with a discussion of the barriers to acquisition of environmentally sound technology by developing countries.

In the third issue of *Clime Asia*, the focus was on the NGOs, demand for equitable climate rights. Equitable climate rights and effective participation became the major concerns in the issue distributed during INC 4. These demands grew out of the efforts to formulate a draft convention during the meeting of NGO members of CAN held in Paris in December 1991. The Paris meeting of NGOs expressed great concerns over the inequitable relationship between North and South, a balance of power that seemed likely to be exaggerated further in a future world confronted by the impacts of rapid global climate change.

Issue four focused further on developing countries. It highlighted India's activities in limiting CO_2 emissions. This issue also criticized "Joint Implementation," a concept floated originally in the negotiation by the Norwegians as a mechanism for achieving emissions reduction through bilateral cooperation between an industrial and a developing country. The Southern NGOs were concerned about joint implementation and raised doubts as to various aspects of its implementation—such as beneficiaries, role of forests and other sinks and difficulties of honest monitoring and evaluation. The NGOs also criticized the GEF as being a limited and unrepresentative institution, dominated by the World Bank. This issue of *Clime Asia* also observed that rich, i.e., Northern, countries essentially get "free lunches" through their consumption patterns and demonstrated that the current pattern of resource transfer from the South was a threat to achieving a sustainable world.

Successful outcomes depend not just on "participation" but "effective participation," i.e., the capacity to participate in an informed way and be able to affect policy and protect one's own interest. *Clime Asia* regularly highlighted the lack of effective participation on the part of Southern governments and expressed concern that many were coming to the negotiating table without adequately preparing themselves to discuss the very complex issues that were being negotiated.

In the April–June 1992 issue of *Clime Asia* that was distributed during INC 5, great concerns were expressed as to the future of the negotiation. The lead story noted ". . . one of the success stories that UNCED can and must have is the (successful conclusion of a) convention by the Intergovernmental Negotiating Committee on Climate Change. INC 5 held from 18 to 28 February 1992 in New York wasted a lot of time due to lack of commitments from the powerful economies. A new session termed as the second part INC 5 will be held in May 1992 at the UN in New York. This is the last hope for any meaningful global conventions to be signed by the heads of states present in the 'Earth Summit' in June 1992."

The April-June 1992 issue also illustrated the impact of sea-level rise on Bangladesh. Furthermore, in another article it was argued that the financial questions cannot be used as excuses for non-activity.

This issue compared and reviewed three independent studies from the United States, Europe and Australia and showed that it is possible to make CO_2 emission control economically viable and sometimes even profitable.

This last issue of *Clime Asia* before UNCED also reported on the results of the South Asian summit of NGOs held in New Delhi in February 1992. The Summit Communiqué highlighted the impacts and consequences of global climate change and concluded that the climate convention should be signed only when the equal rights of all to the atmosphere are accepted. This demand again brought to the surface the question of equity and the plight of the more vulnerable states to the risks of rapid climate change.

Wider Access to Newsletters

All the regional CAN newsletters were circulated in the INC meetings and the subsequent GEF meetings. They were widely read not only by people who participated in these meetings but also by government agencies and NGOs who could not participate. These publications have strengthened the Climate Action Network—helping it to become not only a community of scientists but a community where science policy analysis produces policy prescriptions that can assist and influence the formal negotiations among governments.

Analyzing NGO Concerns, Involvement and Impact on the INC Process

The presence of a large number of NGOs, with their organizational capacity, information gathering, dissemination and lobbying potential, affected the debate on many issues and helped to maintain the momentum of the INC. These same factors encouraged the cultivation of a consensus towards a climate convention. The NGOs raised many issues, succeeded in achieving several of their goals but were less successful with others.

The following is an analysis of some of the issues raised by NGOs in the negotiations. Successes are set against failures to give a flavor of the breadth of the issues covered and of the intensity of the debates that characterized them. Rather than an exhaustive list of themes or a detailed analysis of each of the issues, this is an attempt to demonstrate the complexity of the debates and to present clear examples of the role and dynamics of NGOs in the INC process.

Ecological Targets

The most general NGO goal was to establish the importance of ecological limits to GHG buildup and thus to ensure the minimization of risks from anthropogenic intervention in the global climate system. Achieving this goal required having an effective national action plan, and a clear idea of national responsibilities for past emissions.

Throughout the INC process, the NGOs worked towards enshrining the concept of ecological limits in the text of the Convention. This proved quite difficult owing to the attempts to dilute the principle by some country delegations. Furthermore, NGOs maintained a consistent position in attributing the responsibility for the present and historical accumulations of GHGs where it belonged—on the industrialized countries of the North. The Convention reflected this attribution of responsibilities. However, in their attempts to elicit binding commitments to rapid, effective, and meaningful reductions in GHG emissions, the NGOs were utterly disappointed.

The NGO position on ecological limits was consistently affirmed throughout the INC process by groups from both the North and the South. While in INC 5, the EC countries were prepared to commit to global emissions limits, the US consistently opposed any such concrete steps. The consensus for binding commitments to emissions reductions broke down in the EC when the UK did not actively support the EC commitment and chose instead to remain very discrete and inactive. This shift destroyed any semblance of Northern resolve in support of a serious commitment to emissions reductions. It was evident at that point that a powerful or effective Climate Convention would not be forthcoming. But the political imperative to have a convention, however weak, ready for signature at UNCED remained a priority for the NGOs too. Whenever the NGOs expressed their frustration at the lack of "teeth" in the convention, several of the more seasoned national negotiators reminded them that this was only the natural working of the UN system. The NGOs had little choice but to accept that the current convention document was only the beginning of a process that would ultimately lead towards a more effective international agreement— if national negotiators wanted it to do so. Comparing it to the Convention on Biodiversity, NGOs took some solace in the fact that at least there was a global consensus on a convention on climate change, however weak the initial instrument might be.

Financial Mechanisms and the Global Environment Facility (GEF)

The GEF, brought into existence by a directive of the Board of the World Bank, started in April 1992 to present itself as the leading candidate to be the financial mechanism under the Climate Convention. A different set of NGOs involved in "Bank Watch"—but not active in the INC—had systematically analyzed World Bank projects and programs and demonstrated the disastrous role the World Bank had been playing in developing countries. In particular, the World Bank's recent role in structural adjustment had systematically crippled developing economies, burdening them with an insupportable debt responsibility. Anticipating the major role played by World Bank in the GEF pilot phase (since both the chairman and the GEF administrator are Bank employees) NGOs were extremely critical of the GEF becoming the major financial mechanism.

To the Southern NGOs, the World Bank embodied everything that was wrong with the Western-oriented development paradigm. Hence, the Southern NGOs were extremely vocal in the intra-NGO debate on the role of the GEF and of the World Bank as the dominant agency in the GEF. The Northern governments (especially France and Germany) were concerned about the capability of the UN system, which they viewed as a financially inefficient mechanism, to handle efficiently the large sums of "new and additional funding" that were expected to be forthcoming after the Convention was signed. Informally they indicated that the World Bank was the only organization that could be trusted to handle the fund of potentially billions of dollars that would be available to implement the Convention.

Many governments and NGOs from the South noted that the four priority areas set by the GEF in the pilot phase already demonstrated a Northern bias. Ozone depletion and global warming, for example, resulted from emissions caused largely by the Northern industrial system. The Southern NGOs argued that large World Bank projects that were typically funded were both anti- environment and anti-people. Thus, the World Bank style of industrial development underpinned by mega-projects would be in constant conflict with the sustainable development paradigm that was the essence of UNCED. To avoid or minimize this conflict, the Southern governments—acting through the Group of 77—consistently opposed the identification of any single financial mechanism as the vehicle for distributing funds made available through the convention. These countries—and the NGOs allied with them—favored a new fund and the creation of a new financial entity.

Northern donors were committed to ensure the GEF was the dominant and single financial mechanism and that it remain safely in the hands of the World Bank. The UN Development Programme (UNDP) and the UN Environment Programme (UNEP) could be allowed to participate only as junior partners, helping to give the entire enterprise a semblance of legality and balance. For a long time, negotiators for the Southern countries insisted that institutional options be kept open, allowing more than one contender to be considered. Some NGOs believed that the North would use the GEF as a new strategic device to shift responsibility for past and present damage to the climate system onto the Southern countries. The NGOs condemned the approach to naming the GEF in the Convention as the "interim financial mechanism," regarding this as an attempt to let in the GEF "through the back door." Southern NGOs went as far as to call this diplomatic finesse "an immoral act."

Northern views often reinforce the GEF's role, describing it as the existing reality, arguing that now it must be made to work and modified where appropriate. Nonetheless, well after the convention had been signed in February 1993, attempts were still being made to define this "reality" for the first time. Developing countries expressed grave concerns about any definition which will limit their entitlement to the purported "new and additional funding"—since the GEF does not take into account their greatest environmental concern: poverty.

Southern NGOs still feel strongly that history will be the judge of whether this will give life to the Convention or continue to reinforce the existing and unequal trade system. Some fear that using the World Bank as the medium of transferring resources will continue to promote a net flow of funds from the South to the North rather than from North to South.

A key compromise in the final period of negotiations of the financial mechanism was the introduction of the concept of "agreed full incremental costs" to describe the obligations of the Northern countries to finance measures taken by developing countries to achieve the objectives of the Convention. The enshrining of incremental costs in the Convention without any prior analysis, understanding, definition or clarity, was a means of maintaining ambiguity toward the role of the funding mechanisms. Because of concerns about the implications of this clause in Article 4, Paragraph 3, NGOs—particularly from the South—are still debating whether the GEF has any future for transferring new and additional funds from North to South.

Joint Implementation

NGOs were often divided on new concepts that arose during the negotiations, one important example being the concept of "joint implementation." Many Southern NGOs consistently opposed the concept of joint implementation. Floated originally by Norway, the idea was strongly criticized by NGOs from the South on several grounds.

Joint implementation was viewed with some skepticism by Southern NGOs who saw it as a potential "policy vent" through which culpable Northern governments and enterprises would be able to continue their environmentally damaging activities while appearing to contribute directly to meeting the GHG crisis through a formalized—and therefore respectable—facility. The main concern of the Southern NGOs was that joint implementation would give rise to a new form of colonial enterprise where defaulting Northern countries or companies that produce large amounts of GHGs could transfer their obligations under the Convention to weak governments in the South—which would accept these obligations in exchange for money regardless of the terms.

Global Equity and the Per Capita Concept

Developing countries insisted that the central focus of the Convention should be on per capita entitlement to the sink capacity of the atmosphere. Despite the fact that this concept was pushed by, among others, Ghana and Pakistan—as leaders of the Group of 77—and India in particular, the words "per capita" still appear only once in the Convention—and then only in its preamble. The concept of rights to the atmospheric sink has been "forgotten."

The large amount of time spent in debating the per capita concept left no imprint on the Convention. It was the NGOs in the *ECO* newsletter who mentioned that per capita entitlement reflected the UN charter—which enshrines equality of all human beings past and present. If the per capita entitlement to future GHG emissions and historical responsibility for past emissions were accounted for, there would be no real need for "new and additional funding." These rights would have resulted necessarily in an automatic transfer of resources from industrial to developing countries.

This was not uniquely a North–South debate: there were also so-called Southern countries which, on the last day of the debate, raised objection to the presence of per capita entitlements even in the preamble. These objections were raised by several of the of the oil-exporting developing countries, including Saudi Arabia, which has one of the world's highest GNP per capita but is entirely dependent on non-renewable resources. The role of oil-producing countries (particularly Saudi Arabia and Kuwait) in trying to delay any convention at all and especially obstructing agreement on the per capita concepts was not easy to understand initially. Some NGOs and government delegates saw this anomalous behavior as a reflection of the newly found wealth of these countries and a concern for their disproportionate number of expatriate workers. Since these workers do not form part of the population count and therefore do not form part of the calculation in terms of per capita entitlement, the interests of the oil-producing countries diverged from those of other Southern countries. Furthermore, some Latin American countries with low population densities and high forestry coverage saw the per capita concept as having a potentially negative effect on their short-term economic interest.

The essential issue of equity in development—which could have been addressed through the concept of the per capita entitlement—was thus lost. As a consequence, the Convention missed an opportunity to commit itself to a more equal world in the future.

Economic Crisis in the North and New Funding

The timing of the INC negotiations coincided unfortunately with a slump in the economic performance of many Western countries. The Western governments were therefore preoccupied with addressing their domestic ills. Whatever meager aid already existed was being re-evaluated, and was often viewed as a burden on their own societies.

The sense of a global moral obligation to address global environmental issues was at a low ebb during 1992 in the United States as well as elsewhere in the North. Hence, few governments were in a position to make firm and effective commitments to new and additional funds. The opinions of environmentally literate leaders such as Al Gore did not have a strong constituency in their respective national electorates during the Bush presidency. The question of using the Climate Convention as a means of resolving global inequity never appeared high in the priority of Northern leaders. By contrast, many Southern leaders and NGOs clamored to use UNCED and particularly the INC process as a means to do so.

Consumption, Lifestyle, and the Population Issue

Southern NGOs insisted that excessive consumption and lifestyle were critical parameters which had to be brought into the center of the INC debate. Many Northern NGOs indicated that neither their governments nor their people were yet prepared to make any major shift in lifestyle or consumption patterns, although many Northern NGOs recognized the need to do so. Several Northern NGOs insisted that energy efficiency would be able to redress part of the over-consumption problem in the North. The bitter pill of an absolute reduction in consumption in order to reduce GHG emissions to the atmosphere was never explicitly prescribed in the Climate Convention.

At Nairobi during INC 3, one Northern NGO member was worried, reflecting his delegation's concern, that the question of present population and the potential of future population growth in developing countries had not been adequately addressed. The Southern NGOs retorted by saying that as far as the risks of rapid climate change are concerned, the population problem was essentially in the North, where each child born is likely to consume much more, causing the emission of more GHGs. Hence, the Northern bias toward considering GHG emissions alone must be contested. The NGOs reiterated that per capita consumption and population together are responsible for increases in GHGs.

This debate was initiated by the publication entitled "Exploding the Population Myth: Population versus Emissions, Which is the Time Bomb?" (Rahman, Robins and Roncerel, eds.) The more forceful of the Southern NGOs emphasized that it is the North's historical emission for which they themselves must now take responsibility and thus, must pay appropriate compensation. Potential GHG emissions in the future by unborn generations are impossible to assess now.

Other Initiatives: Pledge and Review, Prompt Start, and Scientific Seminars

Pledge and Review

NGO criticism of some of the initiatives that arose in the INC debates had direct and specific impacts. One example concerns the

NGO response to the Japanese initiative of "pledge and review." During INC 3, Japan floated the idea of "pledge and review" as a means of softening any commitment to emissions reductions and building in a review process by which soft targets could be set with no legally binding targets. The Japanese suggested that this was one possible way of keeping the US "in the game" of negotiation and making some commitment to emissions reductions, however limited.

The NGOs, particularly several from the US with contacts in the US Administration, knew that the Japanese proposal would have, at best, a limited impact. Many NGOs felt totally opposed to this ill-defined approach and lampooned the Japanese proposal as "hedge and retreat," reflecting their perception of a generally weak Japanese position toward emissions reductions. *ECO* observed that pledge and review had no serious potential to achieve lasting emissions reductions. *ECO* argued that the concept could not work for technical reasons either as a mechanism or as a strategy to achieve greater gains in the future. This strong attack by NGOs on the pledge and review idea may have contributed to its demise after only a short life in the negotiations.

Prompt Start

The NGOs played a supportive role in promoting the concept of "Prompt Start." NGOs argued that a prompt start was need to implement some specific actions by the INC members following the signing of the Convention and before its ratification. Several members of the INC Bureau had formal and informal meetings in which NGO experts gave concrete suggestions on how to make the prompt start concept a reality.

Informal Consultations and Scientific Seminars

Aside from NGOs being given their opportunity to deliver a short statement at the plenary session in each INC session, the INC Bureau members frequently had formal and informal meetings where NGO representatives were given opportunities to air their views and to represent the views of their colleagues.

Another important role played by the NGOs was to keep the most up-to-date information on the science of climate change in the forefront of the INC negotiations. NGOs brought independent scientists to deliver seminars during the INCs. One such example is the seminar conducted in New York by Jim Hansen of NASA. These seminars raised the profile of the science in the diplomatic process and helped to keep attention focused on the objectives of the negotiations rather than just the details of the negotiating process.

Unequal Participation

NGOs made a serious attempt to encourage equitable participation in the INC process. But owing to lack of personnel or expertise, small countries with small delegations found it difficult to follow the multiple sessions of the INC that were held in parallel working groups. Many poor countries with small delegations often did not even participate in the plenaries and were thus considered by some to be "marginal," with little or nothing to contribute to the debate.

NGOs, particularly those from the South, frequently reiterated their concerns about the effects of unequal participation of national governments in the process. They expressed a common fear that the concerns of developing countries would be forgotten by default rather than through an orchestrated movement against them. These NGOs expressed anger and resentment that many of the same small countries were likely to be early victims of rapid climate change, but because of other domestic concerns this issue had not emerged on their political agenda. In light of this situation, NGOs made a distinction between participation and effective participation. Nonetheless, with the assistance of friendly NGOs, in some cases, these small delegations were able to gather information from the proceedings and track the discussions which they could not attend.

One important exception to the problem of limited or unequal participation was the Association of Small Island States (AOSIS). AOSIS was well represented at all the INC sessions and was well supported by several specialized advisers, seconded to AOSIS by NGOs.

In addition, during the negotiations, NGOs were meticulous in ensuring that there was no session in which at least one NGO was not present to brief the other NGOs on the discussions and on the personalities involved. NGOs established the procedure of a daily lunchtime briefing and review that was held throughout the INC sessions. This daily meeting also served as a vehicle in which to delegate responsibilities for the upcoming sessions that had to be covered. The individuals responsible for specific sessions often wrote up the proceedings in *ECO*. This procedure ensured a regular and detailed coverage of the full range of activities in the negotiation process. The resulting capsule summaries often became a dominant vehicle of communication among negotiators themselves and between negotiators and their headquarters.

The daily briefing and review process led Northern and Southern NGOs to undertake parallel debates in the same room. In these debates, some of the more critical issues were confronted directly and frankly; any agreements reached amongst NGOs were communicated to their respective governments.

Environment–Development Linkages and the Universal Right to Development

The debates among NGOs reflected the formal negotiations on the linkages between environment and development. The Southern countries insisted that the right to development be enshrined in the Climate Convention. Many Northern countries were prepared to accept this moral argument in principle but were nevertheless in no mood to accept it into the Convention. The Northern NGOs had been convinced that this was a right which all Earth's citizens shared. The Northern NGOs were more amenable to accept this right than were Northern governments which took the view that enshrining this right in a binding legal text might open the door to a larger transfer of resources than they were willing to commit.

The NGOs made special efforts to encourage some participation in the negotiation process from representatives of the economies in transition. In every session, there were some NGOs from Eastern Europe and the former Soviet countries, but the number was never large. Similarly, the voices from Eastern Europe were muted in the plenary sessions.

The Way Forward: Issues and Concerns

In the months after Rio, we are facing the task of giving practical effect to the Climate Convention. As finally agreed, the Convention falls far short of the original expectations held by the international community. The Convention deals with complex interlocking commitments, provisions and institutional arrangements, and reflects considerable tension between developed and developing countries, through the numerous escape clauses in the final text. The Convention is replete with caveats, qualifications and provisions for special circumstances. The concept of ecological targets is mentioned, but not quantified; commitments from industrialized countries are weak, obscure, and confused. Joint implementation is allowed and the GEF has been adopted as the interim funding mechanism. All of these compromises are measures that the NGOs originally opposed.

Bridging the Gap between Scientific Analysis and Political Necessity

In a report published by the Stockholm Environment Institute in 1990 called "Responding to Climate Change: Tools for Policy Development," the working group on targets and indicators of climatic change said:

> Based on an analysis of changes in the past and the adaptive capabilities of terrestrial and marine ecosystems, the working group proposes the selection of quantitative targets for changes of the global mean temperature and sea level . . . a maximum rate of change in temperature of 0.1 degrees C per decade for mean global temperature and a maximum rate of rise of between 20 and 50mm per decade for sea-level rise.

Article 2 of the Convention established a powerful ecological goal, stating that the "ultimate objective" is the "stabilization of greenhouse gas concentrations in the atmosphere at a level that would prevent dangerous anthropogenic interference with the climate system"; and more importantly, it states that this should be achieved "within a time-frame sufficient to allow ecosystems to adapt naturally to climate change, to ensure that food production is not threatened and to enable economic development to proceed in a sustainable manner." But a cap on greenhouse gas concentration in the atmosphere which is necessary to have such an effect could not be agreed in these negotiations.

Moreover, the latest scientific research points towards more alarming connections between increased CO_2 emissions and the potential occurrences of an Arctic ozone hole. This research suggests that atmospheric processes may dramatically intensify the ozone-depleting effect in the Northern Hemisphere that would be expected from CFCs alone. Some of the benefits of a phase-out of CFCs may be lost if there are no concurrent steps to reduce CO_2 emissions.

Therefore, environmental NGOs suggested in Geneva during INC 6—the first negotiating meeting after Rio—that the IPCC takes up the issue of examining the implications of ecological limits as a priority in order to assist the Conference of the Parties in meeting the objectives of other Convention.

It is also clear, in a general terms, that the IPCC will have to deliver the best scientific guidance to politicians on these risky areas. IPCC scientists and negotiators must now have the political courage to go back to these issues and confront the issue with sound scientific expertise.

What Do the "Commitments" Really Mean?

At a minimum, NGOs had expected that the industrialized countries would agree to stabilize their emissions of CO_2 and other GHGs at 1990 levels by the year 2000. Instead, only a confused commitment by industrialized countries to "bring emissions back to earlier levels" has been agreed, reflecting the US refusal to agree to the stabilization target. The text of Article 4(2)(a) and (b) on specific emission reduction commitments by industrialized countries reflects the tense battle between most industrialized countries, who wished to have an emission stabilization commitment, and the US, who adamantly refused to agree to "targets and timetables."

"We are creating a baby and babies are not born with teeth," said the Saudi Arabian delegate at INC 5 (February 1992, New York). US intransigence on specific targets and timetables has meant that the industrialized countries, as a group, were unable to agree on stabilizing CO_2 and other GHG emissions at 1990 levels by 2000. This despite the Intergovernmental Panel on Climate Change stating in February that "more far-reaching efforts are required than are now being contemplated."

Led by Saudi Arabia, Kuwait and Iran, the oil-producing countries also played an obstructive role. Many INC delegates believe that the OPEC countries had no wish for a convention. Other fossil fuel interests, particularly the coal industry (through the World Coal Institute for example), hindered the process by feeding delegates misinformation on the greenhouse issue. Whilst the twelve EC countries, as well as Finland, Austria, Sweden, Switzerland, Australia, New Zealand, Japan and Canada, maintained their commitment to the stabilization target, none was prepared to have a convention without the US. As well as being the world's largest GHG emitter, the US is also the largest single source of funds for any UN program.

Legally, Article 4, sub-paragraphs 2(a) and 2(b) do not constitute a binding commitment to stabilize emissions. However, they do constitute a political signal that developed countries must adopt measures that can stabilize emissions by the end of the century. Ultimately, NGOs believe that even the US will be forced to agree to this.

The text of these two paragraphs created a legal quagmire. The legal problems with these two sub-paragraphs are essentially as follows: in sub-paragraph 4(2)(a) Parties may "return" emissions to "earlier levels" by 2000, depending on the need for "strong and sustainable economic growth" or their "economic structures and resource structures and resources bases." What happens to emissions levels after 2000 is left open. Emissions may, for example, rise again, after briefly returning to an "earlier" level such as 1999.

Sub-paragraph 4(2)(b) refers to reporting on projected emissions, taking into account the effects of the policies adopted under 4(2)(a), adding that the aim is to return emissions to 1990 levels at some unspecified time in the future. Legally, there is no requirement that the timetable for return to 1990 levels is that of 4(2)(a), i.e., 2000.

Developed countries may therefore stabilize emissions at some unspecified time in the future at 1990 levels, after perhaps returning to some previous level, however briefly. All this, however, is still qualified by, for example, the need for "strong" economic growth.

Where are the Limits to Flexibility?

Other concepts of "flexibility" are built into this already weak Article on commitments. All these provisions offer industrialized countries significant, "flexible" opportunities to avoid taking substantive action. The most significant of these include the following:

(1) Comprehensive approach: The specific emission commitments refer to all GHGs (except those ozone-depleting gases being phased out under the Montreal Protocol) rather than those for which scientific knowledge urges immediate action, such as CO_2.

(2) Net approach: The convention allows the parties to use a "calculation" of changes in GHG sinks (e.g., forests which absorb CO_2) to offset emissions in determining whether or not emissions limits are being met. This approach could lead to failure to act on primary emissions and is open to scientific uncertainties and to deliberate abuse.

(3) Joint approach: Developed countries will be able to meet their commitments by taking action (probably projects) in developing countries on whom no emissions limit is placed. Thus, in aggregate, increases in global emissions may be accelerated and not slowed down.

Michael Oppenheimer, a senior scientist with the Environmental Defense Fund warned the delegates in May 1992 about Paragraph 1(d) of Article 4. It calls upon all Parties, inter alia, to "promote and cooperate in the enhancement, as appropriate, of all sinks and reservoirs of greenhouse gases, in particular, biomass, forests and oceans."

Michael Oppenheimer warned that this formulation could encourage dangerous tampering with marine ecosystems:

> The apparent motivation here is that the ocean's capacity to absorb carbon dioxide could be enhanced by human intervention. This notion draws on earlier and more sensible proposals to protect and expand forests for the purpose of sequestering CO_2. The difficulty with manipulation of marine carbon capacity lies in our relative ignorance of the implications and consequences of this manipulation. Large-scale forest manipulations carry risks as well as benefits. But we are at least in a position to understand and evaluate these dangers. Our ignorance of the oceans is profound, he noted. The Climate Convention ought to limit its attention to the ocean to encouraging research. As it stands, Article 4(1)(d) is a dangerous invitation to manipulations which could have catastrophic consequences.

Unfortunately, the original wording was adopted by the negotiators despite Oppenheimer's warning.

Green Business or a Fair Share for Development?

The first implicit element of the Convention, which is fundamental to the implementation and funding mechanisms, is in the preamble—and only there—where the Parties are "noting that the largest share of historical and current global emissions of greenhouse gas has originated in developed countries . . . and that per capita emissions in developing countries are still relatively low."

This paragraph sets out the situation quite clearly: industrialized countries have contracted an ecological debt by increasing the concentration of greenhouse gases in the atmosphere and still contribute to this phenomenon by their high per capita consumption patterns. It is in these terms that the responsibility of the North should be described, and not simply—as is often the case in the treasuries and development ministries of the North—as the comfortable position of a generous donor country.

The financial mechanism of the Convention is described under Article 11. This article was fiercely negotiated by the countries and does not mention the GEF. It describes a mechanism which should function under the guidance of, and be accountable to, the Conference of the Parties, whose function is to decide policies, program priorities, and eligibility criteria related to this Convention.

It is only in a separate later article of the Convention, Article 21, that the GEF is mentioned as an interim arrangement. It is accepted as the financial mechanism only on the condition that it be appropriately restructured and its membership made universal.

A signal was given by the Convention on the governance of the GEF, which must have an equitable and balanced representation of all Parties within a transparent system of governance. A signal must also be given on membership: universality is agreed under this Convention and should be integrated into the structure of the GEF.

One last point is also essential: eligibility criteria for projects have to be established by the Conference of the Parties, and not by the GEF. The Conference of the Parties will formally exist only after entry into force of the Conventions. Projects of the fourth tranche have still to be evaluated according to the criteria of the GEF Pilot Phase. The current Pilot Phase criteria are clearly unsatisfactory if one is thinking beyond macroeconomic terms, but aware that sustainable development contains other parameters. New eligibility criteria for GEF measures have to be established by the Conference of the Parties. Present GEF eligibility criteria cannot generate the right answers because global benefits cannot be measured without local concerns. This has to change for the GEF to become more than an interim solution for developing countries under the Climate Convention.

Developing countries who have negotiated and signed the conventions have made commitments to introduce global concerns in their policies, but they have also established their own priorities. These priorities are identified under Article 4(7). They emphasize economic and social development and poverty eradication.

In December 1992, NGOs concluded their statement by stating that they expected the accomplishment of the following:

> First, industrialized country commitments should be the primary focus of one of the Working Groups. We call upon the industrialized countries to set 1994 as the target date (no later than the first Conference of the Parties) for adopting a protocol to bring about a 25% reduction in CO_2 and other greenhouse gas emissions by the year 2005, at the latest. We were pleased to hear the EC and German statements call for negotiations on protocols to reduce CO_2. These efforts should be complemented by the implementation of the EC commitment to stabilization of CO_2 emissions by 1990. Such a target date was set in a Resolution adopted with the Vienna Convention which created the international momentum to negotiate and adopt the Montreal Protocol before the convention entered into force. Therefore there is no reason for parties to wait until the climate convention enters into force to start negotiations on protocols.
>
> Second, we call upon developed countries to publish their national plans in accordance with agreed criteria by June 1, 1993. The

Secretariat should assist in preparing a paper on suggested criteria for the next INC meeting, including common methodologies
for emissions inventories. At a minimum, these national plans
should ensure that developed countries' greenhouse gas emissions
are capped at 1990 levels by the year 2000 and that the primary
focus of these plans is on countries taking responsibility for reducing their own emissions.

Third, developed countries, in coordination with the INC Secretariat, should assist developing countries in compiling these country plans. We encourage them to do it promptly taking into account
relevant initiatives of other international bodies.

Fourth, the convention states that the financing mechanism should
function under the guidance of, and be accountable to, the Conference of the Parties. GEF is for the moment an interim mechanism;
independent review of its Pilot Phase should be completed before
any decisions are made on a financing mechanism. Attention should
also be given to possible alternatives. The World Bank Administrator of the GEF is now struggling to improve its image. Developing countries are risking potential polarization and conflict. They
are opting for short-term economic benefits without considering
the will of the people on whose account they assume the obligations of government. That threat imposed by climate change implies very serious human-rights issues underlining the urgency of
an appropriate financial mechanism. It is in the interest of all governments to have a broad based support from a healthy and pros-
perous constituency assured a livelihood. Accordingly, we offer
the following principles for establishment of a climate convention
financial mechanism:

- Parties and their subsidiary bodies retain effective, equal and
  full control over the operation of the financial facility. This will
  ensure that operations are targeted at meeting the objectives as
  agreed by contracting parties.
- Funds should be disbursed from the facility on a predictable
  and equitable basis.
- There is a clear separation of the administrative function of the
  financial facility from its policy development and project implementation functions.
- The facility's objectives are not frustrated by the agenda of dominant existing institution(s).
- The GEF remains faithful to the principles and the objectives of
  the convention.
- The GEF should be accountable to both contracting parties and
  their citizens. It must provide free and open access to all

information related to its policy directives and project development, implementation and assessment.

- The GEF should select projects based on their ability to meet specific criteria rather than being influenced by current institutional bias toward large projects. These criteria must address local manifestations of global problems by addressing local needs.
- The GEF must involve grassroots organizations at the local and regional level in all stages of the process.
- Acceptance of funds does not impose collateral or unilateral conditions on recipient countries.

Can All This Be Done?

While not as clear and binding as one might have hoped, the Convention's structure must be used as the existing tool for leading the global community into stronger action. As seen in the analysis, the Convention has strengths and weaknesses: it is the result of a series of compromises which were sometimes reached through deliberate ambiguity.

Imaginative delegates now have the opportunity to stimulate new thinking on unsolved issues turning them into well-defined commitments (such as a clear and healthy definition of joint implementation, or the creation of a dynamic funding mechanism offering enough confidence to both the North and the South). In any event, it paves the way for continued participation by non-governmental actors who have demonstrated their willingness and their capabilities. Many of them are already involved in emissions inventories and the evaluation of country programmes. They will certainly continue to play an important role in these types of activities which—in the end—lead to the monitoring of the implementation of the Convention on the ground.

Since Rio, persistent economic recession in leading Northern countries is preventing them from making stronger financial commitments. To this situation must be added the fact that oil countries played an enormous role in slowing down the whole process; their role and influence was perhaps undervalued, by NGOs in particular. Long-term solutions must be negotiated and dialogue must be established with the OPEC nations and the industrial economies of the North in order to make further progress on Climate Change.

PART IV

Prospects for the Future

CHAPTER 13

Towards a Winning Climate Coalition[1]

James K. Sebenius
Harvard Business School
Cambridge, Massachusetts

Introduction

One key legacy of the mammoth June 1992 Earth Summit is the Framework Convention on Climate Change (also referred to as the FCCC or simply the Climate Convention). Signed by 154 nations following sixteen months of negotiation within the Intergovernmental Negotiating Committee (INC), the Climate Convention enters into force ninety days after the fiftieth ratification is received.

Many environmentally concerned citizens regarded the set of principles embodied in this document as essentially meaningless since the agreement lacks specific targets or timetables for reducing greenhouse gas emissions or any concrete commitments for financial transfers from developed to developing countries. Many of these observers argue that the "new and additional" funds provided for in the Convention will be necessary to induce developing countries to pursue more greenhouse-friendly strategies of national economic development.

But other advocates were more circumspect about the result, believing it to be an essential, if cautious, first step toward controlling human actions that may lead to rapid climate change—with unpredictable damages for human societies and natural ecosystems. Still others found grounds for optimism in the unprecedented scale of governmental and non-governmental participation in the Earth Summit and the white-hot glare of publicity generated worldwide on behalf of environmental issues. Whatever the ultimate verdict, the present Climate Convention is likely to be followed by years of on-and-off international negotiations over more specific strategies designed to control the rate of future global warming.

This chapter offers an assessment of the climate talks thus far and asks how best to move them forward. In so doing, it does not seek to evaluate or resolve the considerable scientific and economic uncertainties that surround the global warming issue, nor does it analyze the merits of the many proposed policy responses. In particular, it does not weigh the option of no further international action on the climate issue. Instead, as a reasonable (but contested) assumption for purposes of analysis, this chapter uncritically maintains that the damages associated with a rapid anthropogenic climate change could be quite serious. Furthermore, this analysis adopts the perspective of an advocate who seeks some more stringent sort of coordinated international action to reduce the risks of rapid climate change, but who is agnostic as to the particular form those actions ought best to take (e.g., targets, carbon taxes, tradeable permits, etc.).

In addressing this question, this chapter draws on concepts from the emerging prescriptive field of "negotiation analysis," an approach with roots in game theory, decision analysis, and social psychology.[2] In particular, the analysis is organized around the problem of building and maintaining a meaningful "winning coalition" of countries that will take cooperative concrete measures to slow the rate of climate change. In a parliamentary context, a winning coalition is a group sufficient to enact legislation. Here, the term is used in a more subjective and expansive sense and is defined with respect to the goal of controlling greenhouse gas buildup. From this point of view, a winning coalition is a set of countries whose actions to control the greenhouse gas emissions or to expand the natural sinks for greenhouse gases are sufficient to meet that goal. (Because different goals imply different winning coalitions, this chapter's analysis is framed to apply readily to different goals.) The opposite of a winning coalition is a "blocking coalition," a group able to prevent—passively or actively—a winning coalition from emerging and to keep that coalition from taking effective actions to limit greenhouse gas emissions over time.

There is a temptation among some observers to attribute the apparent lack of progress in the INC talks to the United States, and, in particular, to the notable opposition of powerful Bush Administration figures (such as former Chief of Staff John Sununu, Budget Director Richard Darman, and chairman of the Council of Economic Advisers Michael Boskin). It follows, therefore, that a more environmentally sensitive and committed Clinton–Gore Administration should unblock progress. This chapter argues that such a view, although not wrong, is too narrow. There is a much broader range of scientific, economic, and ideological barriers that combine to obstruct the efforts of those Parties who wish to supplement the Framework Convention with additional emissions control measures that are both meaningful and sustainable. As such, the present analysis will first describe these barriers and then concentrate on how potential and actual "blocking" coalitions of interests opposed to such a regime can be prevented from forming, can be acceptably accommodated, or can be otherwise neutralized.[3]

Given the prominence and publicity surrounding international conferences, getting stronger and more specific treaty provisions (or "protocols") negotiated, signed, and ratified by a suitable number of countries may seem like the obvious route to achieving a winning coalition. Yet, to those who seek to control the buildup of greenhouse gases, the success of a negotiation should not be measured by paper mandates and signatures affixed, but instead by changes in the actual rates of greenhouse gas emissions or rate of increase in atmospheric concentrations. These outcomes may be observed indirectly as programs that, for example, increase energy efficiency more than would otherwise have occurred, slow down deforestation or introduce new programs of afforestation, shift energy use from relatively carbon-intensive fuels such as coal to cleaner ones such as natural gas, and the like.

By analogy, a committed Europhile may have thrilled to the sight of heads of state signing the Maastricht Treaty that promised closer monetary, economic, and foreign policy integration among the European Community Member States. Yet if these leaders prove to have been too "far ahead" of their national constituents—who, like the Danish people, end up voting against ratification—the negotiated instrument may even set back the cause of European unity. In that event, some advocates may wish in retrospect that there had been greater efforts at consensus building early on and a less ambitious treaty that might have proven to be a more solid building block. In short, while an advocate's natural focus may be on formal negotiations and far-reaching treaties, these should be viewed as means to ends rather than as ends in themselves.

Any responsible actor who wishes to slow the rate of future climate change faces an analogous strategic choice. One option would be to devote primary energies to exploring those diplomatic and procedural devices which are most likely to achieve a negotiated instrument that directly mandates strict adherence to greenhouse-friendly policies—perhaps represented by some form of global environmental protection agency posessing powerful capabilities to enforce targets and timetables for future emissions reductions. Another approach would be to think about how the negotiations themselves can be used indirectly to achieve their goals, acting as instruments to strengthen the broad and deep consensus needed to overcome the already formidable obstacles to action. From this latter perspective, the negotiating process can be regarded less as a device whose sole purpose is to reach formal binding agreements and more as a means of generating publicity, raising public awareness, helping to mobilize sympathetic environmental opinion worldwide, forging coalitions across national boundaries, and providing support for broad-based scientific and policy research to resolve the remaining very troublesome questions about climate change. Deciding which of these approaches, or what combination of them, holds the most long-term promise for crafting a winning climate coalition depends on the nature and extent of the obstacles to be overcome. It is to an examination of those obstacles—and the actual and potential blocking coalitions to which they give rise—that this chapter must turn. First, however, it is useful to place the climate negotiations and the barriers they face in some historical perspective.

Sustained and Effective Agreements Will be Difficult to Attain

The widely accepted goal for the climate change negotiations has been to adopt a general "framework convention" and one or more "protocols" on specific subjects. A framework convention was indeed signed in Brazil although specific protocols for the control of greenhouse gases were not agreed to at that time.[4] In part, this step-by-step, framework–protocol approach was a reaction against the years of negotiating the detailed and comprehensive Law of the Sea (LOS) treaty that was ultimately rejected by the United States and opposed by a few other key powers. In part, the present approach to climate negotiations seeks to build on the perceived success of an analogous process that led to international control measures applied to the production and use of chlorofluorocarbons (CFCs), a class of ozone-depleting chemicals that also contributes to global warming owing to the enhanced greenhouse effect. While a large number of other international negotiations have influenced the dominant course of climate change negotiations and contain useful insights, both the LOS and the CFC negotiations concern global resources (like the atmosphere), embody valuable lessons, and serve as especially salient examples for many informed observers.[5]

Many who are concerned about the greenhouse problem were especially disappointed that the United States—virtually alone among industrialized countries—remained resolutely opposed to adopting any language in the Climate Convention that implied agreement to binding national targets or timetables for stabilizing greenhouse gas emissions. Prior to the 1992 Earth Summit, the other members of the Organization for Economic Cooperation and Development (OECD) had taken more forthcoming positions on the control of greenhouse gas emissions. The nations of the European Community (EC) and the European Free Trade Association (EFTA), along with Japan, Australia, and Canada, for example, had adopted national targets for stabilizing or reducing annual emissions of greenhouse gases.[6] However, on April 21, 1993, President Bill Clinton committed the United States to return its greenhouse gas emissions to 1990 levels by the year 2000 and on October 19 of that year unveiled a plan to achieve this goal.[7]

Prior Negotiations and their Relevance to the Climate Convention

The Third United Nations Conference on the Law of the Sea, launched by the General Assembly in 1970, led in 1982 to a comprehensive treaty signed by 159 states (and other authorized parties). The LOS treaty enters into force once the sixtieth instrument of ratification is deposited; there are now over sixty ratifications.[8] On the positive side, this mammoth effort produced a broadly acceptable LOS Convention—a "constitution for the oceans"—the dire predictions of many knowledgeable observers notwithstanding, and despite technical complexity, scientific uncertainty, and deep ideological divisions on many issues. The negotiation process and the LOS treaty have reduced or eliminated much of the conflict over commercial and military uses of the oceans that was burgeoning at the outset of the negotiations. Given these factors—and the widespread recognition that the atmosphere, like the oceans, is a global resource—there were calls from some quarters for efforts to negotiate a loosely analogous, comprehensive "Law of the Atmosphere" to address global warming.

By contrast, many view the Law of the Sea as precisely the *wrong* way to negotiate a climate convention. The process was conducted at a level of detail that many argue should have been unthinkable in a treaty framework; moreover, twenty years after its inception, the resulting treaty will only shortly enter into force. In the views of skeptics, the result of this unwieldy process, especially with respect to deep seabed resources, was fundamentally unworkable, a dangerous precedent, and counter to Western interests. Just as the United States rejected this flawed treaty, goes this line of argument, so should it reject any analogous process or result on climate change.

Concurrent with the later stages of the LOS talks, another relevant set of negotiations got underway.[9] In 1977, the United Nations Environmental Programme (UNEP) and other UN agencies drew up an "Action Plan to Protect Stratospheric Ozone" that strengthened international efforts at research, monitoring, and assessment. Under the auspices of UNEP, a working group was established in May 1981 to try to come up with a global agreement, a "framework convention," to protect the ozone layer from chemical attack by chlorofluorocarbons. After seven rounds of negotiations, a path-breaking compromise, the Vienna Convention for the Protection of the Ozone Layer, was signed in March 1985 by twenty countries and the European Community.

The Convention created a framework for international cooperation on research, monitoring and exchange of information regarding conditions and trends in the stratosphere, and provided procedures for developing "protocols" containing specific control measures. In 1987, twenty-four countries signed the Montreal Protocol on Substances that Deplete the Ozone Layer calling for the consumption of most CFCs to be cut by 50 percent by 1999. By mid-1990, over sixty countries had ratified the Protocol or announced their intended date of ratification. This list included key developed countries, including the United States, the former Soviet Union, Japan, and the Member States of the European Community. However, relatively few of the less developed countries (LDCs) had ratified the Montreal Protocol; holdouts included potentially major CFC producers such as India, China, and Brazil. Following some North–South pyrotechnics, during a June 1990 meeting held in London, ninety-three nations—including some vocal LDC holdouts such as India—signed a much strengthened CFC convention that would virtually ban CFC production and use by the year 2000. The new agreement also promised substantial financial and technical assistance to the developing world.

In direct contrast to the blunt US rejection of the LOS treaty, President Reagan described the 1987 Montreal Protocol as "the result of an extraordinary process . . . of international diplomacy . . . a monumental achievement." In assessing the relevance of this approach for climate change negotiations, some of those involved with the CFC process noted that the complexity of climate issues makes it "impossible to deal with everything at once" but still recommended disaggregating the problem, and following a step-by-step framework–protocol process modeled after the CFC experience. Subsequent official action by both developed and developing countries endorsed the Montreal Protocol model and supported the framework–protocol approach to the climate negotiations.

Factors Contributing to the Difficulty of the Climate Negotiations

Many factors make achieving tangible policy results from the climate change negotiations far more difficult than dealing with ocean resources and preserving the ozone layer. Some of the most important of these factors are described below.

Scope and Complexity of the Climate Issue

To place the climate change negotiations in perspective, especially with respect to the earlier negotiations concerned with the oceans and the ozone layer, it is important to understand the nature of the issue. Consider four complementary dimensions along which to understand the sources of greenhouse gas emissions:
- In conventional scenarios, slightly less than half of the expected warming from emissions during the decade of the 1980s, for example, came from activities associated with the use of commercial fuels (coal, petroleum, and natural gas—used in industry, home heating, transportation, etc.). Non-energy industrial activities delivered about a quarter, and land-use activities (deforestation, rice cultivation, fertilization, etc.) caused the rest.
- About 55 percent of the expected contribution to warming from emissions during this period is due to carbon dioxide, with CFCs (currently estimated at 24 percent, but likely to be somewhat less depending on the effects of the Montreal Protocol and of findings

that the effective warming contribution of CFCs may be small), methane (15 percent), and nitrous oxides delivering the rest.

- About half of the expected warming will reflect global population growth and about half will reflect growth in per capita demand for energy and land.
- Developing countries now account for emissions responsible for 40 percent of the annual global contribution to expected future warming, a figure that may rise to 60 percent by the end of the next century. (Today, by contrast, more than 60 percent of greenhouse gas emissions arise from activities in the industrialized countries; by 2100 these countries will contribute less than 40 percent of global emissions.) Thus, both issues of economic growth for the industrial countries and development in the Third World will be affected by the choice of policy responses to the risks of rapid global warming.

This examination of the present and future causes of the enhanced greenhouse effect reveals the range of causes and policies that could make some difference in the amount or rate of expected future warming. No approach narrowly focused on carbon dioxide, for example, or fossil fuels or energy conservation or deforestation can fully solve the problem. More important, this look at the vast scope of the greenhouse problem underscores just how deeply its causes are embedded in the central aspects of the world's economic and social activity: across transportation, industrial, agricultural, and forestry practices; from the developed to the developing world; and in the very growth of populations and economies. This carries an important implication: Although some expected a full solution to the climate change problem to emerge from the negotiations that culminated in the Earth Summit, it is realistic only to consider these talks at best as a first step in a series of greenhouse negotiations that will likely stretch over decades.

Thus, one can only conclude that negotiating and sustaining serious actions to mitigate greenhouse gas emissions will entail great difficulties. While the Montreal Protocol model has generally been seen as appropriate, the number of significant CFC-producing countries was small. The economic costs, required institutional changes, and affected industries were relatively limited. Those firms that expected to be able to produce CFC substitutes could benefit compared with their competitors and thus could even gain from the treaty. Few of these conditions apply to limits that might be negotiated in the future on emissions of carbon dioxide and other greenhouse gases. Thus relative to a "successful" treaty for ozone layer protection, the greenhouse problem poses far greater challenges.

A Climate Convention Will Impose Economic Limitations on the Parties

Further, negotiating a broad-scale convention on the apparent causes of global warming will be much more difficult than reaching an agreement on the Law of the Sea. Much of the LOS accord granted or legitimated a series of new claims to ocean resources that had previously been made by many states. Devising an LOS "convention of expansion" involved the relatively easy problem of how to divide what many believed to be an expanding pie.[10] By contrast, negotiations to reduce the risks of rapid climate change must focus on working out conventions of limitation, of shared sacrifice, and of painful transfers and compensation. The resulting agreements will probably require curtailments in energy use, more expensive LDC development paths, changes in agricultural patterns, cessation of currently profitable deforestation, and other such activities. To the extent that climate change negotiations are perceived as allocating sacrifices, they will be fundamentally more difficult than the less politically difficult problem facing negotiators of the LOS, i.e., the problem of allocating "new" resources. Of course, to the extent that the participants focus on the joint gains relative to feared climate disaster, the process will be that much easier. And some groups that will directly benefit—such as vendors of renewable, cleaner, more efficient energy and the technologies that make such energy use possible— may join environmental advocates as vocal proponents of a greenhouse control regime.

Powerful Disincentives Discourage Efforts to Protect the Global Atmosphere

In 1970, the UN General Assembly unanimously declared deep seabed resources (e.g., manganese nodules) to be the "common heritage of mankind." Although this may seem to be a philosophical point, it was in fact a statement about property rights related to newly discovered (and still undiscovered) resources of the world's deep seabeds. By contrast, the global atmosphere is a true commons in the economist's sense, i.e., all greenhouse gas emissions from a single country eventually mix and adversely affect the entire world. Resources involving the global commons generate economic disincentives for individual initiatives to reduce collective risks since the full costs of efforts to mitigate harmful actions by one state can be borne fully by that state—while the benefits of such actions are diffused throughout the global community.[11] Moreover, any benefits of actions now that would slow the present rate of growth of greenhouse gases would only be felt decades hence by the inhabitants of a *future* world. Thus, facing the full costs of abatement today but enjoying only a fraction of any future benefits, individual entities have powerful incentives to continue emitting and to take a "free ride" on any costly actions others might take to mitigate the problem. As such, strong political and economic forces can lead states and private parties to postpone any action absent a broad international agreement.

Many Plausible Bases Exist for Forming "Blocking Coalitions"

As presently contemplated, the Framework Convention on Climate Change–which sets forth an agreement on the definition of the problem, joint research, monitoring, coordination, national reporting—is to be followed by specific protocols on actions to restrict future emissions of greenhouse gases and to enhance greenhouse gas sinks in various sectors. *In such an approach, the choice of which specific protocols to pursue singly, in combination, or in sequence (e.g., transportation, energy, tropical forestry, etc.) will heavily determine which interests will arise to oppose action; in choosing one's issues, one chooses one's opponents.* As elaborated below, this "choice of potential opponents"— which can be expected to be both private and sovereign, and located both in the developed and developing worlds—should be a conscious and strategically sophisticated decision.

In contemplating a strengthened climate convention, attention turns naturally enough to the identities of those who will oppose future actions on the basis of economic self-interest. Yet this is too constricted a view; as the LOS and CFC experiences attest, scientific disagreement, ideology, and opportunism may also animate blocking coalitions—or the elaboration of non-joining but opposing entities—whose joint actions prevent agreement on or implementation of an otherwise desirable treaty.

Lessons from the LOS Experience: Blocking Coalitions Based on Economic Interest, Ideology, and Opportunistic Linkage

The negotiations on the Law of the Sea offer a number of important lessons for those wishing to promote cooperative international action on global environmental issues. The following sections highlight some of those broader lessons.

Genuine Dependence of the North on the South. It is perhaps sobering to recall how the LOS treaty's burdens on seabed mining—for all intents and purposes a non-existent industry segment—engendered tenacious and ultimately effective opposition, for both pragmatic and ideological reasons. Major maritime establishments, especially in the former Soviet Union and the United States, were powerfully motivated in the 1960s by the desire to stop "jurisdictional creep." This phrase refers to the tendency, especially by coastal and island developing countries, to expand claims of national sovereignty over offshore territory and to cast an ever-widening net of restrictions on submarine, ship and aircraft mobility in what had traditionally been the international waters of the "high seas." With the likely acceleration of this powerful trend—to extend territorial sea claims from three to twelve and outward to 200 miles offshore—more than one-third of formerly open ocean could have the same sovereign status as the land-based territory of states. To the world's "blue water" navies, this was an intolerable prospect. Thus the developing world threatened to influence something of high value to the maritime powers.

Attempted Use of this Southern Leverage. Emboldened by this genuine maritime interdependence, many developing countries effectively pressed for a seabed regime modeled on the precepts of the New International Economic Order (NIEO). This new regime might include significant wealth redistribution, greater LDC participation in the world economy, and greater Third World control over global institutions and resources. Real LDC leverage meant that the maritime powers could not costlessly reject NIEO demands and just walk away. This perceived vulnerability to LDC coastal state power kept the United States and other maritime powers at the LOS bargaining table for years but ideological disagreements ultimately spurred the treaty's rejection.

Northern Reaction Creates Blocking Counter-coalitions. As the LOS seabed regime took on more of an NIEO-like character, industry opposition grew. Industrial interests discovered that their most persuasive rationale for opposing the treaty was not so much its potential effect on their own private economic self-interest but rather the ideological cast of the emerging regime. Potentially frightening elements included the declaration that seabed resources were the "common heritage of mankind" (seeming collectivist), seabed production controls (OPEC-like cartelization), mandatory technology transfer (seeming to ride roughshod over intellectual property rights), financial requirements (that functioned as globally-levied taxes), new voting schemes (more like the UN General Assembly), and the creation of international mining enterprises (worse even than state-owned enterprises). A number of these issues—such as LDC demands for technology and resource transfers, and demands for new institutions—were very similar to those now animating climate change negotiations.

Richard Darman, once the vice-chairman of the US LOS delegation and subsequently a senior policy advisor in the Reagan White House and Director of the Office of Management and Budget during the Bush years, contended in an influential article in the journal *Foreign Affairs*, that "the most important issues at stake in the deep seabed negotiations, however, are not merely questions of manganese nodule mining. What is fundamentally at stake is a set of precedents with respect to systems of governance." In particular, he distinguished between the "precedential elements of the *seabed regime* (as distinguished from *seabed mining*)."[12] The Reagan Administration generally concurred. Seabed mining was only a small part of the LOS treaty, but the blocking coalition of seabed miners and policy skeptics that it engendered in the United States was ultimately successful, prevailing over the defense and environmental interests that were the strongest supporters of the LOS convention.

In sum, this dynamic of the Law of the Sea negotiations derived from three factors that may also play out in the climate talks: (1) genuine Northern dependence on the South, (2) Southern realization and attempted use of this potential leverage on issues with economic and ideological components, and (3) a counter-reaction involving the formation of potent blocking coalitions in the North organized around economic and ideological interests.

US negotiating behavior throughout the deliberations of the INC and into the Earth Summit was likewise animated by a sizeable and parallel ideological component: suspicious of multilateral institutions and proposed measures of a non-market nature, hostile to actions that might seem to kowtow to a demanding Third World, and negative on perceptions of "environmental extremism." Much of this ideological animus was ascribed by environmental advocates to the personality and convictions of Bush Administration figures such as John Sununu, Richard Darman, and Dan Quayle, but the concern is larger than a few individuals or a single Administration.

Future Dependence of the North on the South.   Undoubtedly, the ideological tenor of US participation in further climate negotiations will shift in a Clinton Administration from its expression during the

Reagan–Bush years, but such considerations will very likely continue to play a significant role in these talks for a very basic reason: like the Law of the Sea or the Montreal Protocol, long-term success is impossible without the cooperation of the developing world. Greenhouse gases in the atmosphere are now mainly due to past and present activities of developed nations. However, with projected population and economic growth in the developing world, the source of the greenhouse problem will rapidly shift over time, especially if India and China choose their least-cost development paths and rely for fuels primarily on their vast coal resources. China, for example, now plans to expand its coal consumption five-fold by the year 2020, a result that would add nearly 50 percent to current worldwide carbon emissions.[13] Any emissions control measures agreed and taken by the developed world alone could be heavily offset over time by inaction in the developing countries; by the year 2050, projected warming without developing country cooperation would be 40 percent higher than with it.[14] Thus, the developed world cannot solve the climate problem in the long run without the cooperation of the LDCs.

Dangerous Temptations for the South. Especially given prevailing levels of distrust—not to mention the steep energy requirements of industrialization and economic development—a threat by key developing nations not to cooperate with an emerging climate regime could have a clear rationale and a measure of credibility. This remains true even if the required actions would be ultimately mutually destructive, and even if their full effects might be more severe in the developing world. No wonder that, in the words of a recent discussion of climate change and overall Third World concerns, "The problems presented by climate change also present opportunities to reexamine and correct many of the underlying problems of development that have led to the current dilemma . . . including trade issues, debt, technology transfer, technical assistance, and financial assistance."[15] To Southern diplomats with this view, the climate change issue may prove to be a very potent bargaining lever with application well beyond the climate context. According to another observer, "this group sees environment as the same kind of issue in the 1990s that energy was in the 1970s. They hope that the developed countries' interest in the environment can be used over time to wring concessions on development issues from the North."[16]

Economic and Ideological Opponents Increase the Risk of Gridlock.
Fundamentally conflicting North–South agendas and the reality of mutual dependence have found and will continue to find

expression in climate negotiations.[17] The underlying ideological template, also present in the LOS and Montreal negotiations, is that of the New International Economic Order (NIEO). A great deal of the preparatory negotiations for the 1992 Conference focused on generalized North–South concerns expressed in well-worn NIEO terms; the Earth Summit itself was barely able to find common ground in principle between a highly negative United States and the "North–South cold warriors" from the developing countries.[18] The risk is that attempted use of real Southern leverage on behalf of NIEO precepts might meet enduring Northern intransigence based on antipathy to the underlying ideology. Parties on either or both sides of this divide could block sustained and effective action to reduce the risks of rapid climate change.

Lessons from the Ozone Negotiations: Blocking Coalitions Based on Science and Economic Interest

Although the CFC accords indeed represent important international coordinating steps, they further illustrate bases of potential blocking coalitions—scientific disagreement and economic interest—complementary to those explored above in the LOS context. Despite periodically intense public concerns dating from fears that the SST and aerosols would deplete the ozone layer, the actions of certain policy skeptics in the major countries along with a relatively small number of industry leaders, e.g., Dupont and Allied in the United States, ICI and others in Europe, were able to delay action on an ozone convention for a number of years. To understand why, it is critical to focus on "internal" (domestic) considerations along with considerations in the "external" (international) negotiating forum.

It is both instructive and sobering to see how CFC industry opposition had been overcome by 1987. In part, it was a matter of science. Though predictions of individual scientists varied greatly, consensus estimates of the extent, likelihood, and danger of ozone depletion had declined from the early 1980s prior to the surprise discovery of the Antarctic ozone hole in 1985; thus industry opposition to regulation during this period had a scientific basis. However, by 1987, public statements to the Congress by officials of the Dupont Corporations had committed the company to a policy which said that, if scientific evidence conclusively showed adverse health effects, Dupont would no longer produce CFCs; this was a key factor in its "conversion."

But two other special dynamics may have been at work in overcoming Dupont's early blocking actions. First, though it put the work on hold for a time in the early 1980s, Dupont had been intensively engaged in the search for CFC substitutes, and appeared to be well ahead of its global competitors in this regard. If this were so, international limits on the amount of the most dangerous conventional CFCs that could be produced and consumed would both permit the price of the allowed production to be raised, and place Dupont in a favorable competitive position vis-à-vis the safer alternatives it had made efforts to develop. Second, as public concern culminated in tremendous attention focused on the Antarctic ozone hole, prospects grew substantially for US legislation that would have restricted CFC production and use unilaterally. From Dupont's point of view, while no regulation would have been the preferred alternative, international rules that constrained the entire global industry were far preferable to a US law that singled out domestic companies. Thus, the unusual confluence of several distinct factors—scientific evidence coupled with prior public statements by the company, competitive dynamics within the industry driven by the development process for CFC substitutes, and the threat of domestic legislation—were sufficient to turn Dupont around and open a split in the ranks of a global industry.

Varied Forces Will Animate Blocking Coalitions in Future Negotiations

Successfully achieving follow-on agreements that can strengthen the Climate Convention will require skillful negotiations. Knowing the opposition and, in particular, understanding the dynamics of the varied forces which may align to form blocking coalitions will be critical to that success. The following paragraphs highlight some elements of those dynamics.

Fear of High Costs of Collective Response

These LOS and CFC accounts illustrate how potent the arguments raised by the opponents of the Climate Convention may be—on scientific, economic, and ideological bases. After all, the LOS treaty was scuppered in the United States and in other important industrial nations by the economic and ideological concerns of an industry segment (seabed mining) that did not even exist. With respect to the ozone experience, the 1990 *Economic Report of the President* estimated the US costs of compliance with the Montreal Protocol at $2.7 billion—one measure, since reduced, of the costs motivating skeptical policy-makers and corporate opponents of the treaty.[19] Despite public concern over the ozone layer, the Montreal treaty was effectively delayed for several years by these groups until the scientific consensus shifted.

Now $2.7 billion is certainly a high cost, but the same report cites the costs of reducing US carbon dioxide emissions by 20 percent at between *$800 billion and $3.6 trillion.*[20] If these figures are even remotely accurate, they suggest that many of those concerned about the prospect of large-scale reductions in greenhouse gas emissions (e.g., policy skeptics, coal and oil companies, auto makers, etc.) would have an economic motivation for opposition—regardless of the level of environmental benefits—that could be literally hundreds of times stronger than that of the CFC industry. The battle throughout the 1980s over amendments to the Clean Air Act, with annual costs "merely" in the $25–35 billion range, gives another sobering point of comparison. Cost estimates of the magnitudes mentioned above are by no means universally accepted; respectable analyses suggest that some reductions may be achieved at low or even negative cost.[21] Yet it is the credible *prospect* of burdensome costs that will engender opposition, especially among risk-averse firms that fear that they will bear the costs. Further, since the benefits are uncertain, diffuse, and will mainly accrue to future generations, today's opponents are likely to speak with the clearest—and the loudest—voices.

Indeed, the powerful coalitions that will arise to resist major international action on the climate problem are now fairly quiet (although coal and oil producers have certainly made their views known). Yet they will certainly awaken to the extent that the prospect of such action becomes more likely and the *feared* costs are large. Look for example to Canada, a country in the rhetorical vanguard of greenhouse concern. If serious actions are proposed, however, will the Canada that pumps oil, cuts forests, and builds cars really just go along? And are those Brazilians who profit from burning rainforests today readily going to buy arguments about future world benefits? More broadly, blocking coalitions are just as likely to arise in so-called "Southern" countries, whose development could be impeded by emissions control measures, as in developed countries, whose industries and consumers could face heavy cost burdens. Oil-producing states have, of course, powerfully lined up against a climate deal. Likewise, the imperative for Eastern Europe to grow economically and to consolidate its political gains will weigh against major greenhouse gas control action. Such coalitions will likely be composed not only of traditional nation-states but also of domestic interest groups and transnational alliances. In short, the potentially huge costs that are feared to result from significant emissions control policies offer one measure of the economic motivation for opposition to action—and a partial guide to the strength of likely blocking coalitions.

Opposition May Grow as Target Emission Limits Become More Specific

This implication has particular force with respect to allocating national "targets," or reductions from given emission levels that would collectively be within an overall world reduction target. Emission targets and timetables have been the dominant theme in international discussions over a greenhouse gas control regime; environmental advocates and media observers have generally judged the seriousness of national governments by their willingness to endorse binding targets and timetables—and judged the Climate Convention a failure given its absence of binding commitments. Especially given the high level of public concern about the greenhouse issue, many environmental advocates expected quick negotiations and decisive agreement on targets. The significant number of industrial countries that by early 1992 had committed unilaterally or in small groups to stabilizing greenhouse gas emissions was in line with this optimistic view (although there is a long road between target and result). Yet US (and OPEC) opposition to an overall target—e.g., limiting greenhouse gas emissions in the year 2000 to 1990 levels—effectively kept such targets out of the climate change agreement that was signed at the Earth Summit, although the United States later agreed to a greenhouse gas emissions target.

US opposition to targets may appear anomalous, especially given that all other OECD countries except Turkey had agreed to stabilization of emissions by mid-1992 (and the United States had concurred by mid-1993). *Yet as the effects of targets become more specific and stringent, the more resistance from those affected will generally grow.* This implies that negotiating meaningful policies for emissions control is likely to take considerable time. The above analysis spells out the extent to which climate change negotiations could seriously impinge on a range of vital activities—far more than the twelve-year Law of the Sea process. The much simpler CFC negotiation process—from which specific country obligations emerged—took over *five* years from the start of negotiations and over *ten* years from the announcement of UNEP's 1977 Action Plan to Protect the Ozone Layer. Similarly, the twelve-nation European Community Large Combustion Plant Directive to limit acid rain took five years of what were often twice-weekly negotiations among a relatively homogeneous group to agree on targets.

More recently and ominously, although the European Community *as a whole* agreed to stabilize its *overall* greenhouse gas emissions at 1990 levels by the year 2000, its internal negotiations over which nations would be required to make what reductions—the "target-sharing" problem—have utterly broken down. This should be especially sobering to proponents of strict emissions targets given the EC's high level of greenhouse concern and its relative homogeneity (especially compared with the broader UN membership that is charged with negotiating the next phase of a global climate treaty). With this failure to negotiate country-specific targets, EC attention then shifted to imposing a carbon-related tax. As this alternative was being developed, the *Economist* observed that "the proposed carbon tax has been subject to some of the most ferocious lobbying ever seen in Brussels."[22] Carlo Ripa di Meana, then the EC Environment Commissioner, charged that the EC faced "a violent assault from industrial lobbies and the [oil-producing] Gulf countries, which even threatened to break off diplomatic relations" following the announcement of the energy tax.[23] Largely as a result of industry opposition, before the carbon tax was even proposed as a directive to the Council of Ministers, both energy-intensive industries and major exporters were pre-emptively exempted from the tax. Further, rather than apply the tax *unconditionally* as a means of reducing EC carbon emissions (as environmental advocates urged), the tax was made *conditional* on comparable action by the Community's main trading partners.

In some cases, not only industrial stakeholders but also the wider electorate have raised objections to policies that might control greenhouse gases. For example, a mere three days after Britain's Environment Minister announced that Her Majesty's Government could meet the carbon dioxide emission reductions called for at the Rio Earth Summit, the Conservative Party suffered its worst by-election defeat since the end of the Second World War. "The leading factor causing the defeat . . . [was] plans to add a value added tax on domestic heating fuel . . . aimed at cutting CO_2 emissions. . ."[24]

These episodes stand as testament to the power of potential blockers in the realm of climate negotiations—and should be far more worrisome than the image of one or more powerful individuals single-handedly preventing climate action.

Targets May be Relatively Easy to Adopt but Difficult to Implement

At first blush, the acceptance of emissions stabilization targets by all the OECD countries except Turkey and the United States might seem to contradict the above analysis arguing for the extent and power of potential blocking coalitions. Yet another interpretation is possible: as illustrated by the EC experience, *targets may be relatively easy to adopt but difficult to implement.*

For example, prior to the Earth Summit and amid some fanfare, Japan accepted targets to cut its carbon dioxide emissions to 1990 levels by the year 2000 and to eliminate ozone-depleting substances. Yet by June 1993 a different picture was emerging. By that time, legislation designed to discourage industries from generating large amounts of greenhouse gases had been watered down in response to strong opposition from Japanese business leaders and the Ministry of International Trade and Industry. In particular, it would not require environmental impact assessments, penalties, or taxes to discourage polluters.[25]

One might even draw the analogy to the Gramm-Rudman anti-deficit law, which eerily resembles a "framework" climate convention in that it contained targets and timetables for reducing the American budget deficit but left specific agreement on spending cuts and tax increases for later. As such, this law served for years as an expedient political "solution"—at a time of intense public concern about the deficit—allowing executive and legislative officials to declare the problem "solved" and return to budgetary chicanery. Similarly, following the second "oil shock" in 1979, member nations of the International Energy Agency agreed on targets and timetables for dramatically reducing their oil imports from OPEC. Despite this pious-sounding agreement, little has changed; the results are hard to discern today. It is quite possible that the significant number of unilaterally adopted greenhouse gas control targets or a very weak framework convention that was politically touted as the "solution" to global warming could have similarly limited practical effects. In short, the more clearly identified the objects of measures to control greenhouse gas emissions—such as targets and timetables or carbon-related taxes—the greater the likely opposition from the affected parties and the more likely that, if adopted, the measures will not be implemented. Practical and effective implementation of controls cannot be achieved without a far broader and deeper scientific and public consensus on the problem than has emerged so far.

* * *

In sum, *although economic reasons are most often cited as the basis for opposition to green collective action on the climate problem, that is too narrow a view; scientific disagreement, ideological clash, and opportunistic use of apparent bargaining leverage are also likely to play roles.*

In principle, each type of blocking coalition might be dealt with according to its basis; in practice, the bases are likely to be intertwined. (These are not the only bases for opposition; for example, conflicting values or different attitudes toward risk or the passage of time may engender opposition.) The seabed mining industry appealed to economic interest and ideology in opposing the LOS treaty; science and self-interest played complementary roles in delaying a CFC accord; ideological clash and opportunism may well combine in further global climate talks. Opposition for one set of reasons will often masquerade behind another, perhaps more politically palatable, rationale. Opposition may manifest itself in the negotiations themselves, but may also rise up and block action at the implementation stage rendering prior treaty agreements impotent.

Strategies to Overcome Blocking Coalitions in the Climate Negotiations

In light of this exploration of potential blockers, I return to the basic choice posed at the beginning of this chapter. One option would be to focus negotiating energies directly on achieving binding restrictions on greenhouse-unfriendly policies—a frontal attack, so to speak, on the forces that thwarted universal adoption of targets and timetables in the last session of the INC. In particular, the goal would be to tighten the Climate Convention by the addition of new agreements or restrictive protocols that specifically limit future emissions of greenhouse gases.

Another option would follow a more indirect route, using the negotiating process itself and agreements reached as instruments for nurturing a far more broadly based and deeper consensus among scientists, policy-makers, and the public on the need for control of greenhouse gas emissions.

Of course, these options are not mutually exclusive, but the second approach is more suited to a view that suggests the scale of the climate change problem is unprecedented. In this approach, the role of negotiations themselves and any agreements they produce is important but only one measure of the success of the process. In this interpretation, any solution to the problem of global warming will be the accumulation of many disparate influences operating at international, national, regional, and local levels. The task of negotiators would be to ensure that their process and any agreements they reach will enhance public discussion and education about climate issues, mobilize decentralized responses—especially among nongovernmental organizations. It must also increase public support and stimulate worldwide scientific involvement in the search for a deeper understanding of the phenomenon and the nature of effective responses.

Strategy 1: Build on Public and Private Actions Short of Internationally Mandated Emissions Limits

Instead of immediately seeking a traditional control regime, other approaches can partly sidestep and prevent the problems of blocking coalitions as well as some of the time-lags and sovereignty difficulties characteristic of formal protocol negotiation, ratification, and implementation. For example, former UNEP Deputy Executive Director, Peter Thacher, has argued against the conventional wisdom of waiting for a negotiated framework convention as a "first step," to be followed by specific negotiated protocols. Instead, in line with the experience of the Mediterranean and Ozone Action Plans, he suggested that as many countries as are now willing should first agree on a greenhouse "Action Plan" that contains no formal obligations, but that offers the willing sponsors a vehicle within which to promptly commence valuable research, monitoring, and assessment programs as well as to offer developing countries needed assistance to participate in technical and negotiating fora.[26] Such voluntary actions would support and may well speed up the protocol negotiations.

A slightly more difficult option has been suggested by Abram Chayes by analogy to the launching of the International Monetary Fund.[27] By creating a post-Second World War "transition" period during which treaty members could simply "maintain" various forbidden restrictions until they *voluntarily* relinquished them, the necessary institutional apparatus could be developed, professional staffs and reporting practices established, and momentum built toward the result which was ultimately widely accepted. Applied to the greenhouse problem, this would permit further collection of detailed statistical series on global emissions, facilitate technical assistance to environmental agencies especially in the developing world, permit the development and empirical validation of more specific performance criteria, and help develop a technically competent and credible monitoring and compliance capability.

Peter Haas and Emmanuel Adler recently edited and contributed to an issue of *International Organization* (Winter 1992) that contained numerous case studies of the formation of so-called "epistemic communities". These loose-knit transnational coalitions of experts form a worldwide cadre of advisors and analysts that are reasonably close to the decision processes in various countries, who have common views of a problem and who come to exercise great influence on the international response. One view of a very useful function for subsequent climate negotiations would be to foster the development of an increasingly self-conscious epistemic community around issues of climate science. At present, there is consensus on much of the basic science of the greenhouse problem, but issues of timing, magnitude, and distribution still loom large. The two key ingredients that might be provided by upcoming negotiations are resources, especially to enable the sustained participation of scientists from developing countries, and a public spotlight.

Absent a natural climate catastrophe, it is unlikely that future climate negotiations will have anything like the public salience of the Earth Summit, which brought together more than 150 nations, 1,400 nongovernmental organizations, and 8,000 journalists. Nevertheless, future talks should explicitly seek to build on this history of widespread public exposure. Given the potential of global communications technologies and the efforts of concerned governments and interested non-governmental organizations, future climate negotiations themselves and the public awareness they stimulate can help to spur "informal" control regimes, in part by building on and influencing domestic opinion—often led by the actions of nongovernmental organizations. (The most striking example of this phenomenon probably occurs in the area of human rights.) The "national reporting requirements" contained in the current climate convention—if beefed up and properly funded—could provide a natural vehicle for involving and mobilizing citizens and advocates. In turn, stronger informal regimes may come to be embodied in more potent formal instruments that might earlier have been blocked by opposing coalitions.

Arguably, enough countries and environmental organizations are already sufficiently supportive of actions to control greenhouse gas emissions that they should not have to wait for the conclusion of protocol negotiations to take meaningful action. In effect, the Climate Convention signed at the 1992 Earth Summit—which contained no binding greenhouse gas reduction requirements for signatory nations—adopted a version of the approach sketched above, postponing negotiations over actual restrictions to a later, protocol stage. A negotiating process and outcome that did *not* build on the framework now in place would be publicly invisible and exclusive. A negotiating process and outcome that *did* would be designed for public visibility and inclusiveness as well as extension of the scientific consensus on the problem both to new areas of the science and to a broader range of scientists worldwide. In short, if successful, it would aim to channel resources and energy into activities that would broaden and deepen the scientific/political coalition in favor of substantial anti-greenhouse action.

* * *

A second group of strategies returns to the conventional protocol negotiating process and offers a number of ways of enhancing the prospects for success.

Strategy 2: Choose the Subject and Nature of Later Protocols with Great Care

Evidently, the choice of protocols and the negotiating relationship that is envisioned among them is of central importance; after all, with the choice of a protocol comes a set of opponents (as well as supporters). Protocols have been suggested, in some cases without much explicit analysis of their implications for negotiating success, on a virtually endless number of potential subjects: targets for reducing national greenhouse gas or carbon emissions; credits for providing carbon "sinks;" efficiency improvements in automotive transportation and industrial energy use; limitations on tropical forestry, reform of agricultural practices, sea-level rise, technology transfer, international funds to aid LDCs, a carbon tax, tradeable emission permits, methane, and so on.

While it is beyond the scope of this chapter to develop and justify a specific agenda for this process, the choice of protocols should maximize substantive desirability and the potential of the chosen issue to contribute joint gains to a broad-based group of adherent countries—while reducing the likely opposing interests that will be stimulated. Following substantive value, a prime consideration in the choice of protocols should be a clear-eyed view of the likely opposition. Is a proposed target concentrated or diffuse? Politically influential in key countries or not? Are the necessary changes inexpensive or very costly?

Strategy 3: Minimize the Risk of Energizing and Unifying Disparate Interests into a Large Blocking Coalition

A good way to guarantee an endless negotiating impasse would be to handle all the above-mentioned subjects in a single protocol or "Law of the Atmosphere" package to be agreed by consensus. Comprehensive emissions control efforts that affect a number of potentially powerful interests risk energizing and unifying other-wise independent, blocking forces. A protocol that, for example, explicitly targeted oil companies, coal mining interests, or automobile manufacturing firms, as well as various agricultural concerns—let alone one that affected the full range of human activities that result in greenhouse gases—would almost certainly take a very long time to negotiate and might never surmount the solid wall of opposition it could raise.[28]

In the greenhouse case, therefore, to avoid creating a potent unified opposing coalition, it may be wise to proceed *sequentially* with protocols. Not entirely tongue in cheek, it may be best to pick "easy" subjects first—protocols directed at greenhouse contributors that are politically weak, morally suspect, and concentrated in highly "green" countries—to generate momentum, with strategically chosen later protocols building on early successes.

In this connection, a widely discussed option for reducing the risks of emissions growth involves allocating a number of "tradeable emission permits" such that the overall level of greenhouse gas emissions could be limited worldwide. Beyond the initial allocation, the ultimate distribution of the permits would not have to be negotiated or bureaucratically determined since these permits could be bought and sold. In theory at least, they would end up in the hands of those entities that could reduce emissions most efficiently. An ongoing question with respect to such a tradeable permits regime is whether it should only cover carbon dioxide emissions or should extend to other greenhouse gases such as methane and nitrous oxides (in order that the *overall* least cost control actions be chosen). A full answer to this question depends on issues such as the ability to identify and monitor various sources of emissions and the expected complexity of the associated negotiations. Yet from the standpoint of blocking coalitions, it is clear that seeking to negotiate a more comprehensive regime would also risk unifying a much wider set of disparate, opposing interests. Analogous reasoning about negotiating tactics and institutional dynamics applies as well to other proposed regimes, including outright emission limits and various forms of "carbon taxes."

Strategy 4: Exploit the Potential of Incremental Approaches

Beyond measures to prevent the formation of blocking coalitions in the first place, a number of other approaches can be characterized as incrementalist. The idea behind them is to gain agreement on a relatively weak or nonspecific treaty or plan of action in the expectation that over time the resulting instrument will be progressively strengthened. This approach may be a conscious, initial choice or it may simply reflect the strength of opposing forces in the early negotiations. Advocates may "settle for what they can get" or take "half a loaf" and hope that the stage is set for another round that will conclude more in line with their preferences. This section considers a few such incrementalist approaches in rough order of the specificity and weight of the commitments that would be undertaken.

Develop a "Baseline Protocol"

Even in the best of circumstances, a great deal of valuable time may be lost as countries wait until the international process concludes before taking actions to mitigate greenhouse problems. Some domestic opponents of action in different countries will cynically argue for delaying domestic action until all countries have agreed on reductions. Others will merely regard arguments for avoiding domestic actions as a prudent bargaining technique to hold off any unilateral action until an international accord is reached. Either way, their blocking (and delaying) potential can be damaging.

One approach to this problem would be the early negotiation of a protocol specifying a starting point or baseline date—perhaps in the past—after which emissions control measures taken by individual countries would be credited against the requirements of a later international agreement.[29] With such a date agreed, states could promptly undertake unilateral or small group initiatives to reduce greenhouse gas emissions in the confidence that these measures would "count" toward the reductions required by any subsequent regime. Such a "baseline year" agreement, perhaps negotiated as a protocol, could help to neutralize a major argument of domestic opponents of emissions control measures who hold that action absent overall international agreement is either unwarranted or foolish.

Given the time likely to be required for negotiating an overall agreement embracing substantive measures such as binding targets and timetables, a preliminary "baseline" protocol of this sort should prove far easier to negotiate quickly. Incidentally, such a baseline protocol need only assure states that their actions subsequent to the agreed baseline year would count; the question of the status of actions taken before the agreed date could be explicitly left for future negotiation. (A rough US analogue to this approach is contained in the 1992 energy strategy bill that permits companies to report emissions control measures they take today and to count the resulting emissions reductions against future regulatory requirements.)

Design Flexible "Ratchet Mechanisms"

Suppose that greenhouse gas reduction targets were set at extremely modest levels in an initial protocol. Likewise, imagine that an international tax on carbon emissions was initially set at a very low rate—for example to collect resources for an international environmental fund. Given its low rate, this tax (or set of reduction targets) might not trigger concentrated opposition. Later with the monitoring and collection structures in place, if the state of the science merited it and broad-based support existed for such a move, the tax rate might be "ratcheted" up or the emissions limits made more stringent.

Indeed, a review of the history of the ozone negotiations suggests the potential value of such a "ratcheting" device. When an agreement to set CFC limits proved unreachable in 1985, the United States and others pressed for the Vienna "framework" convention that collectively legitimated the problem, set in motion joint efforts at monitoring, coordination, and data exchange, and envisioned the later negotiation of more specific protocols.[30] After scientific consensus on the problem had solidified and industry opposition was largely neutralized, the Montreal Protocol was agreed in 1987. Although this agreement will cut CFC production and use 50 percent by the year 2000, many environmental activists harshly criticized these targets as inadequate.

Yet the institutional arrangements set up by the Montreal Protocol included provisions to facilitate a review of the agreed limits in the face of new evidence (or, effectively, with shifts in public opinion). In effect, these provisions functioned as a "ratchet." Scientific findings subsequent to the signing of the Convention—such as the discovery of a direct link between CFCs and the seasonal occurrence of the Antarctic ozone hole—stimulated treaty parties to tighten the limits beyond the 50 percent reductions originally agreed to. As UNEP's Mostafa Tolba put it, "By aiming in 1987 for what we could get the nations to sign . . . we acquired a flexible instrument for action. If we had reached too far at Montreal, we would almost certainly have come away empty-handed . . . [The] protocol that seemed modest to some . . . is proving to be quite a radical instrument."[31] This assessment was borne out by the 1990 London negotiations that converted a 50 percent reduction into a virtual CFC ban. This model of settling for relatively modest restrictions on which early agreement can be reached, together with arrangements that facilitate reconsideration, may well be emulated in the climate context.

Yet there is a danger to partial agreements that is exemplified by experiences with the 1963 Limited Test Ban Treaty. A number of observers have criticized these accords as stopping too soon and bleeding the intense public pressure for change—when, arguably, intensified negotiating efforts might have achieved a *comprehensive* test ban treaty. By addressing the concerns about Strontium-90 from atmospheric testing in the food chain (and its appearance in mothers' milk in particular), this argument goes, the broader dangers of nuclear testing were not addressed and a valuable opportunity was squandered. Rather than acting as a stepping stone to a larger accord, the Partial Test Ban Treaty became a stopping place. (Recall also the analogy to the US Gramm-Rudman anti-deficit law that was drawn above.)

With respect to climate change negotiations in particular, it is quite likely that public concern will be cyclic, in part as a result of natural climate variability as well as unrelated environmental events (such as medical waste on beaches and the *Exxon Valdez* oil spill). Arguably a naturally occurring period of climate calm, including milder summers and normal rainfall, will lead to reduced public concern and pressure for action. Moreover, scientific understanding will change over time. These prospects argue for more limited agreements that included analogs to the ratchet mechanism in the Montreal Protocol—ratchets that can be tightened down if and as more stringent action appears warranted. Such agreements could constitute a "rolling process of intermediate or self-adjusting agreements that respond quickly to growing scientific understanding."[32] And an even more fundamentally adaptive institution might be envisioned, better matching the rapidly changing science and politics of this set of issues.

Strategy 5: Establish Limited Linkages among Subsets of Issues

In the face of the substantial challenges, a successful accord on climate change calls for a process designed to achieve results that can be sustained over time and modified as appropriate. In particular, a Vienna/Montreal-like process with independent protocols to be negotiated on a step-by-step basis was thought to have the advantage of speed and relative simplicity over a comprehensive LOS-like approach. This raises the more general question of how to deal with greenhouse issues (or protocols): singly, comprehensively, or in intermediate-sized linked packages. The answer, explored below in the LOS context, has a direct implication for ensuring enough gains in an agreement to attract a winning coalition.

Problems of Comprehensiveness and Universal Participation

Many factors contributed to the lengthy LOS process but four procedural cornerstones virtually guaranteed its duration and if adopted in the future negotiations on the climate issue, could easily do the same to global warming negotiations. These included (1) virtually universal participation, combined with (2) a powerful set of rules and understandings aimed at making all decisions by consensus (if at all possible), (3) a comprehensive agenda, plus (4) the agreement to seek a single convention that would constitute a "package deal." [33] The rationale for each of these components was understandable; however, *a universally inclusive process with respect to both issues and participants, together with the requirements of consensus on an overall package deal would be very time-consuming—holding the ultimate results hostage to the most reluctant party on the most difficult issue.* In practice, the LOS conference was less constrained by absolute adherence to these procedural choices, but the powerful bias toward a snail's pace was very real.

Reacting against the LOS approach (i.e., a comprehensive agenda with the requirement of a package deal), climate change negotiators aimed for a framework convention to be followed by specific protocols. In line with the CFC experience, this retained the aims of universality and consensus, but dropped comprehensiveness and the goal of a package deal—in favor of single, separable protocols on limited subjects. This alternative has attractive negotiating features, but it is worth noting that it was the failure of precisely this approach—negotiation of separate "mini-conventions," analogous to protocols—in earlier LOS conferences (in 1958 and 1960) that indirectly led back to the comprehensive package approach of the 1973 LOS conference.

The Problem of Single-Issue Protocols and Selective Adherence

The LOS experiences in 1958 and 1960 suggest that sometimes issues must be linked. By 1958, for the first UN LOS conference, the International Law Commission had suggested a negotiating structure with four separate conventions, concerning different issues such as the breadth of the territorial sea and the extent of the continental margin. With respect to the comprehensive agenda of the 1973 LOS talks, conference president Tommy Koh observed, "A disadvantage of adopting several conventions is that states will choose to adhere only to those which seem advantageous and not to others, leaving the door open to disagreement and confrontations. The rationale for this [comprehensive] approach was to avoid the situation that resulted from the 1958 conference which concluded four [separate] conventions."[34]

Such an uneven pattern might also result from a framework/protocol structure on climate change. Imagine Libya signing a forestry convention while Nepal agreed to a transportation and automotive protocol. For individual countries or groups of similar ones, a single issue often represents either a clear gain or a clear loss. As with the early LOS conferences (with independent mini-conventions), countries sign the gainers and shun the losers. In a climate context, for example, China may resist a specific fossil-fuel protocol that would place restrictions on the development of its extensive coal resources but sign a protocol that encouraged the transfer of advanced technologies to developing countries.

The Benefit of Cautious Linkages among Carefully Chosen Issues: Breaking Single-Issue Impasses

Such single-issue protocols may prove non-negotiable unless they can be combined with agreements on other issues that offset the losses (or at least seem to distribute them fairly). A package deal may offer the possibility of "trading" across issues for joint gain—thus breaking impasses that would result from treating issues separately.

For example, following the 1958 and 1960 LOS experiences, two *separate* negotiations were attempted; until linked, each proved fruitless. With deep seabed resources the "common heritage of mankind," the "Seabeds Committee" undertook a negotiation over the regime for seabed mining. Developing countries wanted this convention to offer meaningful participation in deep seabed mining and sharing of its benefits. Yet the developed countries whose companies had the technological potential, the capital, and the managerial capacity ultimately to mine the seabed, saw no reason to be forthcoming, and these negotiations were inconclusive. At about the same time, strenuous efforts by the United States, the Former Soviet Union and other maritime powers—greatly concerned about increasing numbers of claims by coastal, strait, island, and archipelagic states to territory in the oceans—sought to organize a set of negotiations that would halt such "creeping jurisdiction." In effect, the maritime powers were asking coastal states, without compensation, to cease a valuable activity (claiming additional ocean territory). Not surprisingly, these discussions over limits on seaward territorial expansion in the ocean yielded scant results.

Seen as separate "protocols," these two issues taken independently were not susceptible to agreement. Yet together with concerns over the living resources and outer continental shelf hydrocarbons, it was ultimately the linkage of these two issues, navigation and nodules in the LOS bargaining—together with concerns over the living resources and outer continental shelf hydrocarbons—that came to be at the heart of the final LOS conference negotiations. With respect to climate change negotiations, it is easy to imagine that separate protocols calling on different groups to undertake painful and costly measures will similarly be rejected unless they can be packaged in ways that offer sufficient joint gains to key players. Since any action on climate change will largely involve shared and parallel sacrifice, it is probably only by linking issues such as technological assistance and various forms of financial or in-kind compensation that many developing countries will be induced into joining the international effort to reduce the risks of greenhouse gas buildup. As such, one should expect great pressure toward combining issues for purposes of negotiation that might initially be conceived as separate protocols.

Certain classes of issues, of course, have already been inextricably linked in the negotiations. In particular, the question of how greenhouse gas control measures—among the many other items in the Earth Summit's agenda—are to be financed was the subject of furious negotiation in the INC and at the Earth Summit. While no specific commitments were made by the North, the prospect of substantial resources being funneled through the World Bank proved to be a critical "sweetener" in moving the negotiations forward. Almost independently of how the Global Environmental Facility negotiations proceed, donor country support for an additional "earth increment" to World Bank financing resources (the "tenth replenishment" of the International Development Association) will be seen as critical.[35] Such linkages, perhaps external to the actual climate negotiations, need special attention.

Just as in the LOS experience, mutually beneficial "manageable packages" of protocols under a framework climate convention might be cautiously extended to other environmental issues that arose in the context of the 1992 Conference. Expanding the scope of negotiations in this way might have the effect of bringing on board potential "blockers" from the developing world. For example, desertification and soil erosion issues may be more pressing to key developing countries than greenhouse questions. Many developed countries that are unwilling to make "bribes" to induce developing country participation may nonetheless be genuinely concerned about and more willing to be forthcoming on these regional issues in the context of a larger agreement that promised global climate benefits. ⌐Similarly, more expansive versions of so-called "debt-for-nature" swaps may be explored↓ One of the most potent long-term steps that could be taken by developing countries to combat global warming (as well as a host of other environmental issues) would be significantly stepped-up population control programs.[36] Unlike, say, energy use restrictions, this course of action has the virtue of helping rather than hindering economic development objectives. For cash-strapped LDCs, relatively modest aid from developed countries in this area could considerably enhance domestic population control efforts.

Given this analysis, a central problem in the design of future greenhouse negotiations would seem to be finding a constructive path between the Scylla of a comprehensive package agenda that risks LOS-like complexity, and the Charybdis of independent, single-issue protocols that may lack sufficient joint gain, and risk very selective adherence.[37] Rather than trying to predict the appropriate linkages, the conference should be designed in such a way as to facilitate them as they become evident and necessary. It is generally preferable to deal with issues on their separate substantive merits as much as possible, yet be alert to potential linkages to break impasses. This suggests a conference design with independent working/ negotiating groups with a higher-level body seeking to integrate issues across groups and facilitate valuable, but limited, "trades."

Yet issues should be linked with caution. It can be extraordinarily difficult to "unpackage" them once they have been combined for bargaining purposes. For example, the United States was generally in favor of the navigational portions of the LOS treaty, but had problems with the concessions demanded on a seabed regime. It exerted strenuous efforts at unlinking or separating these topics into "manageable packages," but to no avail. The "package deal" was too strong in the minds of many delegates, and ultimately the LOS convention contained both elements.

Strategy 6: Give Special and Specific Attention to the North–South Dimensions of the Negotiations

Although the climate negotiations focus ostensibly on a global environmental issue, many other factors will complicate future efforts to supplement the Framework Convention on Climate Change. Achieving additional agreements will require paying close attention to the dynamics of North–South relations, especially in the areas of development assistance and trade policy.

Take Active Steps to Avoid Triggering a North–South Impasse

As discussed above, there is an acute risk that a larger North-South agenda—some of it only loosely related to climate change and much of it highly contentious—will occupy center stage in future greenhouse negotiations. Indeed, these talks have already been characterized by aggressive LDC demands for technology transfer and large resource commitments from the industrial world. It is clear that finance and technology transfer, for example, are legitimate interests, but the extent to which developed countries will be forthcoming on them in the context of future climate change negotiation is far less clear—especially given ideological reservations about what could be seen as resurgent demands for a "discredited" New International Economic Order (NIEO). Moreover, despite the keen concern in many nations about climate change, the greenhouse problem is still viewed by many observers as speculative, contested, far in the future, and very costly to address now merely on its own terms—absent additional resources to mitigate the generalized problems of developing countries. The uncertain prospect of global warming may not be a strong enough hook on which to hang a larger North–South agenda. The specter of a breakdown in communications here is all too real—with the North speaking "environmentalese" and the South speaking "developmentese," and each side talking past the other.

With the crumbling of socialist ideology in Eastern Europe and the former Soviet Union, many Europeans are also becoming less receptive to formerly attractive NIEO precepts. Thus, if the language negotiated as part of a climate change convention invokes images such as central command, heavy-handed international bureaucracy, forcible technology transfer, blame-casting ideological declarations, guilt-based wealth transfers, and the like, the results of any such negotiation run substantial risk of being overturned. Indeed Northern, especially US, opponents of an expanded climate change convention may well base their negative stand on the perceived ideological cast of the proposed regime.

Like the Law of the Sea, therefore, real mutual interdependence means that climate change talks have the ingredients for an inescapable, long-term, North–South engagement: Southern insistence on NIEO-like measures that would meet with Northern resistance. Given that Southern commitment to the NIEO *per se* has moderated considerably since the 1970s, the risk of an ideologically driven impasse is probably manageable with some conscious effort. Both the INC and the Earth Summit itself offer grounds for optimism although the issue may only be in temporary "remission." As will be discussed below, creative steps are essential to meet legitimate LDC interests while reducing the risks that such an engagement would result in endless delay and damaging ideological confrontation— with no action to address either the greenhouse problem or development imperatives as a result.

Build Trust and Address North–South Concerns Through an Informal Negotiating Process Parallel to the Formal One

A number of well-publicized regional workshops in advance of the negotiations—presented by regional scientists and policy figures that focused on possible local impacts—could help spread the conviction that global warming is a common threat from a shared problem. Joint research and study by mixed teams of experts from developing and developed countries should likewise be encouraged, perhaps building on the work of the Intergovernmental Panel on Climate Change that has been jointly sponsored by UNEP and the World Meteorological Organization.

During the negotiations themselves, similar informal educational events could be helpful. One extraordinary element of the LOS experience (that has been detailed by outside observers) consisted of the influence of a computer model of deep ocean mining developed at the Massachusetts Institute for Technology. This model came to be widely accepted in the face of the great uncertainty felt by the delegates about the engineering and economic aspects of deep seabed mining. A critical point in the negotiations occurred during a Saturday morning workshop—held outside UN premises, under the auspices of Quaker and Methodist non-governmental organizations—in which developed and developing country delegates were able to meet and extensively query the MIT team that had built the model. Indeed, over time the delegates came to make frequent use of the model for learning, mutual education, invention of new options— and even as a political excuse to move from frozen positions.[38]

Similarly, the Montreal Protocol process was aided by a series of informal, off-the-record workshops where diplomats and politically active participants in the negotiation gathered informally. These events greatly increased mutual understanding, improved relationships, and contributed to a successful treaty. Despite its potential abuse by advocates, outside scientific information—when it can be seen to be objective and is accessible to the participants—can help move a complex negotiation, even one that is highly politicized and ideologically controversial, in the direction of mutual cooperation. (On the other hand, improved science also poses a risk to the negotiations. New data might instead clarify winners and losers, thus polarizing the issue.)

As a broader proposition, negotiations that take place entirely through formal diplomatic means and at arm's length have a diminished prospect of success relative to those that encourage informal interactions, the buildup of trust, and the enhancement of personal relationships. Such a parallel, informal process can be "hosted" by any number of sympathetic participants and observers, from delegations themselves to non-governmental organizations of all sorts. A wise Secretariat will seek to provide occasions and venues for such events to flourish.

Enhance the Role of Advisory Groups in Promoting Cross-cutting Coalitions

Given the actual and feared adverse impacts of measures under discussion, conference leadership would be wise to continue to make extensive use of broadly constituted advisory groups. These can be composed of business and other multinational interests as well as environmentalists, in order to understand concerns, anticipate emerging problems, correct misapprehensions, and communicate about the issues and evolving negotiating responses. Not only could the two-way communication be useful in such settings, but cross-cutting coalitions might form. For example, industries that could gain from substantial anti-greenhouse action in the developing world (for example, by supplying critical technology for energy efficiency) might make common cause with key LDCs and environmental advocacy organizations in arguing the case for more developed country financial assistance to the South. The potential value of such groups as the Commission on Sustainable Development and the Business Council for Sustainable Development is thus very high indeed.

Elaborate and Diffuse a New Ideological "Template"

The North–South conflict has been a staple of recent global negotiations—from the UN Conference on Trade and Development to debt talks and the debate on codes of conduct for transnational corporations—although the overt NIEO focus has moderated in the years since the LOS discussions.[39] Joint development of a new "ideological template" within which the climate question could be negotiated would offer another means to escape this traditional impasse. Such a new conception could avoid lumping together countries with vastly different *climate* interests into catch-all categories such as "North" and "South," simply because they share certain characteristics. Without this new ideological template, we risk simplistically grouping together coal-rich developing countries such as China and India into one category, all of sub-Saharan Africa into another, the so-called "Second World" of Central Europe into another, and industrial countries as disparate as Norway and the United States into yet another.

The most promising candidates to date for such a template are the principles of "sustainable development"—insisting on development that meets the needs of the present without compromising the ability of future generations to meet their own needs. These principles were best articulated by the Brundtland Commission in *Our Common Future* and elaborated in the discussion of the Earth Charter and Agenda 21.[40]

Though in need of clearer definition, these widely discussed principles call for tight links between environment and development, for institutions that integrate environmental and economic decision-making, for international cooperation on global issues, and for major efforts toward more sustainable paths of population, energy, and resources. At a minimum, such a template needs explicitly to incorporate notions of *fairness* that loom large in this kind of negotiation. In particular, fairness concerns should address the issues of who can and should bear the burden of implementing more environmentally friendly policies as well as adapting to the consequences of climate change.[41] Whether such principles can come to have the acceptance, weight, and specific implication needed to steer climate negotiations safely away from stale North–South rhetorical exercises of the recent past remains to be seen, but they are a promising possibility.

Strategy 7: Direct Negotiating Energies Toward a Small-scale, Expanding Agreement

The complexities of a universal process may still threaten endless delay or impasse. But suppose that a smaller group of industrialized states—with potent domestic interests keenly interested in anti-greenhouse measures—were to negotiate among themselves an emissions reduction regime including timetables and targets, either voluntary or mandated. Presumably the core group would include major contributors to the greenhouse problem in which there was substantial and urgent domestic sentiment for action. A natural starting core would be the twelve nations of the European Community, the six member states of the European Free Trade Association, plus Japan, Australia, and Canada—all of which by 1992 had unilaterally or collectively adopted greenhouse gas stabilization or reduction targets.[42] With the advent of the Clinton–Gore Administration, the prospects for more active US participation, support, and leadership in such an effort certainly improve.

Agreement among such a group would likely prove far easier to achieve than a global accord, as a function of the smaller number of states involved as well as their greater economic and political homogeneity. The obvious umbrella for such an effort is the Climate Convention signed at the 1992 Earth Summit, but existing institutions (such as the UN Economic Commission for Europe or the OECD) might also facilitate the process. And while there would clearly be substantial negotiating difficulties involved, this smaller-scale process might avoid the kind of protracted, inconclusive North–South clash that could characterize a larger forum.

To be effective in the longer term, of course, a smaller-scale agreement would have to be expanded later to include key developing countries such as China, India, Brazil, Indonesia, and Mexico, as well as additional developed nations, especially in Eastern Europe. In this sense, an agreement explicitly designed for an increasing number of adherents has strong parallels to agreements that "ratchet" down to become increasingly stringent. The design of the smaller negotiation could anticipate and facilitate such an expansion in several ways.

First, the smaller agreement should seek to build on the present Framework Convention on Climate Change, in which the general problem will have been legitimated and accepted to the largest extent possible. Second, it should be cast not as an alternative to the global process over protocols, but as a complement to it—in which those nations most responsible for the present greenhouse gas problem are taking early actions to mitigate emissions. This would give the smaller group that had agreed to cuts a higher moral standing in soliciting later reductions from others. Third, the smaller-scale group should structure its accord with the explicit expectation of collectively negotiating incentives for key developing nations to join the accord. For example, the smaller group might agree to tax its members on their carbon emissions. All or part of those tax proceeds could be transferred to other key countries and used encourage their participation in emissions control measures. The smaller group could create an entity that would carry out the negotiations with these key countries, rather than leaving such negotiations to ad hoc and bilateral efforts by individual member countries. And the smaller group might be especially effective in soliciting support for the next replenishment of international development assistance.

Negotiations between the smaller treaty group and, say, China, could, for example, set a schedule of emission targets and offer China significant incentives to reach them. Or it could address a range of China's environmental and other concerns in return for less climate-damaging development (e.g., assistance with greater exploration for Chinese natural gas reserves; Chinese agreement to use CFC substitutes in refrigeration; agreement to make its coal development more greenhouse-friendly, perhaps by the transfer of more efficient electrical generating equipment). Such "customized" small-group negotiations—with China, India, Brazil, and others—should be more conducive to environmentally desirable results than would be the expected North–South clashes that might occur in a full-scale UN conference.

Fourth, and considerably more ambitious and contentious, as the group of adherents to the smaller convention grew in size, it might choose to impose a tax on products imported into member countries from non-adherents, perhaps based on the direct or indirect carbon content of those products. While this would provoke an angry reaction from the General Agreement on Tariffs and Trade (GATT), the carrot (providing individually tailored negotiated incentives for non-adherents to join) and the stick (raising such a "carbon fence" around the group of greenhouse activists), might together lead to a much larger number of countries jointly taking measures to prevent climate change. Evidently, a price to be faced, deliberated, and accepted by the smaller group would be a substantial number of free-riding countries. With a large enough group of adherents, however, the smaller group could still be preferable to no agreement at all.

Ironically, although a number of developing countries have joined the Montreal Protocol, it is quite possible to interpret this, after the fact, as strongly analogous to the smaller-scale convention just discussed. While carried out in the context of a widely accepted framework (the Vienna Convention), the relatively small number of key CFC-producing countries ultimately acceded to the CFC reductions in the Montreal Protocol in 1987. However, important LDCs (India, China, Brazil) did not agree until 1990. India, for example, demanded $2 billion—a number related to its cost of using more ozone-friendly technology in the future—as its price to join the 1987 protocol.[43] In 1990, a number of developed nations agreed to provide assistance up to $240 million. This proved sufficiently attractive to representatives of states such as India and China that they indicated their willingness to join the Protocol. Yet as a result of the "smaller-scale" Montreal Protocol, extremely significant ozone-protection measures are now underway even before the full resolution of important issues concerning financial aid and technology transfer to the developing world.[44]

Summary and Conclusions

The problem of negotiating the next phases of a regime to control global warming amply illustrates powerful barriers to agreement, versions of which apply in a large number of contexts. This chapter has sought to clarify the nature of these barriers and suggest constructive responses to them. Environmental diplomats have largely taken negative lessons from the LOS negotiations and positive ones from the CFC accords in envisioning a framework/protocol process for global warming. Yet gaining significant future action to curb future greenhouse gas emissions will be a far more difficult task than either dealing with ocean resources or protecting the ozone layer. Despite the apparent appeal of the step-by-step approach, a review of the evolution of the LOS process from separate "mini-conventions" into a comprehensive treaty illustrates the powerful forces that will likely operate on a climate change negotiation to combine protocols and to collapse what is seen as a many-stage process into a more unified effort. The challenge will be to find smaller, more manageable packages that embody enough mutual gains to attract key players.

The power of the coalitions that will arise to block international action to reduce the risks of rapid climate change must be taken into account in designing an effective future negotiating process. These forces will come into play not merely for reasons of economic interest, but also for reasons of science, ideology, and/or opportunism. Preventing and overcoming these forces could be aided by sophisticated choice and sequence of protocols, as well as innovative devices such as "ratchet" mechanisms, negotiated "baselines," and voluntary actions short of negotiated targets. Even if these hazards are avoided, the possibility of a North–South impasse looms; a number of actions could mitigate it, including workshops, negotiation process choices, creative linkages, and advancement of new ideological "templates." If these measures are unsuccessful, attention may shift to a smaller-scale, expanding convention that could use incentives and penalties to later bring other states into its fold. Good candidates to start this process include those countries that have unilaterally committed to greenhouse targets.

Yet advocates of international cooperation on climate change should bear in mind the distinction between success measured by the ratification of diplomatic instruments and actual policy shifts implemented over time. The obvious focus of current energy is on the former. Given the sheer scale of the factors contributing to the climate problem, however, negotiated results will at best be one of many factors that accumulate to change widespread and deep-seated behaviors that generate greenhouse gases. *Thus, rather than conceptualizing negotiations primarily as potential direct producers of restrictions on greenhouse gas emissions, this chapter has argued that negotiations themselves should also be understood as a potential contributor to broader-scale awareness, scientific development, and consensus.* The latter effect may be far more powerful than the former. A climate negotiating process and its outcome should be measured by the extent to which they stimulate public visibility and education, mobilize non-governmental action, and foster widespread involvement of scientists globally in the development of a fuller consensus. The opposite of success in these terms would be a process and result that were diplomatically correct but proceeded invisibly and exclusively. One scarcely looks forward to the international equivalent of the US Gramm-Rudman anti-deficit targets and timetables that were passed in the late 1980s.

In sum, to an advocate of a new cooperative international regime to reduce the risks of rapid climate change, the fundamental negotiating task is to craft and sustain a meaningful winning coalition of countries backing such a regime over time. Two powerful barriers to this fundamental task are (1) that each member of the coalition fails to see enough gain in the regime, relative to the alternatives, to adhere, and (2) that potential and actual "blocking" coalitions of interests opposed to the regime are neither prevented from forming, acceptably accommodated, nor otherwise neutralized. The negotiation design recommendations developed in this chapter suggest that, over time, as the science and politics warrant, there are many ways to surmount these daunting barriers.

Notes

1. This chapter has evolved from an earlier article of mine, "Designing Negotiations Towards a New Regime: The Case of Global Warming," that appeared in *International Security*, vol. 15, no. 1 (Spring 1991), pp. 110–148; it contains an extensive set of background and supporting citations that are incorporated by reference into the present chapter. I am indebted to the same people and organizations acknowledged

therein, especially the Negotiation Roundtable at Harvard and the Salzburg Environmental Initiative, as well as to Kenneth Arrow's subsequent helpful suggestions on a related paper. Support of the Office of Policy and Evaluation of the United States Environmental Protection Agency, the Charles Stewart Mott Foundation, and the Stockholm Environment Institute is gratefully acknowledged.

2.　"Negotiation analysis" is a prescriptive approach to negotiating situations that draws on game-theoretic concepts but does not presuppose the full "rationality" of the participants or "common knowledge" of the negotiating situation. For expositions, see, e.g., Howard Raiffa, *The Art and Science of Negotiation* (Cambridge, MA: Harvard University Press, 1982); David A. Lax and James K. Sebenius, *The Manager as Negotiator* (New York: The Free Press, 1986); James K. Sebenius, "Negotiation Analysis: A Characterization and Review," *Management Science* vol. 38, no. 1 (January 1992), pp. 19–38; or H. Peyton Young, ed., *Negotiation Analysis* (Ann Arbor: University of Michigan Press, 1991).

3.　In the climate case, "blocking" coalitions may include non-joiners and free riders. However, peculiarities in the rules of conference diplomacy may allow such non-joiners actually to block agreements that are widely desired. For traditional discussions of these coalitional concepts see R. Duncan Luce and Howard Raiffa, *Games and Decisions* (New York: Wiley, 1957) or William Riker, *The Theory of Political Coalitions* (New Haven: Yale University Press, 1962). Here, "winning coalitions" are only defined with respect to a set of policy measures from the point of view of a particular actor or actors; such coalitions consist of sufficient numbers of adherents to render the policy effective (again from the point of view of the specific actor or actors). "Blocking" coalitions are those opposing interests that could prevent a winning coalition from coming into existence or being sustained. The term "actor" should be contextually obvious and can include states, domestic interests, and transnational groupings of either as appropriate. Though the "necessary" conditions described above are extremely important, "sufficient" conditions do not in general exist for an agreement to be reached and impasse or escalation avoided. See Lax and Sebenius, *The Manager as Negotiator*.

4.　Climate change was but one of the many subjects for the 1992 conference, which was timed to take place on the twentieth anniversary of the initial UN environmental conference held in Stockholm. The vast agenda of the 1992 conference also included other atmospheric issues (ozone depletion, trans-boundary air pollution), land resource issues (desertification, deforestation, and drought), biodiversity, biotechnology, the ocean environment, freshwater resources, and hazardous waste. United Nations General Assembly, "United Nations Conference on Environment and Development" (General Assembly Resolution 228/44 UN GAOR Supp. (No. 49) at 300, UN Doc. A/44/49 (1989)).

5.　Other useful precedents include the Limited Test Ban Treaty and nonproliferation agreements, the Basel convention on hazardous wastes, the Convention on International Trade in Endangered Species, the Antarctic Treaty, and various regional environmental accords such as the Mediterranean Action Plan. For useful distillations of some of the lessons from these and many other related accords, see Oran R.Young, "The Politics of International Regime Formation: Managing Natural Resources and the Environment," *International Organization*, vol. 43, no. 3, pp. 349–375; Peter S. Thacher, "Alternative Legal and Institutional Approaches to Global Change," *Colorado Journal of International Environmental Law and Policy*, vol. 1, no.1 (Summer 1990), pp. 101–126; and, especially, Peter H. Sand, *Lessons Learned in Global Environmental Governance* (Washington, DC: World Resources Institute, 1990).

6. For a summary of the unilateral and small-group greenhouse gas reduction and stabilization targets adopted worldwide, see the *Global Environmental Change Report*, vol. 2, no. 19 (Arlington, MA: Cutter Information Corporation, November 9, 1990), pp. 1–5, as well as subsequent issues.

7. "Clinton Commits U.S. to Greenhouse Target, Biodiversity Convention," *Global Environmental Change Report*, vol. 5, no. 8 (Arlington, MA: Cutter Information Corporation, April 23, 1993), p. 1; see also vol. 5 no. 20, October 22, 1993, of the same source describing the US Climate Action Plan.

8. The following LOS discussion generally relies on Ann L. Hollick, *U.S. Foreign Policy and the Law of the Sea* (Princeton, NJ: Princeton University Press, 1981); James K. Sebenius, *Negotiating the Law of the Sea: Lessons in the Art and Science of Reaching Agreement* (Cambridge, MA Harvard University Press, 1984); Bernard Oxman, David Caron, and C. Buderi, *Law of the Sea: U.S. Policy Dilemma* (San Francisco: ICS Press; Council on Ocean Law, 1983); E.L. Richardson, *The United States and the 1982 UN Convention on the Law of the Sea: A Synopsis of the Status of the Treaty and Its Expanded Role in the World Today* (Washington, DC: Council on Ocean Law, 1989); E.L. Richardson, "Law of the Sea: A Reassessment of U.S. Interests," *Mediterranean Quarterly: A Journal of Global Issues*, vol. 1, no. 2 (Spring 1990), pp. 1–13. The recent report on ratifications to the LOS Convention can be found in *Ocean Policy News*, vol. X, No. 7 (November 1993), p.1 (published by the Council on Ocean Law, Washington, DC)

9. The following account draws generally on Richard E. Benedick, *Ozone Diplomacy* (Cambridge, MA: Harvard University Press, 1991); Benedick, "Ozone Diplomacy," *Issues in Science and Technology*, vol. 6, no. 1 (Fall 1990), pp. 43–50; Benedick, "The Montreal Ozone Treaty: Implications for Global Warming," *The American University Journal of International Law and Policy*, vol. 5, no. 2 (Winter 1990), pp. 227–234; David D. Doniger, "Politics of the Ozone Layer," *Issues in Science and Technology*, vol. 4, no. 3 (Spring 1988), pp. 86–92; and Peter M. Haas, "Ozone Alone, No CFCs: Ecological Epistemic Communities and the Protection of Stratospheric Ozone," Conference on Knowledge, Interests and International Policy Coordination, Wellesley College, MA 1989.

10. There were, of course, limitations on various activities (e.g., coastal state seaward territorial claims, marine scientific research) negotiated in the LOS context. Not surprisingly, they were among the most difficult aspects of the conference.

11. Garrett Hardin, "The Tragedy of the Commons," *Science,* vol. 162, no. 3859 (December 13, 1968), pp. 1243–1248.

12. Richard G. Darman, "The Law of the Sea: Rethinking U.S. Interests," *Foreign Affairs,* vol. 56, no. 2 (January 1978), pp. 373–395.

13. Michael Grubb, "The Greenhouse Effect: Negotiating Targets," *International Affairs,* vol. 66, no. 1 (1990), p. 75.

14. D. Lashof and D. Tirpak, *Policy Options for Stabilizing Global Climate* (Washington, DC: United States Environmental Protection Agency; Office of Policy, Planning, and Evaluation, 1989), pp. 40–43.

15. Christopher D. Stone, "The Global Warming Crisis, If There Is One, and the Law," *The American University Journal of International Law and Policy,* vol. 5, no. 2 (Winter 1990), pp. 497–511.

16. Richard H. Stanley, *Environment and Development: Breaking the Deadlock,* Report of the 21st UN Issues Conference (Muscatine, IA: The Stanley Foundation, 1990), p.8.

17. See, e.g., Stephen D. Krasner, *Structural Conflict: The Third World Against Global Liberalism* (Berkeley: University of California Press, 1985).

18. The phrase is from Richard N. Gardner, *Negotiating Survival* (New York: Council on Foreign Relations, 1992), p. 25.

19. US Council of Economic Advisors, *Economic Report of the President* (Washington, DC: Government Printing Office, February 1990).

20. See *Economic Report of the President,* p. 234, based on A. S. Manne, and R.G. Richels, "Global CO_2 Emission Reductions - the Impacts of Rising Energy Costs" (Menlo Park, CA: Electric Power Research Institute, 1990).

21. For a critique of the Manne–Richels estimates, see R.H. Williams, "Low Cost Strategies for Coping with Carbon Dioxide Emission Limits," Center for Energy and Environmental Studies (Princeton University, 1989). More generally, for a sophisticated review of various cost estimates, see William R. Cline, *Global Warming: The Economic Stakes* (Washington DC: Institute for International Economics, 1992).

22. *Economist,* May 9, 1992, p. 19.

23. *Financial Times,* May 15, 1992, p.3.

24. "Election Defeat Hurts British Conservatives' Global Warming Strategy," *Global Warming Network Online Today* (Alexandria, VA: Environmental Information Networks, August 12, 1993), p. 1.

25. "Japan Backing Down from Large-Scale Global Warming Reduction Plans," *Global Warming Network Online Today* (Alexandria, VA: Environmental Information Networks, June 3, 1993), p. 1.

26. Thacher, "Alternative Legal and Institutional Approaches." (See n. 5 above)

27. Abram Chayes, "Managing the Transition to a Global Warming Regime or What to Do Until the Treaty Comes," in Jessica Mathews,

ed., *Greenhouse Warming: Negotiating a Global Regime* (Washington, DC: World Resources Institute, 1991), pp. 61–68.

28. An unlikely but illustrative US domestic parallel involving the creation of an unusual and potent blocking coalition may be found in Michael Pertschuk's stewardship of the formerly sleepy Federal Trade Commission (FTC) in the late 1970s. The FTC had recently launched a number of rule-making efforts directly affecting a range of small business interests in the United States, such as funeral homes, used car dealers, and optometrists. Further, the FTC decided to take on the issue of children's TV advertising, which not only threatened major media advertising revenues, but also smacked of First Amendment restrictions. In effect, having energized and unified an enormous coalition of large and small business and media companies—many of whom had been bitter rivals before—the FTC engendered a hail of protest, had its budget and authority slashed, and was even shut down for a while. In part, Pertschuk's unintended legacy was a far more unified and politically effective business community. See Philip B. Heymann, *The Politics of Public Management* (New Haven: Yale University Press, 1987).

29. William R. Moomaw, "A Modest Proposal to Encourage Unilateral Reductions in Greenhouse Gases," unpublished paper (Medford, MA, Tufts University, 1990).

30. Indeed, the legal discussions that led to the Vienna Convention began in 1981, four years after UNEP had formulated a World Plan of Action on the Ozone Layer. See Thacher, "Alternative Legal Regimes," pp. 108–9.

31. Mostafa Tolba, "A Step-by-Step Approach to Protection of the Atmosphere," *International Environmental Affairs*, vol. 1, no. 4 (Fall 1989), p. 305.

32. Jessica T. Mathews, "Redefining Security," *Foreign Affairs*, vol. 68 (Spring 1989), pp. 162–177.

33. Tommy T.B. Koh and Shanmugam Jayakumar, "The Negotiating Process of the Third United Nations Conference on the Law of the Sea," in M.H. Nordquist, ed., *United Nations Convention on the Law of the Sea 1982: A Commentary* (Boston: Martinus Nijhoff, 1985), pp. 29–134.

34. Ibid., p. 41.

35. See Gardner, *Negotiating Survival*, pp. 24–33 (See n. 18 above).

36. See, generally, Paul R. Ehrlich, and Anne H. Ehrlich, *The Population Explosion* (New York: Simon and Schuster, 1990).

37. For a general treatment of the underlying theoretical issues of issue linkage and separation, or "negotiation arithmetic," see James K. Sebenius, "Negotiation Arithmetic: Adding and Subtracting Issues and Parties," *International Organization*, vol. 37, no. 1 (Autumn 1983), pp. 281–316; or Chapter 9 of Lax and Sebenius, *The Manager as Negotiator* (*See n. 2 above*).

38. James K. Sebenius, "The Computer as Mediator: Law of the Sea and Beyond," *Journal of Policy Analysis and Management*, vol. 1, no. 1 (Fall 1990), pp. 77–95.

39. See, e.g., Robert L. Rothstein, *Global Bargaining: UNCTAD and the Quest for a New International Economic Order* (Princeton, NJ: Princeton University Press, 1979), for an account of an earlier such engagement.

40. World Commission on Environment and Development, *Our Common Future* (Oxford: Oxford University Press, 1987).

41. For an extensive discussion of these issues, see Michael Grubb, James Sebenius, Antonio Magalhaes, and Susan Subak, "Sharing the Burden," Chapter 21 in Irving M. Mintzer, ed., *Confronting Climate Change: Risks, Implications and Responses* (Cambridge: Cambridge University Press, 1992), pp. 305–322.

42. At present, the OECD countries account for approximately 45 percent of carbon emissions; with the addition of the Former Soviet Union and Eastern Europe, the total would rise to 71 percent. Manne and Richels, "The Costs," p. 15 (See n. 20 above).

43. Stone, "The Global Warming Crisis." (See n. 15 above.)

44. The experience of the Long-Range Transboundary Air Pollution Convention, in which groups of expanding size acceded to the later sulfur and nitrogen oxides protocols, is also generally in accord with this "small-scale" approach. For a summary, see C. Ian Jackson, "A Tenth Anniversary Review of the ECE Convention on Long-Range Transboundary Air Pollution," *International Environmental Affairs*, vol. 2, no. 3 (Summer 1990), pp. 217–226.

CHAPTER 14

Visions of the Past, Lessons for the Future

Irving M. Mintzer
Stockholm Environment Institute
Washington, DC

and

J. Amber Leonard
Stockholm Environment Institute
Washington, DC

The negotiations of the United Nations Framework Convention on Climate Change involved a long and dramatic process. Many players took the stage by turn, but none appeared in all the scenes nor dominated the action throughout. There were many heroes and few villains.

By examining the action in this drama through the lenses of more than a dozen witnesses, we have learned how different the perceptions of a shared event can be. Drawn from their personal recollections in late 1992 and 1993, each reconstruction of the events stands on its own for the reader to judge. Each participant sees the action in terms of how it affects the interests and institutions that he or she knows best. Each vision of the events is colored by the background and experience of the observer, which filters out some elements of the picture that may be readily seen by others. Studying this process can be instructive for future international negotiations.

Nearly all observers agree on certain factors that led to the success of the climate negotiations. We begin this chapter with an overview

321

of some of those areas of agreement, then highlight a number of unresolved issues that must be addressed in the second phase of the climate negotiations. Based on the recollections presented in this volume, the third section of this chapter summarizes a set of proposals for shaping the structure of the next phase of negotiations. In the final section we summarize the lessons learned from the climate debate that may be applicable to other international negotiations.

Building Consensus

Most of the observers whose views are reflected in the present volume give special emphasis to the *structure and process* of the debate as a key contributor to successful negotiations. They emphasized, for example, the importance of keeping the process open, transparent, and highly participatory.

Second, all observers emphasize the significance of broad participation from the outset by scientists, legal experts, economists, and representatives of non-governmental organizations (including both business and environmental groups). Each felt that such participation was critical to maintaining the momentum and the sense of political urgency that characterized these negotiations. They conclude that while the scientists and economists brought an element of technical rigor to the negotiating table, the non-governmental participants brought a freshness of vision, a concern about equity, and a measure of pragmatism that may not otherwise have been available to the diplomats and ministers who participated on behalf of their governments. In addition, they note that the broad array of interest groups represented by the non-governmental observers provided a diversity of perspectives and a range of information that enriched the negotiations in ways that would have been exceptionally difficult to achieve if participation had been strictly limited to those who were nominated by governments.

Third, nearly all of the observers who contributed to this volume emphasized that the timing of the negotiations was critical. Most indicated that a looming, politically important deadline (in this case the implacable deadline represented by the opening of the Earth Summit) gave a heightened sense of urgency to the climate negotiations. Unlike the dialogue which led to the signing of the Convention on the Law of the Sea, the politics of the Earth Summit ensured that the climate negotiations simply *could not* be allowed to continue for ten years. And, perhaps—as Hyder points out—equally important, no government was eager—or even willing—to be seen as the principal reason why the deadline was missed for a meeting which more than 100 heads of state would attend. This suggests that, at least in some circumstances, presenting the diplomatic community with *an apparently impossible* deadline (and keeping up the pressure to meet it) is necessary to break the bonds of diplomatic tradition and to overcome widespread political inertia.

Fourth, most observers share the general consensus concerning a need to approach issues that are as complex as climate change with more than the usual degree of flexibility. The situation in this sense is similar to the one encountered by the international community when seeking to control the manufacture and use of substances which deplete the stratospheric ozone layer. In the climate case, the effort began without a universally perceived need to reduce the risks of rapid global warming and during a period of continuing scientific uncertainty about the potential impacts of a greenhouse gas buildup. Given the range of views on the seriousness of the problem, the diplomats, scientists, and economists engaged in the climate negotiations felt compelled to develop an international legal regime that was both flexible and subject to periodic scientific review.

By not considering the first agreement as final, the INC established a continuing process of research and dialogue that will keep attention focused on the climate issue for decades to come. This iterative process of scientific assessment and diplomatic review will allow the international community periodically to re-evaluate the adequacy of the specific commitments in the Convention. Following these regular assessments, the Parties may choose to strengthen their commitments if the early projections of future impacts appear to have understated the likely damages or to relax the controls if new scientific evidence suggests that the peak risk has already passed.

The decision to establish this requirement for periodic reassessment complements the decision to incorporate the Precautionary Principle into the Convention. The agreed objective of the Convention is stated not in terms of controlling a particular technology or activity, but in terms of stabilizing *the concentration of greenhouse gases at a level that would prevent dangerous anthropogenic interference with the climate system*. Achieving this objective while promoting sustainable economic development will require all countries to monitor the implications of their economic, environmental, agricultural, energy, and social policies that affect the rate of future release and uptake of greenhouse gases. In some cases, trade-offs must be evaluated explicitly in order to balance the short-term economic benefits of traditional policy choices, which ignore impacts on the environment, against the benefits of alternative policies that place greater emphasis on preserving the environment and promoting economic development for the long term.

To further "hedge the bet," the Climate Convention also encourages the use of the Precautionary Principle as a guide to formulating domestic policies. Thus, it gives added impetus to the preferential consideration of long-term benefits over short-term gains at both the national and international levels.

Finally, the Climate Convention suggests that complex problems like climate change, in which all countries bear an uncertain future share of the damages, cannot be solved by the wide distribution of some new technology or simplistic answer. Solving these problems will require a careful and systematic consideration of the entire pattern of future economic development—in both industrial and developing countries—as well as a lasting commitment to continuing international cooperation.

Living with Unresolved Differences

In addition to these areas of consensus, it is not surprising that the first stage of the climate negotiations produced many areas of disagreement. A number of important issues remain unresolved. The recollections presented in this volume highlight some of these areas of continued dispute:

(1) the extent and implications of responsibility for historical emissions of greenhouse gases by industrial countries;

(2) the responsibility of countries with historically low levels of per capita emissions to take measures designed to limit the future rate of growth in emissions of these gases;

(3) the balance of responsibility for future buildup of greenhouse gases between the impacts of continuing population growth in the South and of "excessive consumption" in the North;

(4) the need for co-development of new technologies and the formation of new partnerships to promote their rapid deployment;

(5) the appropriate role for multilateral banks and other development assistance institutions in support of measures taken by developing countries that simultaneously reduce the risks of global environmental damage while advancing national objectives for sustainable economic development and local environmental protection; and

(6) the relationship between cooperative international efforts to protect the global environment and simultaneous efforts to enhance

economic efficiency and increase equity in matters of international trade.

These issues will surface again in a variety of new forms during the next phase of the climate negotiations. They will also surface periodically in other international negotiations. In the climate negotiations, they will come up in the discussions of the "incremental cost clause" and the role of the Global Environment Facility and will also be part of the discussion of "joint implementation." They will be a factor in the discussion of national reports and the development of methodologies for other Convention-related communications. Simultaneously they will percolate into other seemingly unconnected international debates on the implications of the Uruguay Round of the General Agreement on Tariffs and Trade (GATT), on the role of the industrial countries in the economic development of formerly Communist countries, and on a wide variety of North—South negotiations that will take place during the next decade.

We cannot resolve any of these questions now, nor is it likely that they will be resolved in the next phase of the climate negotiations. But we can, perhaps, be guided by the success of the first phase of the climate negotiations that led to the Framework Convention on Climate Change and thus take advantage of precedent the next time we sit down at the negotiating table. It may be most useful, in fact, to draw some conclusions from our experience about what kind of arrangements and structures promote successful negotiations and to ignore for the time being those substantive issues that will remain in dispute during the coming months. The following section highlights some of the observations of our contributors about how to structure the negotiating process so as to increase the likelihood of reaching supplemental future agreements that are both practical and equitable.

Looking Ahead

As Sebenius points out in Chapter 13, in situations where the parties are trying to negotiate limits on the abuse of the global commons, it is always easier to form "blocking coalitions" that can obstruct the path to effective agreements than it is to form "winning coalitions" that share a broad consensus. Blocking coalitions can form for many different reasons, including common economic interests, perceptions of scientific uncertainty, considerations of ideology, and shared values concerning the management of environmental risks.

While blocking coalitions need only to agree that they object to a particular outcome (and may take objection to very different aspects of that outcome), the formation of winning coalitions requires the development of a consensus on an overall strategy or approach to a problem. This often involves a multistage decision process. For complex issues like climate change, both a strong scientific consensus and firm expressions of national political will seem to be essential ingredients. The negotiations themselves must be open to participation by non-governmental organizations, including representatives of industry, environmental and other public interest groups. All of the countries that may be affected either by the uncontrolled situation or by proposed control measures must be invited to participate in the discussion. Special efforts must be made to support the participation of those from poor countries who would not otherwise be able to afford the cost of joining in these meetings and debates.

Given the current state of climate science and the uncertain economics of the impacts of climate change, we believe that recommendations for future climate change negotiations should focus initially on the structure of the process. Therefore, we offer a set of recommendations for that structure, drawing on the contributions presented earlier in this volume. The implementation of these recommendations may, we hope, expedite progress toward practical and equitable agreements at the upcoming sessions of the INC and the first Conference of the Parties.

Recommendations

(1) *Participation:* The next phase of the INC process should continue to be open to observers from international and non-governmental organizations as well as to official representatives of governments. Both industry and environmental NGOs should be invited to participate. Observers should be recognized to speak during the plenary sessions, at the discretion of the chair. An expanded effort should be made to encourage and support the participation of developing country representatives in the INC.

(2) *Timing:* The next phase of the negotiations should be carried out on a fixed timetable, with a deadline for completion linked to an international event with high political visibility, such as a major conference or summit meeting.

(3) *Schedule:* As Faulkner has suggested, the next phase of the negotiations might usefully be organized in two stages. The first stage could include a series of workshops that might be organized by the Intergovernmental Panel on Climate Change (IPCC). These workshops might analyze some of the unresolved technical

issues in the negotiations and provide a variety of inputs to the negotiators. The second stage should include formal negotiations of agreements that would supplement the Framework Convention. In the meantime, developed country Parties should be encouraged to begin implementing voluntary, cooperative measures to reduce emissions immediately and not to wait until further agreements are reached.

(4) *Organization:* The Climate Change Secretariat should be given the permanent responsibility for coordinating the logistics of the climate negotiations. The Secretariat should continue to work closely with the Bureau of the IPCC and with the Administration of the GEF to implement the terms of the Framework Convention. In particular, the Secretariat should work with the Administration of the GEF to develop operational definitions for a number of key issues related to the activities of the financial mechanism. (See Recommendation (5), *Unresolved Technical Issues* below.) Finally, the Secretariat should explore alternative ways of coordinating its work with the efforts of the United Nations Commission on Sustainable Development. The Secretariat should be requested to report back to the INC plenary on the status of these organizational questions at the beginning of each plenary session.

The INC should request that the IPCC organize a sub-group to sit as the Subsidiary Body on Scientific and Technological Advice (SBSTA) which is called for in the Convention. This will provide a practical vehicle for the IPCC to provide technical advice to the INC on specific questions that arise during the negotiations.

(5) *Unresolved Technical Issues:* There are a number of scientific and technical issues that affect the views of the Parties on the need for additional agreements or protocols to strengthen the Convention. Suggestions for addressing these issues follow:

(a) The INC should direct the SBSTA to:

- determine what constitutes a maximum safe level of concentration for greenhouse gases in the atmosphere,
- explore the technical requirements and financial implications of scenarios in which the atmospheric buildup of greenhouse gases is stabilized at a safe level while continued economic development is stimulated in both developed and developing countries,
- develop methodologies for calculation of emissions by sources and removal by sinks of greenhouse gases, and
- identify ecological indicators of climate change that can be used to signal when a significant change of state in the

global climate system is about to occur or when the global
climate system is approaching some threshold of non-linear
response to further injection of greenhouse gases.

(b) The INC should:

- direct the GEF to conduct extensive analyses of the key
  concepts of incremental cost and joint implementation, as
  well as the criteria and guidelines for funding of GEF
  projects under the Convention, and
- encourage the GEF to supplement these analytic studies
  with extensive, well documented, and operationally oriented case studies in a large number of countries in order
  to address these issues in a practical way.

(c) The INC should establish an ad hoc advisory panel to:

- work with global business NGOs and representatives of
  leading environmental organizations, and
- explore the implications of "joint implementation" and
  other schemes to accelerate the transfer of environmentally sound technology.

(6) *National Action Plans:* The INC should encourage all developed
countries (i.e., those listed in Annex 1 of the Convention) to prepare and to submit quickly national action plans illustrating how
they intend to meet their obligations under Article 4, Paragraph
2 of the Convention. In addition, the INC Secretariat should
develop a methodology for the required periodic review of these
national action plans. This methodology should be designed to
determine the accuracy of the baseline estimates and the
adequacy of measures outlined in the national plans prepared
by Annex 1 countries. In addition, developed countries and
international organizations should be encouraged by the INC to
assist developing countries in the fulfillment of the Convention's
requirements for all countries to provide national reports on
sources and sinks of greenhouse gases. This support should
include both new and additional financial resources as well as
technical assistance and advice.

(7) *Encourage Regional Discussions and Cooperation:* The INC should
encourage participating governments to engage in regionally
based discussions of these issues in parallel to the global negotiations. Regional workshops attended by representatives of
states in the region as well as by scientists and other technical
specialists can help the participants to understand better their
common interests and the special circumstances of the region.
The cooperation resulting from such meetings can only advance
the overall INC process.

Implications of the Climate Negotiations for Other International Debates

Climate change may be the archetype for a growing series of international environmental problems. These problems are defined by a shared set of scientific and economic characteristics. These structural characteristics and their implications include the following:

(1) *Scientific uncertainty concerning the dynamic processes of global environmental change will make it difficult or impossible to predict the regional distribution of physical impacts from those changes.*

This continuing uncertainty will last for decades and will make it extremely difficult, in turn, to predict the magnitude of economic damages resulting from the observed changes in the physical and biological environment.

(2) *Large separations over space and time between the events generating insults to the environment and the realization of observable impacts will make it difficult or impossible for the victims to identify the individuals or activities that have caused their problems and to gain compensation for the associated damages.*

Greenhouse gases, for example, can remain in the atmosphere for centuries and thus become uniformly mixed worldwide. The impacts of GHG emissions on climate will not become visible for decades after the emissions occur. Because the pollutants become well-mixed in the atmosphere, the region bearing the impacts of climate change can not identify the specific area which was the source of the damaging emissions. Furthermore, the long lag times between the events leading to the emissions and the observable damage due to climate change creates an intergenerational inequity. These structural difficulties make it impossible to establish an economic equilibrium in which the beneficiaries compensate the victims through a series of side payments.

(3) *Linkages and synergisms between different aspects of global environmental change create a "joint cost" accounting problem and make it difficult to establish legal and financial "liability" for observed damages.*

For example, tropical deforestation in developing countires may decrease natural sinks for CO_2 on a global basis (thus increasing the rate of CO_2 buildup for any given rate of emissions). The dynamic relationship between loss of sinks and the buildup of CO_2 creates a question of how to allocate responsibility. Should the responsibility for the increased risk of rapid climate change fall to those who cut the trees (thus reducing the sinks for CO_2) or to those who drive cars (and thus generate a disproportionate share of the offending emissions of carbon dioxide)?

There is a similar North–South allocation problem involving future use of conventional chlorofluorocarbons (CFCs) in developing countries. Ozone depletion due to the buildup of conventional CFCs may cause a reduction in the oceanic sink for CO_2. This decline in the sink would occur to the extent that ozone depletion resulted in an increase in ultraviolet radiation and an associated decline in the population of single-celled organisms (i.e., marine plankton) that carry out much of the world's primary photosynthesis. If this population decline causes a subsequent atmospheric buildup of carbon dioxide (because the ability of the ocean to remove CO_2 from the atmosphere is reduced), how much of the responsibility for that buildup should rest with those in developing countries who increase their use of the conventional CFCs (as they are now allowed to do under the Montreal Protocol to the Vienna Convention)? On the other hand, how much of the responsibility should fall to nations which already emit large quantities of CO_2 (due to excessively high levels of per capita energy use)?

These difficulties in assigning responsibility are endemic to discussions of global environmental change and reflect the limited ability of modern science to understand the working of the Earth as a system. They are not part of some temporary environmental hysteria or international conspiracy to avoid responsibility. As dialogue, debate, and eventually negotiations proceed on these problems, the international community will have to adapt to the structural characteristics of this class of problems and resolve them to the extent possible. Their implications are profound but the difficulties which they raise need not be irresolvable.

Lessons for the Future

The successful climate negotiations suggest several lessons which might be brought to bear in future international negotiations on climate change and other similar problems. The principal lesson to be learned from the climate negotiations is that the climate problem is linked to many other challenges of economic development and environmental protection. The interdependence of modern economics, the interdependence of the elements of the global ecosystem, and the synergies between the two systems have removed the possibility that we can treat any of these problems successfully in isolation. The era of simple solutions has ended.

Linked Issues of Global Environmental Change

The most dramatic problems of global environmental change share three levels of interconnection. These problems are connected *economically* because, in many cases, the same economically important activities contribute simultaneously to several sets of risks. They are linked *chemically* because once the pollutants are emitted they interact in complex and synergistic ways in air, water, and soil. Finally, they are linked *politically* because the actions taken in response to any one of these problems will inevitably affect the timing and severity of the related problems. This class of problems includes: global climate change due to the greenhouse effect, ozone depletion due to the buildup of CFCs and Halons, loss of biodiversity due to tropical deforestation and other forms of land use change, and regional pollution due to trans-boundary movement of conventional pollutants.

In light of these linkages and synergisms, it has become increasingly important to develop complex and sophisticated mechanisms for identifying, testing, and debating response options. The value of an international process that includes complementary mechanisms for cooperative international scientific assessment and continuing diplomatic debate (like that used in both the climate and ozone negotiations) will become obvious as political attention shifts increasingly to problems of this type.

Of particular importance, recent experience in the climate and ozone negotiations demonstrates that such processes are greatly strengthened by broadly based participation. Far from being disruptive and unruly influences, non-governmental organizations representing both business and environmental groups have been absolutely critical to achieving politically successful compromises. As similar processes evolve in the future, extensive efforts should be made to keep them open, transparent, participatory, and inclusive—inviting full participation by NGOs in all relevant fora.

Linked Issues of Population, Consumption, Poverty
and Development

The debate on international responses to the risks of rapid climate change has highlighted the global connections between the equally complex issues of population, excessive consumption, poverty, and sustainable development. Whereas in the past, many believed that the international community could somehow address these issues one at a time, we know now that no such luxury is available to us. Inescapable synergisms among these issues force the world's political leadership to address them all concurrently.

Population growth increases the stresses on modern societies and clearly complicates all other issues which must be faced by governments. But increasing population size does not necessarily drive all environmental problems. Because some populations consume (and waste) so much less than others on a per capita basis, a large increase in the population of a poor country may have a smaller negative effect on the global environment, for example, than does the slowly increasing (or stable) population of a rich country. Thus, the number of people *per se* is not the problem. What determines the scale of the environmental impact generated in each country is how the subject population meets its most basic human needs and organizes its economic activities.

Consumption by itself is not necessarily a problem. But excessive and wasteful consumption is a problem that contributes to negative global environmental change. The Earth has a limited natural capital asset base for some key resources. These include fresh and potable water, certain kinds of energy resources, and some key minerals. The challenge is not to give these up, but to exploit them in ways which minimize long-term, irreversible damage to the planet.

All living things consume resources and convert them to other forms in order to meet their basic needs. The issue is how to organize consumption in a sustainable way so that, as the Brundtland Commission warns, consumption today does not compromise the quality of life for those who follow our own generation. Current patterns of resource consumption in most OECD countries are excessive and unsustainable. It is these familiar and widespread patterns of consumption that generate the greatest insults to the global environment.

But poverty is an equally important problem. Poverty denies human beings options concerning how to organize their lives and their time. Extreme poverty degrades the human spirit and is the principal cause of environmental degradation in many countries. Poverty-stricken human beings make short-sighted decisions that may sacrifice important elements of the natural environment in the interest of immediate survival. Poor people, desperate for resources, often over-harvest their environment, ignoring (or unaware of) the long-run impacts of their actions. And although the individual burden on the environment of each poor person is small, the aggregate impact of the actions of the world's poor is large and profoundly negative in many developing countries.

The principal lesson from the Climate Convention concerns the critical linkage between global environmental protection and sustainable development. The success of the work of both the IPCC and the INC is based on both groups' early recognition of the unbreakable link between environment and development. For perhaps the first time in the history of international environmental diplomacy, the Climate Convention places sustainable development and protection of the global environment on an equal footing. The science of global change and the diplomatic success of the Climate Convention supports the view that protection of the global environment can only be achieved in the context of systematic, continuing, and comprehensive efforts to ensure the success of sustainable development. Attention to this conclusion will greatly strengthen the prospects for other international environmental negotiations.

We are All in This Together

Another lesson of the climate negotiations concerns the context of future negotiations, both on climate and on other issues. Success in the climate debate only became possible when the participants were able to forgo the need to assign blame for the current situation, and focus on finding a path through these difficult challenges.

A key factor in the success of the climate negotiations was the conclusion of the IPCC Scientific Assessment that all countries contribute to global emissions of greenhouse gases and to changes in the strength of greenhouse gas sinks. Similarly, every country will be vulnerable to the impacts of rapid climate change, but to an uncertain (and currently unknowable) degree. Although nations do not contribute equally to the rate of increase in the risks, no country can be sure that it will escape all of the damages.

The implication of this simple result is profound. It provides an unshakable scientific basis for the recognition that whatever the future brings, it will be a shared fate for all humanity. The uncertainty about the scale, timing, distribution, and severity of the damages argues that all must participate in the effort to reduce the risks of rapid climate change. The recognition of differing contributions allows and encourages countries to find the strategies that best fit their own local needs and conditions, rather than to try to impose some singular, universal prescription on all countries at the same time. Aware of the inability of any one country or any coalition of countries to insulate itself from damages, participants in the negotiations must avoid endless diatribes designed to distribute blame. Only in this way can the negotiations channel our various capabilities into collective progress toward sustainable development. If the political courage to continue facing the emerging issues of global environmental change persists, future discussions on global environmental issues may be able to replicate and even to surpass the success of the first phase of the climate negotiations.

The Challenge Ahead

The Framework Convention on Climate Change is an important early step on the path to sustainable development. It represents one of the few instances in which national governments have agreed to take near-term measures with potentially large economic consequences in order to avoid uncertain but possibly catastrophic environmental damage in the future.

The Convention contains no binding commitments to emissions reductions in the North and no pledges of a fixed increase in official development assistance to the South. But the Convention does commit the developed countries to finance the "agreed full incremental costs" of measures taken by developing countries to achieve the overall objective of the Convention. Developing countries, while not required to place any specific limits on their national development strategies, commit themselves to report regularly on the status of domestic sources and sinks of greenhouse gases.

The current commitments are wholly and unequivocally inadequate to meet the stabilization objective outlined in Article 2 of the Convention. Although the specific commitments are limited and weak, the Convention sets up a mechanism to strengthen the climate control regime and to support economic development on a sustainable basis in both the North and the South.

Can the Convention's goal of climate protection and sustainable development be achieved with the limited financial resources available today? We cannot know if nations have the political will to find and allocate the necessary financial resources. But we are hopeful that the lessons learned from the successful climate negotiations can lead us to a better understanding of the challenges of combining climate protection and sustainable development. If these lessons are applied thoughtfully, systematically, and patiently, we believe they will create the conditions in which men and women of good will can work together to develop new political, social, and financial mechanisms that will mobilize human initiative to solve our common global problems.

United Nations Framework Convention On Climate Change

The Parties to this Convention,

Acknowledging that change in the Earth's climate and its adverse effects are a common concern of humankind,

Concerned that human activities have been substantially increasing the atmospheric concentrations of greenhouse gases, that these increases enhance the natural greenhouse effect, and that this will result on average in an additional warming of the Earth's surface and atmosphere and may adversely affect natural ecosystems and humankind,

Noting that the largest share of historical and current global emissions of greenhouse gases has originated in developed countries, that per capita emissions in developing countries are still relatively low and that the share of global emissions originating in developing countries will grow to meet their social and development needs,

Aware of the role and importance in terrestrial and marine ecosystems of sinks and reservoirs of greenhouse gases,

Noting that there are many uncertainties in predictions of climate change, particularly with regard to the timing, magnitude and regional patterns thereof,

Acknowledging that the global nature of climate change calls for the widest possible cooperation by all countries and their participation in an effective and appropriate international response, in accordance with their common but differentiated responsibilities and respective capabilities and their social and economic conditions,

Recalling the pertinent provisions of the Declaration of the United Nations Conference on the Human Environment, adopted at Stockholm on 16 June 1972,

Recalling also that States have, in accordance with the Charter of the United Nations and the principles of international law, the sovereign right to exploit their own resources pursuant to their own environmental and developmental policies, and the responsibility to ensure that activities within their jurisdiction or control do not cause damage to the environment of other States or of areas beyond the limits of national jurisdiction,

Reaffirming the principle of sovereignty of States in international cooperation to address climate change,

Recognizing that States should enact effective environmental legislation, that environmental standards, management objectives and priorities should reflect the environmental and developmental context to which they apply, and that standards applied by some countries may be inappropriate and of unwarranted economic and social cost to other countries, in particular developing countries,

Recalling the provisions of General Assembly resolution 44/228 of 22 December 1989 on the United Nations Conference on Environment and Development, and resolutions 43/53 of 6 December 1988, 44/207 of 22 December 1989, 45/212 of 21 December 1990 and 46/169 of 19 December 1991 on protection of global climate for present and future generations of mankind,

Recalling also the provisions of General Assembly resolution 44/206 of 22 December 1989 on the possible adverse effects of sea level rise on islands and coastal areas, particularly low-lying coastal areas and the pertinent provisions of General Assembly resolution 44/172 of 19 December 1989 on the implementation of the Plan of Action to Combat Desertification,

Recalling further the Vienna Convention for the Protection of the Ozone Layer, 1985, and the Montreal Protocol on Substances that Deplete the Ozone Layer, 1987, as adjusted and amended on 29 June 1990,

Noting the Ministerial Declaration of the Second World Climate Conference adopted on 7 November 1990,

Conscious of the valuable analytical work being conducted by many States on climate change and of the important contributions of the World Meteorological Organization, the United Nations Environment Programme and other organs, organizations and bodies of the United Nations system, as well as other international and intergovernmental bodies, to the exchange of results of scientific research and the coordination of research,

Recognizing that steps required to understand and address climate change will be environmentally, socially and economically most effective if they are based on relevant scientific, technical and economic considerations and continually re-evaluated in the light of new findings in these areas,

Recognizing that various actions to address climate change can be justified economically in their own right and can also help in solving other environmental problems,

Recognizing also the need for developed countries to take immediate action in a flexible manner on the basis of clear priorities, as a first step towards comprehensive response strategies at the global, national and, where agreed, regional levels that take into account all greenhouse gases, with due consideration of their relative contributions to the enhancement of the greenhouse effect,

Recognizing further that low-lying and other small island countries, countries with low-lying coastal, arid and semi-arid areas or areas liable to floods, drought and desertification, and developing countries with fragile mountainous ecosystems are particularly vulnerable to the adverse effects of climate change,

Recognizing the special difficulties of those countries, especially developing countries, whose economies are particularly dependent on fossil fuel production, use and exportation, as a consequence of action taken on limiting greenhouse gas emissions,

Affirming that responses to climate change should be coordinated with social and economic development in an integrated manner with a view to avoiding adverse impacts on the latter, taking into full account the legitimate priority needs of developing countries for the achievement of sustained economic growth and the eradication of poverty,

Recognizing that all countries, especially developing countries, need access to resources required to achieve sustainable social and economic development and that, in order for developing countries to progress towards that goal, their energy consumption will need to grow taking into account the possibilities for achieving greater energy efficiency and for controlling greenhouse gas emissions in general, including through the application of new technologies on terms which make such an application economically and socially beneficial,

Determined to protect the climate system for present and future generations,

Have agreed as follows:

ARTICLE 1

Definitions*

For the purposes of this Convention:

1. Adverse effects of climate change" means changes in the physical environment or biota resulting from climate change which have significant deleterious effects on the composition, resilience or productivity of natural and managed ecosystems or on the operation of socio-economic systems or on human health and welfare.

2. "Climate change" means a change of climate which is attributed directly or indirectly to human activity that alters the composition of the global atmosphere and which is in addition to natural climate variability observed over comparable time periods.

3. "Climate system" means the totality of the atmosphere, hydrosphere, biosphere and geosphere and their interactions.

4. "Emissions" means the release of greenhouse gases and/or their precursors into the atmosphere over a specified area and period of time.

5. "Greenhouse gases" means those gaseous constituents of the

---

* Titles of articles are included solely to assist the reader.

atmosphere, both natural and anthropogenic, that absorb and re-emit infrared radiation.

6. "Regional economic integration organization" means an organization constituted by sovereign States of a given region which has competence in respect of matters governed by this Convention or its protocols and has been duly authorized, in accordance with its internal procedures, to sign, ratify, accept, approve or accede to the instruments concerned.

7. "Reservoir" means a component or components of the climate system where a greenhouse gas or a precursor of a greenhouse gas is stored.

8. "Sink" means any process, activity or mechanism which removes a greenhouse gas, an aerosol or a precursor of a greenhouse gas from the atmosphere.

9. "Source" means any process or activity which releases a greenhouse gas, an aerosol or a precursor of a greenhouse gas into the atmosphere.

ARTICLE 2

Objective

The ultimate objective of this Convention and any related legal instruments that the Conference of the Parties may adopt is to achieve, in accordance with the relevant provisions of the Convention, stabilization of greenhouse gas concentrations in the atmosphere at a level that would prevent dangerous anthropogenic interference with the climate system. Such a level should be achieved within a time frame sufficient to allow ecosystems to adapt naturally to climate change, to ensure that food production is not threatened and to enable economic development to proceed in a sustainable manner.

ARTICLE 3

Principles

In their actions to achieve the objective of the Convention and to implement its provisions, the Parties shall be guided, inter alia, by the following:

1. The Parties should protect the climate system for the benefit of present and future generations of humankind, on the basis of equity and in accordance with their common but differentiated responsibilities and respective capabilities. Accordingly, the developed country Parties should take the lead in combating climate change and the adverse effects thereof.

2. The specific needs and special circumstances of developing country Parties, especially those that are particularly vulnerable to the adverse effects of climate change, and of those Parties, especially developing country Parties, that would have to bear a disproportionate or abnormal burden under the Convention, should be given full consideration.

3. The Parties should take precautionary measures to anticipate, prevent or minimize the causes of climate change and mitigate its adverse effects. Where there are threats of serious or irreversible damage, lack of full scientific certainty should not be used as a reason for postponing such measures, taking into account that policies and measures to deal with climate change should be cost-effective so as to ensure global benefits at the lowest possible cost. To achieve this, such policies and measures should take into account different socio-economic contexts, be comprehensive, cover all relevant sources, sinks and reservoirs of greenhouse gases and adaptation, and comprise all economic sectors. Efforts to address climate change may be carried out cooperatively by interested Parties.

4. The Parties have a right to, and should, promote sustainable development. Policies and measures to protect the climate system against human-induced change should be appropriate for the specific conditions of each Party and should be integrated with national development programmes, taking into account that economic development is essential for adopting measures to address climate change.

5. The Parties should cooperate to promote a supportive and open international economic system that would lead to sustainable economic growth and development in all Parties, particularly developing country Parties, thus enabling them better to address the problems of climate change. Measures taken to combat

climate change, including unilateral ones, should not constitute a means of arbitrary or unjustifiable discrimination or a disguised restriction on international trade.

ARTICLE 4

Commitments

1. All Parties, taking into account their common but differentiated responsibilities and their specific national and regional development priorities, objectives and circumstances, shall:

   (a) Develop, periodically update, publish and make available to the Conference of the Parties, in accordance with Article 12, national inventories of anthropogenic emissions by sources and removals by sinks of all greenhouse gases not controlled by the Montreal Protocol, using comparable methodologies to be agreed upon by the Conference of the Parties;

   (b) Formulate, implement, publish and regularly update national and, where appropriate, regional programmes containing measures to mitigate climate change by addressing anthropogenic emissions by sources and removals by sinks of all greenhouse gases not controlled by the Montreal Protocol, and measures to facilitate adequate adaptation to climate change;

   (c) Promote and cooperate in the development, application and diffusion, including transfer, of technologies, practices and processes that control, reduce or prevent anthropogenic emissions of greenhouse gases not controlled by the Montreal Protocol in all relevant sectors, including the energy, transport, industry, agriculture, forestry and waste management sectors;

   (d) Promote sustainable management, and promote and cooperate in the conservation and enhancement, as appropriate, of sinks and reservoirs of all greenhouse gases not controlled by the Montreal Protocol, including biomass, forests and oceans as well as other terrestrial, coastal and marine ecosystems;

   (e) Cooperate in preparing for adaptation to the impact of

climate change; develop and elaborate appropriate and integrated plans for coastal zone management, water resources and agriculture, and for the protection and rehabilitation of areas, particularly in Africa, affected by drought and desertification, as well as floods;

(f) Take climate change considerations into account, to the extent feasible, in their relevant social, economic and environmental policies and actions, and employ appropriate methods, for example impact assessments, formulated and determined nationally, with a view to minimizing adverse effects on the economy, on public health and on the quality of the environment, of projects or measures undertaken by them to mitigate or adapt to climate change;

(g) Promote and cooperate in scientific, technological, technical, socio-economic and other research, systematic observation and development of data archives related to the climate system and intended to further the understanding and to reduce or eliminate the remaining uncertainties regarding the causes, effects, magnitude and timing of climate change and the economic and social consequences of various response strategies;

(h) Promote and cooperate in the full, open and prompt exchange of relevant scientific, technological, technical, socio-economic and legal information related to the climate system and climate change, and to the economic and social consequences of various response strategies;

(i) Promote and cooperate in education, training and public awareness related to climate change and encourage the widest participation in this process, including that of non-governmental organizations; and

(j) Communicate to the Conference of the Parties information related to implementation, in accordance with Article 12.

2. The developed country Parties and other Parties included in annex I commit themselves specifically as provided for in the following:

(a) Each of these Parties shall adopt national[1] policies and take corresponding measures on the mitigation of climate change,

---

1. This includes policies and measures adopted by regional economic integration organizations.

by limiting its anthropogenic emissions of greenhouse gases and protecting and enhancing its greenhouse gas sinks and reservoirs. These policies and measures will demonstrate that developed countries are taking the lead in modifying longer-term trends in anthropogenic emissions consistent with the objective of the Convention, recognizing that the return by the end of the present decade to earlier levels of anthropogenic emissions of carbon dioxide and other greenhouse gases not controlled by the Montreal Protocol would contribute to such modification, and taking into account the differences in these Parties' starting points and approaches, economic structures and resource bases, the need to maintain strong and sustainable economic growth, available technologies and other individual circumstances, as well as the need for equitable and appropriate contributions by each of these Parties to the global effort regarding that objective. These Parties may implement such policies and measures jointly with other Parties and may assist other Parties in contributing to the achievement of the objective of the Convention and, in particular, that of this subparagraph;

(b) In order to promote progress to this end, each of these Parties shall communicate, within six months of the entry into force of the Convention for it and periodically thereafter, and in accordance with Article 12, detailed information on its policies and measures referred to in subparagraph (a) above, as well as on its resulting projected anthropogenic emissions by sources and removals by sinks of greenhouse gases not controlled by the Montreal Protocol for the period referred to in subparagraph (a), with the aim of returning individually or jointly to their 1990 levels of these anthropogenic emissions of carbon dioxide and other greenhouse gases not controlled by the Montreal Protocol. This information will be reviewed by the Conference of the Parties, at its first session and periodically thereafter, in accordance with Article 7;

(c) Calculations of emissions by sources and removals by sinks of greenhouse gases for the purposes of subparagraph (b) above should take into account the best available scientific knowledge, including of the effective capacity of sinks and the respective contributions of such gases to climate change. The Conference of the Parties shall consider and agree on methodologies for these calculations at its first session and review them regularly thereafter;

(d) The Conference of the Parties shall, at its first session, review the adequacy of subparagraphs (a) and (b) above. Such review shall be carried out in the light of the best available scientific information and assessment on climate change and its impacts, as well as relevant technical, social and economic information. Based on this review, the Conference of the Parties shall take appropriate action, which may include the adoption of amendments to the commitments in subparagraphs (a) and (b) above. The Conference of the Parties, at its first session, shall also take decisions regarding criteria for joint implementation as indicated in subparagraph (a) above. A second review of subparagraphs (a) and (b) shall take place not later than 31 December 1998, and thereafter at regular intervals determined by the Conference of the Parties, until the objective of the Convention is met;

(e) Each of these Parties shall:

    (i) coordinate as appropriate with other such Parties, relevant economic and administrative instruments developed to achieve the objective of the Convention; and

    (ii) identify and periodically review its own policies and practices which encourage activities that lead to greater levels of anthropogenic emissions of greenhouse gases not controlled by the Montreal Protocol than would otherwise occur;

(f) The Conference of the Parties shall review, not later than 31 December 1998, available information with a view to taking decisions regarding such amendments to the lists in annexes I and II as may be appropriate, with the approval of the Party concerned;

(g) Any Party not included in annex I may, in its instrument of ratification, acceptance, approval or accession, or at any time thereafter, notify the Depositary that it intends to be bound by subparagraphs (a) and (b) above. The Depositary shall inform the other signatories and Parties of any such notification.

3. The developed country Parties and other developed Parties included in annex II shall provide new and additional financial

resources to meet the agreed full costs incurred by developing country Parties in complying with their obligations under Article 12, paragraph 1. They shall also provide such financial resources, including for the transfer of technology, needed by the developing country Parties to meet the agreed full incremental costs of implementing measures that are covered by paragraph 1 of this Article and that are agreed between a developing country Party and the international entity or entities referred to in Article 11, in accordance with that Article. The implementation of these commitments shall take into account the need for adequacy and predictability in the flow of funds and the importance of appropriate burden sharing among the developed country Parties.

4. The developed country Parties and other developed Parties included in annex II shall also assist the developing country Parties that are particularly vulnerable to the adverse effects of climate change in meeting costs of adaptation to those adverse effects.

5. The developed country Parties and other developed Parties included in annex II shall take all practicable steps to promote, facilitate and finance, as appropriate, the transfer of, or access to, environmentally sound technologies and know-how to other Parties, particularly developing country Parties, to enable them to implement the provisions of the Convention. In this process, the developed country Parties shall support the development and enhancement of endogenous capacities and technologies of developing country Parties. Other Parties and organizations in a position to do so may also assist in facilitating the transfer of such technologies.

6. In the implementation of their commitments under paragraph 2 above, a certain degree of flexibility shall be allowed by the Conference of the Parties to the Parties included in annex I undergoing the process of transition to a market economy, in order to enhance the ability of these Parties to address climate change, including with regard to the historical level of anthropogenic emissions of greenhouse gases not controlled by the Montreal Protocol chosen as a reference.

7. The extent to which developing country Parties will effectively implement their commitments under the Convention will depend on the effective implementation by developed country

Parties of their commitments under the Convention related to financial resources and transfer of technology and will take fully into account that economic and social development and poverty eradication are the first and overriding priorities of the developing country Parties.

8. In the implementation of the commitments in this Article, the Parties shall give full consideration to what actions are necessary under the Convention, including actions related to funding, insurance and the transfer of technology, to meet the specific needs and concerns of developing country Parties arising from the adverse effects of climate change and/or the impact of the implementation of response measures, especially on:

    (a) Small island countries;

    (b) Countries with low-lying coastal areas;

    (c) Countries with arid and semi-arid areas, forested areas and areas liable to forest decay;

    (d) Countries with areas prone to natural disasters;

    (e) Countries with areas liable to drought and desertification;

    (f) Countries with areas of high urban atmospheric pollution;

    (g) Countries with areas with fragile ecosystems, including mountainous ecosystems;

    (h) Countries whose economies are highly dependent on income generated from the production, processing and export, and/or on consumption of fossil fuels and associated energy-intensive products; and

    (i) Land-locked and transit countries.

Further, the Conference of the Parties may take actions, as appropriate, with respect to this paragraph.

9. The Parties shall take full account of the specific needs and special situations of the least developed countries in their actions with regard to funding and transfer of technology.

10. The Parties shall, in accordance with Article 10, take into consideration in the implementation of the commitments of the

Convention the situation of Parties, particularly developing country Parties, with economies that are vulnerable to the adverse effects of the implementation of measures to respond to climate change. This applies notably to Parties with economies that are highly dependent on income generated from the production, processing and export, and/or consumption of fossil fuels and associated energy-intensive products and/or the use of fossil fuels for which such Parties have serious difficulties in switching to alternatives.

ARTICLE 5

Research and Systematic Observation

In carrying out their commitments under Article 4, paragraph 1(g), the Parties shall:

(a) Support and further develop, as appropriate, international and intergovernmental programmes and networks or organizations aimed at defining, conducting, assessing and financing research, data collection and systematic observation, taking into account the need to minimize duplication of effort;

(b) Support international and intergovernmental efforts to strengthen systematic observation and national scientific and technical research capacities and capabilities, particularly in developing countries, and to promote access to, and the exchange of, data and analyses thereof obtained from areas beyond national jurisdiction; and

(c) Take into account the particular concerns and needs of developing countries and cooperate in improving their endogenous capacities and capabilities to participate in the efforts referred to in subparagraphs (a) and (b) above.

ARTICLE 6

Education, Training and Public Awareness

In carrying out their commitments under Article 4, paragraph 1(i), the Parties shall:

(a) Promote and facilitate at the national and, as appropriate, subregional and regional levels, and in accordance with national laws and regulations, and within their respective capacities:

  (i) the development and implementation of educational and public awareness programmes on climate change and its effects;

  (ii) public access to information on climate change and its effects;

  (iii) public participation in addressing climate change and its effects and developing adequate responses; and

  (iv) training of scientific, technical and managerial personnel.

(b) Cooperate in and promote, at the international level, and, where appropriate, using existing bodies:

  (i) the development and exchange of educational and public awareness material on climate change and its effects; and

  (ii) the development and implementation of education and training programmes, including the strengthening of national institutions and the exchange or secondment of personnel to train experts in this field, in particular for developing countries.

ARTICLE 7

Conference of the Parties

1. A Conference of the Parties is hereby established.

2. The Conference of the Parties, as the supreme body of this Convention, shall keep under regular review the implementation of the Convention and any related legal instruments that the Conference of the Parties may adopt, and shall make, within its mandate, the decisions necessary to promote the effective implementation of the Convention. To this end, it shall:

    (a) Periodically examine the obligations of the Parties and the institutional arrangements under the Convention, in the light of the objective of the Convention, the experience gained in its implementation and the evolution of scientific and technological knowledge;

(b) Promote and facilitate the exchange of information on measures adopted by the Parties to address climate change and its effects, taking into account the differing circumstances, responsibilities and capabilities of the Parties and their respective commitments under the Convention;

(c) Facilitate, at the request of two or more Parties, the coordination of measures adopted by them to address climate change and its effects, taking into account the differing circumstances, responsibilities and capabilities of the Parties and their respective commitments under the Convention.

(d) Promote and guide, in accordance with the objective and provisions of the Convention, the development and periodic refinement of comparable methodologies, to be agreed on by the Conference of the Parties, inter alia, for preparing inventories of greenhouse gas emissions by sources and removals by sinks, and for evaluating the effectiveness of measures to limit the emissions and enhance the removals of these gases;

(e) Assess, on the basis of all information made availabe to it in accordance with the provisions of the Convention, the implementation of the Convention by the Parties, the overall effects of the measures taken pursuant to the Convention, in particular environmental, economic and social effects as well as their cumulative impacts and the extent to which progress towards the objective of the Convention is being achieved;

(f) Consider and adopt regular reports on the implementation of the Convention and ensure their publication;

(g) Make recommendations on any matters necessary for the implementation of the Convention;

(h) Seek to mobilize financial resources in accordance with Article 4, paragraphs 3, 4 and 5, and Article 11;

(i) Establish such subsidiary bodies as are deemed necessary for the implementation of the Convention;

(j) Review reports submitted by its subsidiary bodies and provide guidance to them;

(k) Agree upon and adopt, by consensus, rules of procedure
and financial rules for itself and for any subsidiary bodies;

(l) Seek and utilize, where appropriate, the services and cooperation of, and information provided by, competent international organizations and intergovernmental and non-
governmental bodies; and

(m) Exercise such other functions as are required for the achievement of the objective of the Convention as well as all other
functions assigned to it under the Convention.

3. The Conference of the Parties shall, at its first session, adopt its
own rules of procedure as well as those of the subsidiary bodies
established by the Convention, which shall include decision-
making procedures for matters not already covered by decision-
making procedures stipulated in the Convention. Such procedures may include specified majorities required for the adoption
of particular decisions.

4. The first session of the Conference of the Parties shall be convened by the interim secretariat referred to in Article 21 and
shall take place not later than one year after the date of entry
into force of the Convention. Thereafter, ordinary sessions of
the Conference of the Parties shall be held every year unless
otherwise decided by the Conference of the Parties.

5. Extraordinary sessions of the Conference of the Parties shall be
held at such other times as may be deemed necessary by the
Conference, or at the written request of any Party, provided that,
within six months of the request being communicated to the
Parties by the secretariat, it is supported by at least one-third of
the Parties.

6. The United Nations, its specialized agencies and the International Atomic Energy Agency, as well as any State member
thereof or observers thereto not Party to the Convention, may
be represented at sessions of the Conference of the Parties as
observers. Any body or agency, whether national or international,
governmental or non-governmental, which is qualified in

matters covered by the Convention, and which has informed the secretariat of its wish to be represented at a session of the Conference of the Parties as an observer, may be so admitted unless at least one-third of the Parties present object. The admission and participation of observers shall be subject to the rules of procedure adopted by the Conference of the Parties.

ARTICLE 8

Secretariat

1. A secretariat is hereby established.

2. The functions of the secretariat shall be:

    (a) To make arrangements for sessions of the Conference of the Parties and its subsidiary bodies established under the Convention and to provide them with services as required;

    (b) To compile and transmit reports submitted to it;

    (c) To facilitate assistance to the Parties, particularly developing country Parties, on request, in the compilation and communication of information required in accordance with the provisions of the Convention;

    (d) To prepare reports on its activities and present them to the Conference of the Parties;

    (e) To ensure the necessary coordination with the secretariats of other relevant international bodies;

    (f) To enter, under the overall guidance of the Conference of the Parties, into such administrative and contractual arrangements as may be required for the effective discharge of its functions; and

    (g) To perform the other secretariat functions specified in the Convention and in any of its protocols and such other functions as may be determined by the Conference of the Parties.

3. The Conference of the Parties, at its first session, shall designate a permanent secretariat and make arrangements for its functioning.

ARTICLE 9

Subsidiary Body for Scientific and Technological Advice

1. A subsidiary body for scientific and technological advice is hereby established to provide the Conference of the Parties and, as appropriate, its other subsidiary bodies with timely information and advice on scientific and technological matters relating to the Convention. This body shall be open to participation by all Parties and shall be multidisciplinary. It shall comprise government representatives competent in the relevant field of expertise. It shall report regularly to the Conference of the Parties on all aspects of its work.

2. Under the guidance of the Conference of the Parties, and drawing upon existing competent international bodies, this body shall:

    (a) Provide assessments of the state of scientific knowledge relating to climate change and its effects;

    (b) Prepare scientific assessments on the effects of measures taken in the implementation of the Convention;

    (c) Identify innovative, efficient and state-of-the-art technologies and know-how and advise on the ways and means of promoting development and/or transferring such technologies;

    (d) Provide advice on scientific programmes, international cooperation in research and development related to climate change, as well as on ways and means of supporting endogenous capacity-building in developing countries; and

    (e) Respond to scientific, technological and methodological questions that the Conference of the Parties and its subsidiary bodies may put to the body.

3. The functions and terms of reference of this body may be further elaborated by the Conference of the Parties.

ARTICLE 10

Subsidiary Body for Implementation

1. A subsidiary body for implementation is hereby established to assist the Conference of the Parties in the assessment and review of the effective implementation of the Convention. This body shall be open to participation by all Parties and comprise government representatives who are experts on matters related to climate change. It shall report regularly to the Conference of the Parties on all aspects of its work.

2. Under the guidance of the Conference of the Parties, this body shall:

    (a) Consider the information communicated in accordance with Article 12, paragraph 1, to assess the overall aggregated effect of the steps taken by the Parties in the light of the latest scientific assessments concerning climate change;

    (b) Consider the information communicated in accordance with Article 12, paragraph 2, in order to assist the Conference of the Parties in carrying out the reviews required by Article 4, paragraph 2(d); and

    (c) Assist the Conference of the Parties, as appropriate, in the preparation and implementation of its decisions.

ARTICLE 11

Financial Mechanism

1. A mechanism for the provision of financial resources on a grant or concessional basis, including for the transfer of technology, is hereby defined. It shall function under the guidance of and be accountable to the Conference of the Parties, which shall decide on its policies, programme priorities and eligibility criteria related to this Convention. Its operation shall be entrusted to one or more existing international entities.

2. The financial mechanism shall have an equitable and balanced representation of all Parties within a transparent system of governance.

3.   The Conference of the Parties and the entity or entities entrusted with the operation of the financial mechanism shall agree upon arrangements to give effect to the above paragraphs, which shall include the following:

(a)   Modalities to ensure that the funded projects to address climate change are in conformity with the policies, programme priorities and eligibility criteria established by the Conference of the Parties;

(b)   Modalities by which a particular funding decision may be reconsidered in light of these policies, programme priorities and eligibility criteria;

(c)   Provision by the entity or entities of regular reports to the Conference of the Parties on its funding operations, which is consistent with the requirement for accountability set out in paragraph 1 above; and

(d)   Determination in a predictable and identifiable manner of the amount of funding necessary and available for the implementation of this Convention and the conditions under which that amount shall be periodically reviewed.

4.   The Conference of the Parties shall make arrangements to implement the above mentioned provisions at its first session, reviewing and taking into account the interim arrangements referred to in Article 21, paragraph 3, and shall decide whether these interim arrangements shall be maintained. Within four years thereafter, the Conference of the Parties shall review the financial mechanism and take appropriate measures.

5.   The developed country Parties may also provide and developing country Parties avail themselves of, financial resources related to the implementation of the Convention through bilateral, regional and other multilateral channels.

ARTICLE 12

Communication of Information Related to Implementation

1.   In accordance with Article 4, paragraph 1, each Party shall communicate to the Conference of the Parties, through the

secretariat, the following elements of information:

(a) A national inventory of anthropogenic emissions by sources and removals by sinks of all greenhouse gases not controlled by the Montreal Protocol, to the extent its capacities permit, using comparable methodologies to be promoted and agreed upon by the Conference of the Parties;

(b) A general description of steps taken or envisaged by the Party to implement the Convention; and

(c) Any other information that the Party considers relevant to the achievement of the objective of the Convention and suitable for inclusion in its communication, including, if feasible, material relevant for calculations of global emission trends.

2. Each developed country Party and each other Party included in annex I shall incorporate in its communication the following elements of information:

(a) A detailed description of the policies and measures that it has adopted to implement its commitment under Article 4, paragraphs 2(a) and 2(b); and

(b) A specific estimate of the effects that the policies and measures referred to in subparagraph (a) immediately above will have on anthropogenic emissions by its sources and removals by its sinks of greenhouse gases during the period referred to in Article 4, paragraph 2(a).

3. In addition, each developed country Party and each other d eveloped Party included in annex II shall incorporate details of measures taken in accordance with Article 4, paragraphs 3, 4 and 5.

4. Developing country Parties may, on a voluntary basis, propose projects for financing, including specific technologies, materials, equipment, techniques or practices that would be needed to implement such projects, along with, if possible, an estimate of all incremental costs, of the reductions of emissions and increments of removals of greenhouse gases, as well as an estimate of the consequent benefits.

5. Each developed country Party and each other Party included in annex I shall make its initial communication within six months

of the entry into force of the Convention for that Party. Each Party not so listed shall make its initial communication within three years of the entry into force of the Convention for that Party, or of the availability of financial resources in accordance with Article 4, paragraph 3. Parties that are least developed countries may make their initial communication at their discretion. The frequency of subsequent communications by all Parties shall be determined by the Conference of the Parties, taking into account the differentiated timetable set by this paragraph.

6. Information communicated by Parties under this Article shall be transmitted by the secretariat as soon as possible to the Conference of the Parties and to any subsidiary bodies concerned. If necessary, the procedures for the communication of information may be further considered by the Conference of the Parties.

7. From its first session, the Conference of the Parties shall arrange for the provision to developing country Parties of technical and financial support, on request, in compiling and communicating information under this Article, as well as in identifying the technical and financial needs associated with proposed projects and response measures under Article 4. Such support may be provided by other Parties, by competent international organizations and by the secretariat, as appropriate.

8. Any group of Parties may, subject to guidelines adopted by the Conference of the Parties, and to prior notification to the Conference of the Parties, make a joint communication in fulfilment of their obligations under this Article, provided that such a communication includes information on the fulfilment by each of these Parties of its individual obligations under the Convention.

9. Information received by the secretariat that is designated by a Party as confidential, in accordance with criteria to be established by the Conference of the Parties, shall be aggregated by the secretariat to protect its confidentiality before being made available to any of the bodies involved in the communication and review of information.

10. Subject to paragraph 9 above, and without prejudice to the ability of any Party to make public its communication at any time, the secretariat shall make communications by Parties under this Article publicly available at the time they are submitted to the Conference of the Parties.

ARTICLE 13

Resolution of Questions Regarding Implementation

The Conference of the Parties shall, at its first session, consider the establishment of a multilateral consultative proccss, available to Parties on their request, for the resolution of questions regarding the implementation of the Convention.

ARTICLE 14

Settlement of Disputes

1. In the event of a dispute between any two or more Parties concerning the interpretation or application of the Convention, the Parties concerned shall seek a settlement of the dispute through negotiation or any other peaceful means of their own choice.

2. When ratifying, accepting, approving or acceding to the Convention, or at any time thereafter, a Party which is not a regional economic integration organization may declare in a written instrument submitted to the Depositary that, in respect of any dispute concerning the interpretation or application of the Convention, it recognizes as compulsory ipso facto and without special agreement, in relation to any Party accepting the same obligation:

    (a) Submission of the dispute to the International Court of Justice, and/or

    (b) Arbitration in accordance with procedures to be adopted by the Conference of the Parties as soon as practicable, in an annex on arbitration.

A Party which is a regional economic integration organization may make a declaration with like effect in relation to arbitration in accordance with the procedures referred to in subparagraph (b) above.

3. A declaration made under paragraph 2 above shall remain in force until it expires in accordance with its terms or until three months after written notice of its revocation has been deposited with the Depositary.

4. A new declaration, a notice of revocation or the expiry of a dec-

laration shall not in any way affect proceedings pending before the International Court of Justice or the arbitral tribunal, unless the parties to the dispute otherwise agree.

5. Subject to the operation of paragraph 2 above, if after twelve months following notification by one Party to another that a dispute exists between them, the Parties concerned have not been able to settle their dispute through the means mentioned in paragraph 1 above, the dispute shall be submitted, at the request of any of the parties to the dispute, to conciliation.

6. A conciliation commission shall be created upon the request of one of the parties to the dispute. The commission shall be composed of an equal number of members appointed by each party concerned and a chairman chosen jointly by the members appointed by each party. The commission shall render a recommendatory award, which the parties shall consider in good faith.

7. Additional procedures relating to conciliation shall be adopted by the Conference of the Parties, as soon as practicable, in an annex on conciliation.

8. The provisions of this Article shall apply to any related legal instrument which the Conference of the Parties may adopt, unless the instrument provides otherwise.

ARTICLE 15

Amendments to the Convention

1. Any Party may propose amendments to the Convention.

2. Amendments to the Convention shall be adopted at an ordinary session of the Conference of the Parties. The text of any proposed amendment to the Convention shall be communicated to the Parties by the secretariat at least six months before the meeting at which it is proposed for adoption. The secretariat shall also communicate proposed amendments to the signatories to the Convention and, for information, to the Depositary.

3. The Parties shall make every effort to reach agreement on any proposed amendment to the Convention by consensus. If all efforts at consensus have been exhausted, and no agreement

reached, the amendment shall as a last resort be adopted by a three-fourths majority vote of the Parties present and voting at the meeting. The adopted amendment shall be communicated by the secretariat to the Depositary, who shall circulate it to all Parties for their acceptance.

4. Instruments of acceptance in respect of an amendment shall be deposited with the Depositary. An amendment adopted in accordance with paragraph 3 above shall enter into force for those Parties having accepted it on the ninetieth day after the date of receipt by the Depositary of an instrument of acceptance by at least three-fourths of the Parties to the Convention.

5. The amendment shall enter into force for any other Party on the ninetieth day after the date on which that Party deposits with the Depositary its instrument of acceptance of the said amendment.

6. For the purposes of this Article, "Parties present and voting" means Parties present and casting an affirmative or negative vote.

ARTICLE 16

Adoption and Amendment of Annexes to the Convention

1. Annexes to the Convention shall form an integral part thereof and, unless otherwise expressly provided, a reference to the Convention constitutes at the same time a reference to any annexes thereto. Without prejudice to the provisions of Article 14, paragraphs 2(b) and 7, such annexes shall be restricted to lists, forms and any other material of a descriptive nature that is of a scientific, technical, procedural or administrative character.

2. Annexes to the Convention shall be proposed and adopted in accordance with the procedure set forth in Article 15, paragraphs 2, 3, and 4.

3. An annex that has been adopted in accordance with paragraph 2 above shall enter into force for all Parties to the Convention six months after the date of the communication by the Depositary to such Parties of the adoption of the annex, except for those Parties that have notified the Depositary, in writing, within that

period of their non-acceptance of the annex. The annex shall enter into force for Parties which withdraw their notification of non-acceptance on the ninetieth day after the date on which withdrawal of such notification has been received by the Depositary.

4. The proposal, adoption and entry into force of amendments to annexes to the Convention shall be subject to the same procedure as that for the proposal, adoption and entry into force of annexes to the Convention in accordance with paragraphs 2 and 3 above.

5. If the adoption of an annex or an amendment to an annex involves an amendment to the Convention, that annex or amendment to an annex shall not enter into force until such time as the amendment to the Convention enters into force.

ARTICLE 17

Protocols

1. The Conference of the Parties may, at any ordinary session, adopt protocols to the Convention.

2. The text of any proposed protocol shall be communicated to the Parties by the secretariat at least six months before such a session.

3. The requirements for the entry into force of any protocol shall be established by that instrument.

4. Only Parties to the Convention may be Parties to a protocol.

5. Decisions under any protocol shall be taken only by the Parties to the protocol concerned.

ARTICLE 18

Right to Vote

1. Each Party to the Convention shall have one vote, except as provided for in paragraph 2 below.

2. Regional economic integration organizations, in matters within

their competence, shall exercise their right to vote with a number of votes equal to the number of their member States that are Parties to the Convention. Such an organization shall not exercise its right to vote if any of its member States exercises its right, and vice versa.

ARTICLE 19

Depositary

The Secretary-General of the United Nations shall be the Depositary of the Convention and of protocols adopted in accordance with Article 17.

ARTICLE 20

Signature

This Convention shall be open for signature by States Members of the United Nations or of any of its specialized agencies or that are Parties to the Statute of the International Court of Justice and by regional economic integration organizations at Rio de Janeiro, during the United Nations Conference on Environment and Development, and thereafter at United Nations Headquarters in New York from 20 June 1992 to 19 June 1993.

ARTICLE 21

Interim Arrangements

1. The secretariat functions referred to in Article 8 will be carried out on an interim basis by the secretariat established by the General Assembly of the United Nations in its resolution 45/212 of 21 December 1990, until the completion of the first session of the Conference of the Parties.

2. The head of the interim secretariat referred to in paragraph 1 above will cooperate closely with the Intergovernmental Panel on Climate Change to ensure that the Panel can respond to the

need for objective scientific and technical advice. Other relevant scientific bodies could also be consulted.

3. The Global Environment Facility of the United Nations Development Programme, the United Nations Environment Programme and the International Bank for Reconstruction and Development shall be the international entity entrusted with the operation of the financial mechanism referred to in Article 11 on an interim basis. In this connection, the Global Environment Facility should be appropriately restructured and its membership made universal to enable it to fulfil the requirements of Article 11.

ARTICLE 22

Rarification, Acceptance, Approval or Accession

1. The Convention shall be subject to ratification, acceptance, approval or accession by States and by regional economic integration organizations. It shall be open for accession from the day after the date on which the Convention is closed for signature. Instruments of ratification, acceptance, approval or accession shall be deposited with the Depositary.

2. Any regional economic integration organization which becomes a Party to the Convention without any of its member States being a Party shall be bound by all the obligations under the Convention. In the case of such organizations, one or more of whose member States is a Party to the Convention, the organization and its member States shall decide on their respective responsibilities for the performance of their obligations under the Convention. In such cases, the organization and the member States shall not be entitled to exercise rights under the Convention concurrently.

3. In their instruments of ratification, acceptance, approval or accession, regional economic integration organizations shall declare the extent of their competence with respect to the matters governed by the Convention. These organizations shall also inform the Depositary, who shall in turn inform the Parties, of any substantial modification in the extent of their competence.

ARTICLE 23

Entry into Force

1. The Convention shall enter into force on the ninetieth day after the date of deposit of the fiftieth instrument of ratification, acceptance, approval or accession.

2. For each State or regional economic integration organization that ratifies, accepts or approves the Convention or accedes thereto after the deposit of the fiftieth instrument of ratification, acceptance, approval or accession, the Convention shall enter into force on the ninetieth day after the date of deposit by such State or regional economic integration organization of its instrument of ratification, acceptance, approval or accession.

3. For the purposes of paragraphs 1 and 2 above, any instrument deposited by a regional economic integration organization shall not be counted as additional to those deposited by States members of the organization.

ARTICLE 24

Reservations

No reservations may be made to the Convention.

ARTICLE 25

Withdrawal

1. At any time after three years from the date on which the Convention has entered into force for a Party, that Party may withdraw from the Convention by giving written notification to the Depositary.

2. Any such withdrawal shall take effect upon expiry of one year from the date of receipt by the Depositary of the notification of withdrawal, or on such later date as may be specified in the notification of withdrawal.

3. Any Party that withdraws from the Convention shall be considered as also having withdrawn from any protocol to which it is a Party.

ARTICLE 26

Authentic Texts

The original of this Convention, of which the Arabic, Chinese, English, French, Russian and Spanish texts are equally authentic, shall be deposited with the Secretary-General of the United Nations.

IN WITNESS WHEREOF the undersigned, being duly authorized to that effect, have signed this Convention.

DONE at New York this ninth day of May one thousand nine hundred and ninety-two.

Source: Text and Annexes downloaded from the UNEP/WMO Information Unit on Climate Change (IUCC) as an electronic document.

ANNEX I

Australia
Austria
Belarus[a]
Belgium
Bulgaria[a]
Canada
Czechoslovakia[a]
Denmark
European Economic
  Community
Estonia[a]
Finland
France
Germany
Greece
Hungary[a]
Iceland
Ireland
Italy

Japan
Latvia[a]
Lithuania[a]
Luxembourg
Netherlands
New Zealand
Norway
Poland[a]
Portugal
Romania[a]
Russian Federation[a]
Spain
Sweden
Switzerland
Turkey
Ukraine[a]
United Kingdom of Great
    Britain and Northern Ireland
United States of America

ANNEX II

Australia
Austria
Belgium
Canada
Denmark
European Economic
  Community
Finland
France
Germany
Greece
Iceland
Ireland
Italy

Japan
Luxembourg
Netherlands
New Zealand
Norway
Portugal
Spain
Sweden
Switzerland
Turkey
United Kingdom of Great
    Britain and Northern Ireland
United States of America

---

a. Countries that are undergoing the process of transition to a market economy.

INDEX

A

advisory groups, 184, 310
Africa, 83, 248
Agenda 21
  Atmosphere chapter, 154–157, 172
  Climate Convention and, 17, 155–157, 172
  GEF restructuring and, 218–219, 226
  importance of, 16, 17, 172
  sustainable development and, 311
agreed full incremental costs.
  See financing, agreed full incremental costs
agriculture, 12, 13, 284, 285, 299, 300
Algeria, 28, 108
Alliance of Small Island States. *See* island states (small)
Amazonia forest, 34–35, 175, 176, 178, 180–181
amendments to Climate Convention, 358–360
Annexes to Climate Convention, 365
AOSIS. *See* island states (small)
Arab countries. *See* oil-exporting developing countries
Arum, Gilbert, 253
Association of Small Island States. *See* island states (small)
Australia, 137, 158, 160, 268, 281, 312
Austria, 47–48, 51, 116, 130, 268
automobile industry. *See* transportation

B

Baker, James, 52, 192–193
Bangladesh, 3, 40, 245, 248
  high risk status of, 81, 241–242, 244, 252, 256
Bellagio Conference, 48, 116, 126–127, 130
Bergen Ministerial Conference, 56–57, 100, 116
biological diversity, 16, 17, 33, 101, 182, 245
biota, 10, 11, 15, 178
blocking coalitions
  bases for forming, 286–293, 296, 308, 310–311, 325

blocking coalitions (**cont'd**)
  defined, 278–279, 316*n*3
  and implementing targets, 294–295, 296
  Law of the Sea negotiations and, 287–288, 291, 296, 303–307, 309
  overcoming, 296–315
  ozone negotiations and, 290–292, 296, 302
  winning coalitions vs., 278, 325–326
Bodansky, Daniel, 24, 25–26, 45–74
Bolin, Bert, 51, 84, 92, 151
Borione, Delphine, 24–25, 26–28, 77–96
Boskin, Michael J., 37, 194, 279
Brazil
  blocking coalitions and, 292
  deforestation in, 34–35, 175, 176, 178, 180–181
  energy production in, 191
  environmental policy of, 176
  evolution of climate debate involvement, 34–35, 175–176
  hosting of Earth Summit, 34–35, 175–176
  IPCC and, 58–5
  as key player. *See* key players
  Montreal Protocol and, 282–283
  position of, 35, 176–177, 181–182, 184
  protocol negotiation and, 312, 313
  U.S. and OECD distrusted by, 196–197
Bretton Woods institutions, 203, 216, 219, 220, 226
Brundtland Commission, 183, 311, 332
Bureau (INC), 24–25, 29, 40, 41, 107, 119, 124, 132, 140–141, 222
  critical role of, 29, 32, 102–103, 119, 154
  See also Extended (Expanded or Enlarged) Bureau
Bureau (UNCED PrepCom), 107–108
Bush, George, 48–49, 69, 179, 183
  administration of, 35, 37, 181, 188–190, 192–194, 262, 279, 288
  failure of leadership of, 179, 187–200
business, 229–238
  advisory groups, 310
  climate negotiations and, 22, 39–40, 94–96, 132–133, 229–238
  consulting prior to negotiations, 233–236, 237–238

business (**Cont'd**)
  energy/carbon tax and, 236–237, 294
  future negotiations and, 40, 232–238,
    310, 331
  GEF favored by, 235
  mobilization of, 94–95, 237
  multiple viewpoints of, 39–40, 231
  North-South relations and, 234
  role of, 230–232
  STAP and, 237
  targets and timetables and, 184, 237
  technology transfer and, 236
  trade associations, 231
  See also non-governmental organizations
Business Council for Sustainable Development, 39, 231, 233, 234, 235, 236, 237,
  310
"business as usual" approach, 57

C
Canada, 8, 29–30, 53, 54, 168, 292
position of, 137, 158, 160, 268, 281, 312
CANZ group, 158, 160. *See also* Canada;
  Australia
carbon dioxide
  atmospheric reservoir of, 6, 7, 97
  doubling of, 8, 9, 97
  equivalence concept, 55
  greenhouse effect and, 6–9, 283
  increase in, 7, 9, 46, 97, 129–130
  narrow focus on not useful, 284
  as only measurable greenhouse gas, 168
  ozone hole and, 267
  reducing/stabilizing. *See* greenhouse
    gases, reducing/stabilizing emissions of
  sinks of. *See* greenhouse gases, sinks of
  sources of. *See* greenhouse gases,
    sources of
  temperature and, 8, 9
  See also greenhouse gases
carbon tax. *See* taxes (carbon/energy)
Centre for Science and the Environment,
  244, 248, 250–251
CFCs (chlorofluorocarbons), 7, 71*n*7, 95, 98,
  269, 285–286, 332. *See also* greenhouse
  gases; Montreal Protocol; ozone layer;
  Vienna Convention
Chantilly INC session (INC 1), 60, 61–64,
  85, 117, 131–133, 149, 244, 250, 252
China
  Brazilian delegation to, 181–182
  consensus with Group of 77, 138
  early proposals from, 85, 133, 209

China (**Cont'd**)
  energy production in, 191, 209, 289
  financial mechanism and, 216–217
  fracturing of Group of 77 and China,
    28–29, 35, 66–67, 104–106, 138–139
  future responsibility for greenhouse
    problem, 289
  and historical responsibility of industrial
    countries, 133, 182
  as key player. *See* key players
  Montreal Protocol and, 282–283, 313–314
  New York INC session (INC 5) and,
    215–216
  OECD and U.S. and, 26, 27, 33–34,
    196–197, 198
  Paris INC session (INC 5) and, 211–214
  position of, 35, 66, 85, 138–139, 211–212,
    215
  protocol negotiation and, 312, 313
"clearing house," 158, 161, 167–168, 179, 222
Climate Action Network, 246, 250, 253,
  254, 255–257
climate change. *See* greenhouse effect
Climate Convention
  adoption by consensus, 70, 132, 216
  Agenda 21 and, 17, 155–157, 172
  amendments to, 358–360
  Annexes, 365
  Article 1, 338–339
  Article 2, 17–18, 21, 143, 166, 172, 195,
    204–205, 223, 267, 339
  Article 3
    general principles, 18, 21, 39, 64, 65,
      66, 110, 212–213, 217, 339–341
    Precautionary Principle, 18, 21, 41,
      110, 169, 171, 173*n*3, 323–324, 340
    See also principles
  Article 4
    Paragraph 1, 18–19, 170, 223, 269–270,
      341–342
    Paragraph 2, 19, 145, 165–166, 167–169,
      172, 268–269, 328, 342–344
    Paragraph 3, 19–20, 217, 223, 260–261,
      344–345
    Paragraph 4, 345
    Paragraph 5, 20, 345
    Paragraph 6, 345
    Paragraph 7, 20, 171, 217, 251, 271,
      345–346
    Paragraph 8, 20, 171, 346
    Paragraph 9, 20, 171, 346
    Paragraph 10, 20, 346–347
  Article 5, 347
  Article 6, 347–348

Climate Convention (**Cont'd**)
Article 7. *See* Conference of the Parties
Article 8. *See* Secretariat (INC)
Article 9. *See* Subsidiary Body for Scientific and Technological Advice
Article 10. *See* Subsidiary Body for Implementation
Article 11, 20, 218, 272, 355–356. *See also* financing
Article 12, 20–21, 166–167, 219, 225, 356–358. *See also* reporting; reviewing
Article 13-20, 20, 125–126, 357–361
Article 21, 216, 270–271, 361–362
Article 22-26, 362–364
authentic texts, 364
Biodiversity Convention and, 17
commitments. *See* Climate Convention, Article 4; financing; greenhouse gases, reducing/stabilizing emissions of; targets and timetables; technology transfer; Working Groups (INC), Working Group I on commitments
communications. *See* communications; reporting
Conference of the Parties. *See* Conference of the Parties
cornerstone of, 16, 166
definitions, 338–339
depository, 361
dispute settlement, 125–126, 357–358
education, training, and public awareness provisions, 347–348
entry into force, 90, 109, 165, 217, 239, 277, 363
financial mechanism. *See* financing
further negotiations required, 145–146
historical overview
INC sessions, 60–70, 82–89, 118, 130–131
prior to INC sessions, 45–60, 78–80, 116–118, 129–130
implementing
Bodansky viewpoint, 62–63
Borione and Ripert viewpoint, 84, 91–96
Dowdeswell and Kinley viewpoint, 122–124
Faulkner viewpoint, 237
flexibility in, 169, 171, 269–270
Goldemberg viewpoint, 183, 184
Hyder viewpoint, 202, 217
Kjellen viewpoint, 163, 165–171
resolution of questions, 125–126, 357
See also blocking coalitions; Conference of the Parties; joint implementation

Climate Convention (**Cont'd**)
interim measures, 90, 165, 223–224, 363–364, 171–172. *See also* "prompt start"
multiple interpretations are supported, 23–24
NGO criticisms of, 150–151, 240, 258, 266–267, 268–270, 293
Objective. *See* Climate Convention, Article 2
preamble, 270, 335–338
principal elements, 17–21
principal goals, 4, 17–18, 90
principles. *See* principles
"prompt start," 90, 126–127, 264
ratification of, 89–90, 107, 109, 180, 217–218, 239, 277, 362
reporting. *See* communications; reporting
research and systematic observation provisions, 347
reservations to, 363
Rio Declaration and, 17, 39, 212–213
Secretariat, 20, 351
significance of
action orientation, 127
broad consensus reflected, 17, 31, 128, 258
"buys time" for scientific study, 15
developing countries gained significant concessions, 216–217
differentiated responsibilities reflected, 144
environment and development linked, 16, 17, 21, 31, 81–82, 127
essentially meaningless, 277
financial resources committed by industrial countries, 22
first step. *See* Climate Convention negotiations, success partially attained
historical responsibility of industrial countries recognized, 258
joint implementation supported, 22
landmark achievement, 21
momentum for change created, 17
new model of international cooperation, 31, 226
NGO participation provision, 22, 273, 278
North-South relations, 16, 21, 22, 31
overview of, 4, 21–22
pedagogical role, 96
Precautionary Approach takes precedence, 21, 182
process orientation of, 16, 21–22

Climate Convention (**Cont'd**)
  significance of (**Cont'd**)
    reporting and review obligations, 17, 151
    seriousness of rapid climate change
      recognized, 15, 17, 151, 182
    stabilization of greenhouse gases, 21
    sustainable development focus, 4–5,
      16, 17, 18, 21, 22, 81–82, 93, 113
    universality, 22, 96
    unprecedented scale of participation,
      22, 278
  signing of, 4, 15–16, 89–90, 99, 115, 217,
    239, 277, 361
  Subsidiary Body for Implementation,
    20, 123–124, 353
  Subsidiary Body for Scientific and Tech-
    nological Advice, 20, 100, 122–123,
    171, 221–222, 327–328, 352
  text of, 335–365
  understanding the text of, 201–202
  voting, 360–361
  withdrawal from, 363–364
  See also Climate Convention negotiations;
    Intergovernmental Negotiating Com-
    mittee; protocols; targets and timetables
Climate Convention negotiations
  analyzing, reasons for, 201–204
  bracketing of text, 31, 61, 68, 101–102,
    121–122, 211, 213–214
  CFC negotiations and, 280–281
  complexity of, 23, 114–115, 118, 152–154,
    237, 283–284, 296, 324, 330–333
  consensus vs. voting, 132
  Consolidated Working Document, 67,
    140–141
  contact groups, 67
  difficulty factors, 280–296
    benefits are vague and uncertain, 42,
      292
    blocking coalitions. *See* blocking coa-
      litions
    continual changes in negotiators, 101
    costs, 42, 285–286, 291–293
    dual rotating co-chairs, 102–103
    "free rider effect," 42, 285–286
    future, not present, generations ben-
      efit, 285–286, 292
    multiple meeting sites, 101
    sacrifices must be made, 285
    scope and complexity of climate is-
      sue, 237, 283–284
    United States, 279
    unresolved differences, 107, 324–325
  domestic political events and, 36, 183

Climate Convention negotiations (**Cont'd**)
  drafting and revising vs. bargaining
    and negotiating, 140–141
  dynamic nature of, 21–22
  dynamics of, 42–43, 176–180, 208–210,
    277–320
  on financing. *See* financing
  future success of, factors influencing
    advisory bodies not duplicated, 184
    advisory group role enhancement,
      310
    alternative policy options rigorously
      studied, 234–235
    blame assignment forgone, 333–334
    blocking coalition avoidance, 296–315
    "bottom-up" process, 200
    broad focus maintained, 284, 299, 331
    business mobilization, 94–96
    business participation, 40, 232–238,
      310, 331
    checklist of European Community,
      160
    Commission on Sustainable Devel-
      opment role, 93, 221
    Conference of the Parties role, 93–94,
      100, 108–110, 122–123
    consultations prior to negotiations,
      233–236, 237–238
    core provisions need clarification,
      145–146
    deadlines, 95–96, 183, 233, 326
    developing countries must be helped
      to attend, 326
    dialogue must become more universal,
      203–204, 233
    differentiation of national responsi-
      bilities, 170–171
    economic analyses, 92
    economic policies in industrial
      countries, 92
    energy production transition strategies,
      235–236
    epistemic communities, 297–298
    experts' participation, 95, 200
    fairness, 311
    financial mechanism definition and,
      93–94
    flexibility, 122, 202, 269–270, 302
    GEF restructuring, 216–221, 235
    good faith commitment fulfillment,
      226
    goodwill, 202
    industrial countries must accept his-
      torical responsibility, 96

Climate Convention negotiations (**Cont'd**)
future success of, factors influencing
(**Cont'd**)
industrial countries must help developing countries, 184–185, 226, 326
industrial countries must set
example, 96, 226
informal control regimes, 298
informal interactions, 309–310, 328
innovation, 95, 273
institutional issues and, 171–172
inter-governmental negotiations
directly proceeded to, 183
interim issues that must be handled,
167–170, 221–222
international cooperation, 95, 151,
202, 226
inventories must be produced, 198
IPCC role, 92, 122–123, 171, 221–222,
267–268, 326
"listening" must extend to opposing
viewpoints, 233
national action plans, 198, 328
national interest must yield to global
interest, 235
NGO participation, 42, 95, 183, 232–238,
274, 296, 310, 326, 331
North-South considerations, 307–311
OECD role, 93
participatory, open, transparent process maintained, 22, 203–204, 232,
326, 331
personal relationships, 310
political momentum maintained,
151, 165, 184
political will, 95, 107, 326, 334
priority list of recommendations, 234
protocol negotiation strategies, 183,
299–315
protocols should be avoided, 42–43,
91–92, 145–146, 279–280, 293–294,
296–299, 303
protocols should be negotiated, 27,
166, 183, 271
reporting and review and, 163, 167
resolution by UN General Assembly, 165
SBSTA role, 122–123, 327–328
science must not be politicized, 100,
120, 123
scientific capacities in developing
countries, 120–121
scientific research, 92, 120–121, 171–
172, 296–297, 309–310, 323, 326,
327–328

Climate Convention negotiations (**Cont'd**)
future success of, factors influencing
(**Cont'd**)
Secretariat should coordinate logistics,
327
structure of negotiations, 121–127,
325–328
sustainable development concerns
accounted for, 93
targets and timetables should be set,
166, 184, 271
targets and timetables should not be
set, 43, 91–92, 145–146, 279–280,
296–299, 303
two-phase process, 233–238, 326–327
urgency sense needed, 184
voluntary actions, 43, 280, 297–299, 327
workshops, 183, 235, 309–310, 328
See also Climate Convention negotiations, difficulty factors
historical overview
INC sessions, 60–70, 82–89
prior to INC sessions, 45–60, 78–80
impasses in, 33–34, 39, 102, 103, 150,
208, 311
See also blocking coalitions
implications for international negotiations, 4–5, 16, 21–22, 31, 109, 119, 226,
329–330
indirect goal achievement vs. formal binding agreements, 279–280, 296–299
informal meetings, 210, 215
initial negotiating positions, 26–27, 82–83,
133–136
IPCC role in, 60, 84, 92–93, 100–101,
171, 177, 221–222, 267–268
key players. *See* key players
Law of the Sea negotiations and, 53,
280–282, 285, 286–290, 291, 296, 303–
307, 309
lessons learned from, 23–24, 107–108,
118–121, 182–184, 199–200, 201–204,
330–334
management of. *See* Climate Convention negotiations, structure of
marginalization of small countries, 265
Montreal Protocol negotiation and, 91, 229,
282–283, 284, 290–292, 302, 309, 313–314
multiple interpretations are inevitable,
23–24
negotiation analysis and, 278–279,
316*n*2
and North's dependence on South,
287–290

Climate Convention negotiations (**Cont'd**)
  future success of, factors influencing
    (**Cont'd**)
  ozone negotiations and, 282, 290–292,
    296, 302
  perceived risk of rapid climate change
    and, 5
  politicization of, 28, 57–59, 80, 117, 118, 177
  principles. *See* principles
  prior negotiations' relevance to, 280–283,
    284, 285
  recession and, 179, 273
  Revised Text under Negotiation, 67, 68
  scientific foundation of, 28, 65, 80–82,
    84, 100
  shortcomings of. *See* Climate Convention
    negotiations, success partially attained
  skepticism at beginning of, 84, 85
  speed of, 99–100, 117, 131
  "streamlining" and, 32, 140–141
  structure of
    Bodansky viewpoint, 59, 60–61, 63–64
    Borione and Ripert viewpoint, 85–86
    Dasgupta viewpoint, 139–142
    Dowdeswell and Kinley viewpoint,
      121–127
    Faulkner viewpoint, 230
    future negotiations and, 121–127,
      325–328
    Hyder viewpoint, 203, 210
    Kjellen viewpoint, 150, 152–154
  success of, factors influencing
    appointment of Cutajar as Secretary-
      General, 152, 153
    appointment of Ripert as chairman, 152–
      154
    blame assignment forgone, 333–334
    Bureau, 29, 32, 102–103, 107, 119, 154
    business viewpoints welcomed, 40, 230
    chairman's integrated text, 68–69, 89,
      102, 103, 118–119, 213–216
    common understandings, 34, 150
    competing interests balanced, 27–28,
      31, 35–36, 78, 203–204
    compromise, 34, 89, 103, 119
    deadline set for Earth Summit, 28,
      101, 127, 150, 183, 322–323
    dual rotating co-chairs, 29, 30–31, 64,
      102–103
    flexibility, 34, 121, 323
    global environment concern, 150
    group dynamics, 34, 149, 150, 159
    innovation, 30–31, 84, 121, 124–126
    IPCC preliminary work, 84, 100–101,
      171, 177

Climate Convention negotiations (**Cont'd**)
  future success of, factors influencing
    (**Cont'd**)
  momentum of the negotiations, 183
  mutual cooperation, 101
  NGOs. *See* non-governmental organi-
    zations, critical role of
  OECD Special Working Group, 34
  participatory, open, transparent na-
    ture of process, 22, 40, 64, 82, 120,
    203–204, 215, 230, 232, 265–266, 322
  personal relationships, 34, 202
  political spotlight on climate problem,
    84
  political will, 326
  psychological factors, 150
  public expectations of cooperation,
    183, 202–203
  Rio Declaration principles adopted,
    39, 212–213
  scientific shared understanding, 28,
    100, 120–121, 171–172, 322, 326
  Secretariat, 104, 153
  structure of process, 28, 119, 150, 152,
    203, 322–324
  team effort, 84
  tenacity of negotiators, 84
  travel paid for developing countries,
    28, 203
  Working Group II, 154
  success partially attained
    Borione and Ripert viewpoint, 96
    Dasgupta viewpoint, 32, 33, 144,
      145–146
    Djohhlaf viewpoint, 107
    Hyder viewpoint, 216–217
    Kjellen viewpoint, 150–151, 165
    Mintzer and Leonard, 334
    Rahman and Roncerel viewpoint,
      240, 258, 266–267, 273
    Sebenius viewpoint, 277–278, 284
  suspicions surrounding, 110–111
  traditional convention concepts not
    workable for, 85–86
  unresolved differences, 107, 324–325
  urgency of task, 150
  vagueness of commitments, 18–19, 90,
    126, 143, 144, 145, 188, 202, 268–270
  Vienna Convention negotiations and,
    282, 290–291, 302
  voting vs. consensus, 132
  who should conduct, 59–60, 81–82
  winning coalitions, 278, 325–326
  Working Groups. *See* Working Groups
    (INC)

Climate Convention negotiations (**Cont'd**)
See also Climate Convention; Intergovernmental Negotiating Committee;
protocols; targets and timetables
climate models, 8, 9, 11, 46–47, 114
Climate Network. *See* Climate Action Network
Clime Asia, 255–257
Clinton, Bill, 189, 190, 192, 198
administration of, 198, 199, 226, 279,
288–289, 312
coalitions. *See* blocking coalitions
Collor de Mello, Fernando, 176
Commission on Sustainable Development, 93, 221, 222, 310
commitments. *See* Climate Convention,
Article 4; financing; greenhouse gases,
reducing/stabilizing emissions of;
targets and timetables; technology
transfer; Working Groups (INC), Working Group I on commitments
communications, 18, 19, 20–21, 89, 122,
327, 356–358. *See also* reporting
Communist bloc (former), 29, 104–105,
196, 327. *See also* East European countries; economies in transition
"comprehensive approach," 55, 88, 158,
168, 188, 269, 304–305
Conference of the Parties (COP)
Bureau role in, 119
deadline setting, 108–109
financial mechanism responsibilities
Borione and Ripert viewpoint,
93–94
Climate Convention requirements,
353–354
Dasgupta viewpoint, 144, 147*n*33
Djoghlaf viewpoint, 109–110
Goldemberg viewpoint, 180
Hyder viewpoint, 215, 216–217, 218,
219, 222–223
Rahman and Roncerel viewpoint,
254, 270, 271, 272–273
GEF and, 93, 94, 124, 144, 216–217,
222–223, 271
imbalance in composition of, 109
IPCC and, 221
joint implementation and, 145, 167
"prompt start" and, 126–127
protocols and, 360
resolution of questions, 125–126
reviewing and, 122, 123, 151, 172
role of, 17, 20, 93–94, 100, 108–110,
122–123, 145, 170–171, 348–351

Conference of the Parties (COP) (**Cont'd**)
SBSTA and, 122–123, 352
separation of drafting from negotiating,
100
universal participation in, 203
UN restructuring and, 222
See also Climate Convention negotiations, future success of
conflict, rapid climate change and, 11–12
consumption
excessive, 35, 107, 178, 196, 255, 256,
263, 270, 324, 331–332
sustainable, 332
contact groups (INC), 32, 67
conventions
ambiguity of international, 202
Biodiversity Convention, 16, 17, 101, 182
difficulty in achieving effective, 42
international climate law, 31, 109, 110
"law of the atmosphere" proposal, 53,
72*n*22, 281–282, 299
Law of the Sea, 53, 184, 202, 280–282,
285, 286–290, 291, 296, 303–307, 309
Limited Test Ban Treaty, 302–303, 316*n*5
Maastricht Treaty, 279–280
UN model for negotiating, 119
Vienna Convention, 54, 81, 82, 98, 282,
290–291, 302
See also Climate Convention; protocols
COP. *See* Conference of the Parties
corporations. *See* business
cost-effective implementation, 27, 167, 188
costs. *See* economic costs; financing; social
costs
credits, emission, 64–65, 87, 125, 167–168.
See also "clearing house"
Cutajar, Michael Zammit, 60, 152, 214

D
Darman, Richard G., 37, 194, 279, 288
Dasgupta, Chandrashekhar, 24–25, 31–33,
129–148, 158
debt-for-nature swaps, 307
deforestation
biodiversity and, 33
blocking coalitions and, 286
Brazilian, 34–35, 175, 176, 178, 180–181
cessation of, 285, 299
developing country responsibility for, 178
liability of, 14, 329–330
narrow focus on not useful, 284
rate of, 98
as source of emissions, 176, 178, 283, 329

deforestation
  See also forests
Delors, Jacques, 69
Denmark, 83. *See also* Nordic countries
depasitary for Climate Convention, 361
desertification, 98, 205, 306
developed countries. *See* industrial countries
developing countries
  bargaining power of, 287–290
  binding obligations opposed by, 135
  Clinton administration and, 226
  commitments of, overview of, 18–19,
    20–21, 341–342, 345–347
  Communist bloc breakdown and, 29,
    104–105, 196, 325
  consensus building. *See* key players,
    developing countries as bloc; key
    players, developing country sub-
    groups
  continual changes in negotiators handi-
    capped, 101
  development given primacy by, 38, 58,
    59, 196, 205, 206, 212, 216, 266, 271
  elements considered essential to Cli-
    mate Convention, 215
  equity issues. *See* equity
  evolution of awareness of climate prob-
    lem, 50, 54–56, 58–60, 79–80, 117–118,
    130–131
  financial commitments to. *See* financ-
    ing, for developing countries
  financial mechanism and, 69, 124, 197,
    208, 216–217, 218, 259–260
  financial support for travel, 28, 203
  future responsibility for greenhouse
    problem, 284, 289
  GEF and, 69, 124, 197, 208, 216–217,
    218, 259–260
  greenhouse gas emissions from, 18–19,
    178, 189, 284, 289
  INC Working Group structure and, 63
  industrial countries distrusted by, 196–197,
    207–208
  industrial countries made significant
    concessions to, 216
  industrial country historical responsi-
    bility. *See* industrial countries, histori-
    cal responsibility
  industrial country insistence on ac-
    knowledgment of responsibility by,
    207
  IPCC and, 58, 100–101, 120
  as key player. *See* key players
  local vs. global environmental prob-
    lems and, 38, 177, 205

developing countries (**Cont'd**)
  marginalization of small countries, 265
  Montreal Protocol and, 58, 282–283, 313
  multiple meeting sites handicapped, 101
  national action plans. *See* national
    action plans
  negative flow of resources from, 208,
    256
  NGO recommendations for, 271–272
  population growth and, 263, 289, 307,
    324, 331–332
  poverty issues. *See* poverty
  principal contribution to climate
    change, 178
  principles and, 209, 210, 212, 215
  ratification may be delayed for, 107, 109
  reporting by. *See* reporting
  reviewing and. *See* reviewing
  Special Committee on the Participation
    of Developing Countries, 58, 100–101
  technology transfer and. *See* technology
    transfer
  transparency and universality of rep-
    resentation and, 64, 82
  See also Climate Convention negotiations;
    economies in transition; Group of 77;
    island states (small); Kuala Lumpur
    Group; non-governmental organiza-
    tions; North-South relations; oil-export-
    ing developing countries; targets and
    timetables; *specific countries*
development
  developing countries give primacy to,
    38, 58, 59, 196, 205, 206, 212, 216, 266,
    271
  link with environment, 16, 17, 21, 38,
    41, 81–82, 127, 177, 205–207, 266
  link with other issues, 331–333
  per capita entitlements and, 261–262
  UN General Assembly Resolution 44/28
    on, 205–206
  universal right to, 16, 206, 212, 216, 266
  See also sustainable development
dispute settlement, 125–126, 357–358
Djoghlaf, Ahmed, 24–25, 28–29, 69, 97–111
domestic issues, 36, 183, 294, 298, 324–324
Dowdeswell, Elizabeth, 24–25, 29–31,
  113–128, 154
dual rotating co-chairs, 29, 30–31, 64,
  102–103

E
Earth Summit. *See* United Nations Conference
  on Environment and Development

East European countries, 145, 196, 268, 294, 310. *See also* Communist bloc (former); economies in transition

EC. *See* European Community

ECO, 30, 40–41, 120, 136, 249, 261, 264, 265

ecological limits, NGOs and, 252, 258, 267

economic costs
analyses of, 51, 82, 83, 92, 188
blocking coalitions and, 286–293
cost-effective implementation, 27, 167, 188
disincentives to bearing, 285–286
of eco-tax, 84
estimating future costs, 13, 329
liability and compensation issues, 329–330
NGO viewpoint, 256–257
ocean level rise and, 11
overview of, 12–13, 80–82, 131
rate of climate change and, 10
real and immediate nature of, 42, 285
of reducing carbon emissions in U.S., 189, 192, 198, 257, 292
of sustainable development, 115
of weather extremes, 12
See also financing; poverty

economic development. *See* development

economic interests, blocking coalitions and, 286–293

economies in transition, 169, 171, 189, 268, 347, 367. *See also* Communist bloc (former); East European countries

ecosystem effects, 10–14

education, 309–310, 347–348

EFTA, 83, 158, 167, 281, 312

Egypt, 241–242

emissions. *See* greenhouse gases

energy production
Agenda 21 and, 96, 155, 156, 172
alternative sources, 94, 190, 191, 192, 200
blocking coalitions and, 286, 292, 300
in China and India, 191, 209, 289
curtailments in, 285
efficiency measures, 263
electricity generation, 12, 94, 95, 191
in industrial countries, 177, 178, 189, 190, 191, 192, 200, 283
narrow focus on not useful, 284
negotiating changes in, 285, 299, 300
obstructive role of fossil fuel industry, 268
per capita demand, 284
as principal source of emissions, 81, 131, 177, 178, 189, 191, 283
transition strategies, 235–236

energy resources, limited supply of, 332

energy tax. *See* taxes (carbon/energy)

Enlarged Bureau. *See* Extended (Expanded/Enlarged) Bureau

entry into force. *See* Climate Convention, entry into force

environment
industrial countries give primacy to, 38, 177, 196, 205, 206
link with development, 16, 17, 21, 38, 41, 81–82, 127, 177, 205–207, 266
linked environmental issues, 331
local vs. global degradation of, 38, 177, 205
See also sustainable development

environmental NGOs. *See* non-governmental organizations

environmental protection agency, global, 280

Environmental Protection Agency (EPA), 193

environmental refugees, 11, 12

Environment Ministries, 30, 115

epistemic communities, 297–298

equitable participation. *See* open nature of negotiations

equity, 32, 33, 107, 110, 133–134, 261–262, 324
NGO focus on, 41, 243, 247, 250–251, 255–256, 261–262, 271, 322
Southern focus on, 27, 32, 63, 84, 133–134, 263

Estrada, Raoul, 69

European Community (EC)
common policy approaches and, 27
"comprehensive approach" and, 168
compromise agreed on, 89, 136–137, 142
consensus with other Northern countries, 136–137, 311–312
energy/carbon tax favored by, 27, 83, 84, 96, 163, 294
financial assistance position of, 141
initial negotiating position, 26–27, 83
as key player. *See* key players
meetings of, 158
Montreal Protocol and, 282, 290
north-south problem within, 167
OECD coordination and, 159–160, 163
"pledge and review" and, 66, 136–137
targets and timetables
difficult to implement, 197, 294
favored, 26–27, 34, 57–58, 63, 69, 82–83, 91, 134, 136, 137, 158, 178–179, 258, 268, 281, 294, 312
opposed, 83, 197, 258, 268
United Kingdom withholds support, 258
U.S. conflicts, 26–27, 35, 51, 136, 137, 141

European Community (EC) **(Cont'd)**
  See also East European countries; industrial
    countries; North-South relations; Orga-
    nization for Economic Cooperation and
    Development
European Economic Area, 167
European Free Trade Assocation (EFTA),
  83, 158, 167, 281, 312
excessive consumption. *See* consumption
Extended (Expanded/Enlarged) Bureau,
  70, 154, 163, 164, 215
  significance of, 32, 39, 68–69, 88–89,
    102, 103, 142–144, 154, 162

F
Faulkner, Hugh, 25, 39–40, 229–238
FCCC. *See* Climate Convention
financing
  Agenda 21 and, 157, 218–219, 226
  agreed full incremental costs
    Agenda 21 and, 218
    defining, 93–94, 180, 221, 235, 254
    developing country commitment
      fulfillment and, 144, 170, 180, 194,
      196, 217, 223
    developing country reasons for, 196
    GEF and, 93–94, 180, 218, 221
    industrial country committment to,
      20, 144, 147n33, 196, 217, 223, 334,
      344–345
    intentional ambiguity of, 260
    NGOs and, 254, 260–261
    unresolved differences, 324, 325
  Amazonia forest fund, 180–181
  business sector viewpoint, 235, 236
  Climate Convention Article 11 vs.
    Article 21, 270–271
  Commission on Sustainable Develop-
    ment and, 222
  Conference of the Parties and. *See*
    Conference of the Parties (COP),
    financial mechanism responsibilities
  debt-for-nature swaps, 307
  developing countries
    acceptable terms, 27, 63, 84, 135, 138–139,
      164, 170, 177, 179, 196, 211, 215
    commitment fulfillment and, 144,
      170, 180, 194, 196
    industrial country commitments. *See* fi-
      nancing, industrial country commit-
      ments
    joint implementation and, 161
    reporting requirements and, 90,
      198–199, 207–208, 217, 223, 328

financing **(Cont'd)**
  developing countries **(Cont'd)**
    separate climate fund pressed for, 69,
      124, 197, 216–217, 259–260
    Global Environment Facility. *See* Glo-
      bal Environment Facility
    Global Environment Fund, 90
    INC structure for negotiationg, 62, 63
    India's funding proposal, 63
    industrial country commitments
      Climate Convention requirements,
        19–20, 22, 62, 144, 147n33, 196,
        198–199, 217, 223, 344–345
      positive support of, 135, 141, 206–207
      reticence about, 105, 107, 109–110,
        135, 141, 179–180, 194, 206–207, 211
    liability impossible to establish, 329–330
    mechanism as defined in Climate Con-
      vention, 19–20, 353–354
    "new and additional" resources esti-
      mate, 179
    NGOs and, 245, 246, 253, 254, 256–257,
      259–261, 270–271, 272–273
    Noordwijk Declaration recommenda-
      tions, 56
    Official Development Assistance, 29,
      35, 177, 205
    principles for establishing mechanism,
      272–273
    private investments, 22, 235
    recession and, 179, 262, 273
    travel for developing country partici-
      pants, 28, 203
    UN General Assembly Resolution 44/28
      recommendation, 206
    U.S. position, 135, 141, 194
    World Atmosphere Fund, 49
    World Bank. *See* World Bank
    See also economic costs; joint implemen-
      tation; poverty
Finland, 158, 168, 197, 270, *See also* Nordic
  countries
First Assessment Report (IPCC), 51–52,
  54–55, 57, 58–59, 72n28, 80, 98–99
  scientific consensus contained in, 9, 17–18,
    57, 80, 116
  significance of, 57, 80, 116, 117, 120, 130,
    177, 333
First World Climate Conference, 48, 129–130
Flamingo Park Global Forum, 183
flexibility
  in implementing commitments, 169,
    171, 269–270
  success of negotiations and, 34, 121,
    122, 202, 269–270, 323

forests
  Amazonia forest, 34–35, 175, 176, 178, 180–181
  Climate Convention and Agenda 21 and, 172
  damage to, 98
  Kuala Lumpur Group and, 106
  large-scale manipulation of, 270
  migration of, 13
  "net emissions" and, 168, 269
  Noordwijk proposal on, 55, 56, 110–111
  as sinks, 131, 176, 283, 329
  U.S. proposal on, 178, 181
  See also deforestation
Framework Convention on Climate Change. *See* Climate Convention
France
  differentiation of national responsibilities proposal, 170–171
  Hague Conference sponsorship, 52, 79
  "pledge and review" and, 65
  targets and timetables and, 65, 83, 134, 197
  World Bank and, 259
"free rider effect," 42, 285–286
Friends of the Earth, 245
funding. *See* financing
future negotiations. *See* Climate Convention negotiations, future success of

G
G-7. *See* Group of 7
G-24. *See* Group of 24
G-77. *See* Group of 77
Gayoom, Maumoon Abdul, 244
GEF. *See* Global Environment Facility
General Agreement on Tariffs and Trade (GATT), 83, 110, 313, 325
Geneva INC sessions
  INC 2, 64–65, 133–136, 139–140, 170–171, 209, 253
  INC 4, 33–34, 66–67, 138–139, 140, 158–159, 209–210, 253–254
  INC 6, 222, 254–255, 267
Geneva Ministerial Conference, 99
Geneva PrepCom session (PrepCom 3), 155
Germany, 83, 134, 137, 180, 191, 197, 259
Ghana, 209, 261
Global Environment Facility (GEF)
  Agenda 21 and, 218–219, 226
  agreed full incremental costs and, 93–94, 180, 218, 221
  analytic studies needed, 328
  Bretton Woods framework, 216, 219, 220

Global Environment Facility (GEF) (Cont'd)
  business sector and, 235, 236
  Climate Convention Article 11 vs. Article 21 and, 270–271
  Conference of the Parties and, 93, 94, 124, 144, 216–217, 222–223, 271
  developing countries and, 69, 124, 197, 208, 216–217, 218, 259–260
  Evaluation Committee, 223
  INC relationship with, 222, 226, 328
  industrial countries and, 61, 69, 181, 208, 219, 260
  interim role of, 69, 93, 124, 144, 197, 270–271
  low funding level of, 179–180
  NGOs and, 245, 246, 254, 256, 259–261, 270–271, 272–273
  Participants Assembly, 219–220, 224–225
  replenishment of, 124–125, 223, 224, 313
  restructuring of, 124, 144, 204, 217–221, 222–223, 224–226, 270–271
  Rome meeting, 223–225
  Scientific and Technical Advisory Panel, 123, 184, 221, 237
  Secratariat of, 220, 224
  Secratariat of INC and, 327
  strengths of GEF as financial mechanism, 124
  UNDP and, 197, 218, 219, 260
  UNEP and, 197, 218, 219, 260
  United States and, 181, 195–196
  universal participation and equitable discourse needed, 204, 218–219, 220, 225–226, 270–271, 272
  unresolved differences, 324, 325
  voting system of, 219, 224–225
  workability of, 165, 170
  World Bank and, 196–197, 218, 219–220, 259, 260, 306
Global Environment Fund, 90
Global Forum, 183
global warming. *See* greenhouse effect, as problem
Goldemberg, José, 25, 34–36, 175–185
Gore, Albert, 189, 192, 198, 262, 279, 312
"go slow" approach, 35, 59
Great Britain. *See* United Kingdom
greenhouse effect
  defined, 5–6
  feedback mechanisms, 9–10, 57
  as natural background process, 5–6
  as problem, 6–15
    causal relationships and, 13, 15
    developing country future contribution to, 284, 289

greenhouse effect **(Cont'd)**
    as problem, 6–15 **(Cont'd)**
        development of, 6–9, 14, 114, 177, 283–284
        economic costs of. *See* economic costs
        evolution of awareness of, 45–60, 78–80, 116–118, 129–131
        future causes of, 284
        global effects of, 9–14
        global evidence for past warming, 11, 13
        global warming relationship to, 6–9, 13, 15, 81
        historical responsibility for. *See* industrial countries, historical responsibility
        liability impossible to establish, 329–330
        linked issues of, 331
        long lag times between events and impacts, 329
        natural occurrence of climate change vs., 11, 13, 14
        ocean level rise and. *See* oceans, rise in level of
        rapid climate change and, 10, 11
        rate of warming, 9, 10, 11, 14, 57, 98, 267
        scientific consensus on, 9, 17–18, 26, 45, 46–48, 57, 80–81, 116
        scientific uncertainty about, 13, 14–15, 81, 114–115, 182, 189, 253–254, 329
        scope of, 114
        social costs of, 11–13
    See also greenhouse gases
greenhouse gases
    all countries contribute to, 83–84, 333–334
    "clearing house" for emissions permits, 158, 161, 167–168, 179
    concentrations vs. emissions, 17–18
    defined, 6
    developing country responsibilities, 18–19, 178, 189, 284, 289, 341–342, 345–347
    global effects of buildup of, 9–14
    Global Warming Potential factors (GWPs), 178
    greenhouse problem and, 6–9, 10, 11
    increase in, 5, 7, 9, 14, 46, 97, 129–130, 269
    industrial country responsibilities, 18–20, 286, 291, 343–349. *See also* industrial countries, historical responsibility
    inventories of, 18, 20, 166, 169, 189, 198, 207, 223, 328, 334
    long lag times between events and impacts, 329
    measuring, 26, 46, 161, 168, 178

greenhouse gases **(Cont'd)**
    natural flux of, 6, 9
    NGOs and, 250–252, 258, 261–262, 268–270, 271–1274
    per capita entitlements, 247, 250–251, 261–262
    rate of emissions, 10, 15, 17–18, 129
    reducing/stabilizing emissions of
        caps on emissions, 236–237
        carbon dioxide-equivalence concept, 55
        Climate Convention commitments, 18–19, 165–166, 195, 268–270, 341–347
        Climate Convention Objective, 17–18, 21, 143, 166, 172, 195, 204, 223, 267, 339
        "comprehensive approach," 55, 88, 158, 168, 188, 269
        Conference of the Parties and, 145–146
        credits, 64–65, 87, 125, 167–168
        difficult to foresee what measures to take, 172
        economic impact of. *See* economic costs
        European Community objective, 82
        evolution of approaches to, 45–60
        goals, 17–18, 49–50, 56–58, 66–67, 71n15, 73n33, 78, 134
        immediate reduction, 252, 327
        importance of, 15, 145–146
        inadequacy of commitments, 334
        island states proposal, 66–67
        joint implementation and. *See* joint implementation
        "net emissions," 88, 145, 148n34, 161, 168, 178, 269
        NGOs and, 268–270, 271–272
        Noordwijk Conference recommendation, 80
        "phased comprehensive approach," 64–65
        "pledge and review". *See* "pledge and review"
        targets and timetables. *See* targets and timetables
        Toronto Conference Statement recommendations, 49–50, 78
        tradeable emission permits, 161, 167–168, 179, 236–237, 299, 300
        vagueness of commitments, 18–19, 90, 126, 143, 144, 145, 188, 202, 268–270
    reservoirs, 62, 168
    sinks of
        all countries affect, 333
        carbon-dioxide buildup and loss of, 329

greenhouse gases (**Cont'd**)
  sinks of (**Cont'd**)
    Conference of the Parties and, 169
    dangers of tampering with, 269–270
    forests, 131, 176, 283, 329
    INC structure for addressing, 63, 168
    measurement of biotic uptake, 161, 178
    oceans, 10, 46, 270, 330
    per capita entitlements to atmo-
      spheric sink, 247, 250–251, 261–262
    reduction of, 97–98
    scientific research needed, 168–169
    source relationship to, 168
    strengthening, 19, 27, 62, 64–65
    See also "net emissions"
  sources of
    deforestation, 176, 283, 329
    described, 283
    developing countries, 178, 284, 289
    energy production, 81, 131, 177, 178, 179, 189, 191, 283
    human activities, 7, 9, 14
    industrial countries, 284
    sink relationship to, 168
    soil micro-organisms, 10
  trace gas significance, 7, 47
  See also carbon dioxide; CFCs; methane; nitrous oxide
Greenpeace, 245
Group of 7, 52, 54, 79, 116, 130, 180–181, 203
Group of 24, 66
Group of 77
  Agenda 21 and, 156
  Communist bloc countries (former) and, 29, 104–105
  consensus of, 105, 106, 138, 208, 210
  consensus vs. voting and, 132
  financial commitments expected, 170, 177, 179
  financial mechanism and, 216–217, 259–260
  fracturing of, 28–29, 35, 66–67, 104–106, 138–139
  historical responsibility and. *See* industrial countries, historical responsibility
  as key player. *See* key players
  New York INC session (INC 5) and, 215–216
  OECD ad hoc group text and, 163
  OECD and U.S. conflicts, 26, 27, 33–34, 36, 157, 158, 162
  Paris INC session (INC 5) and, 211–214
  per capita entitlements pushed by, 261
  principles recommended by, 209, 210

Group of 77 (**Cont'd**)
  resumption of meeting as a group, 68
  technology transfer and, 170
  universal participation in COP and, 203
  See also developing countries; island states (small); Kuala Lumpur Group; oil-exporting developing countries; *specific countries*

H
Hague Conference, 52, 79, 100, 130
halons, 14, 331
Hansen, James, 48, 54, 264–265
HCFCs (hydrochlorofluorocarbons), 95
Henningsen, Jörgen, 34, 159
historical overview
  INC sessions, 60–70, 82–89, 118, 130–131
  prior to INC sessions, 45–60, 78–80, 116–118, 129–130
historical responsibility. *See* industrial countries, historical responsibility
Howard, Michael, 194
Hyder, Tariq Osman, 25, 38–39, 201–226

I
Impact, 255
INC. *See* Intergovernmental Negotiating Committee
incremental costs. *See* financing, agreed full incremental costs
India
  Brazilian delegation to, 181–182
  compromises criticized by, 69
  early proposals from, 85, 133–134, 209
  energy production in, 191, 209
  final plenary session and, 143
  funding proposal by, 63
  future responsibility for greenhouse problem, 289
  and historical responsibility of industrial countries, 133–134, 182
  as key player. *See* key players
  Montreal Protocol and, 282–283, 313–314
  new organization for negotiations and, 81
  NGOs in, 245, 248
  OECD and U.S. and, 163, 196–197, 198
  per capita entitlements and, 261
  position of, 35, 66, 136
  protocol negotiation and, 312, 313
Indonesia, 312
industrial countries
  commitments of, overview of, 18–21, 165–170, 341–347

industrial countries (**Cont'd**)
    consensus building among. *See* key play-
        ers, industrial countries as bloc; key play-
        ers, industrial country subgroups
    dependence on developing countries,
        287–290
    developing countries gained signifi-
        cant concessions from, 216
    developing country distrust of, 196–197,
        207–208
    economic policies should combat
        greenhouse effect, 92
    energy production in, 177, 178, 189, 190,
        191, 192, 200, 283
    environment given primacy by, 38, 177,
        196, 205, 206
    evolution of awareness of climate prob-
        lem, 45–60, 78–80, 116–118, 129–131
    example must be set by, 96, 166, 213, 327
    excessive consumption by, 35, 107, 178,
        196, 255, 256, 263, 270, 324, 331–332
    financial commitments. *See* financing,
        industrial country commitments
    financial mechanism and, 61, 69, 181,
        208, 219, 260
    future responsibility for greenhouse
        problem, 284, 289
    GEF and, 61, 69, 181, 208, 219, 260
    greenhouse gas emissions from, 18–20,
        284, 289
    historical responsibility for greenhouse
        problem
        Borione and Ripert viewpoint, 27,
            83–84, 96
        Climate Convention preamble, 270, 335
        Dasgupta viewpoint, 133–134, 135
        Djoghlaf viewpoint, 107
        forgoing assignment of blame, 333–334
        future responsibility vs., 284, 289
        Goldemberg viewpoint, 177, 178,
            179, 182
        Hyder viewpoint, 207
        Kjellen viewpoint, 164
        Mintzer and Leonard viewpoint,
            333–334
        Nitze viewpoint, 196
        Rahman and Roncerel viewpoint,
            240, 258, 261, 263, 270
        rhetoric avoided by Brazil, China,
            and India, 35, 182
        rhetoric opposed by U.S., 134, 181
        Sebenius viewpoint, 289
        UN General Assembly Resolution
            44/28, 206

industrial countries (**Cont'd**)
    historical responsibility for greenhouse
        problem (**Cont'd**)
        unresolved differences about, 107, 324
    immediate action needed by, 252, 301, 327
    as key player. *See* key players
    list of, 365
    local vs. global environmental prob-
        lems and, 38, 177, 205
    narrow positions of, 207–208
    national action plans. *See* national action
        plans
    population growth and, 263, 332
    poverty issue and, 177
    principles and, 164, 209, 212–213
    protocol should be adopted by, 166, 271
    recession in, 179, 262, 273
    technology transfer and. *See* technology
        transfer
    See also Climate Convention negotiations;
        East European countries; economies in
        transition; European Community;
        Group of 7; non-governmental organi-
        zations; North-South relations; Organi-
        zation for Economic Cooperation and
        Development; targets and timetables;
        specific countries
industry. *See* business
Intergovernmental Negotiating Committee
        (INC)
    Agenda 21 and, 155–156
    article cluster groups, 69
    Bureau. *See* Bureau (INC)
    chairperson's power, 32
    Chantilly session (INC 1), 60, 61–64, 85,
        117, 131–133, 149, 244, 250, 252
    "clearing house" for, 222
    Consolidated Working Document, 67
    contact groups, 32
    Decision 1/1, 63–64
    establishment of, 60–61, 84–85, 99, 101,
        105, 130, 152, 203, 242
    Executive Secretary, 222
    GEF relationship with, 222, 226, 328
    Geneva session (INC 2), 64–65, 133–136,
        139–140, 170–171, 209, 253
    Geneva session (INC 4), 33–34, 66–67,
        138–139, 140, 158–159, 209–210, 253–254
    Geneva session (INC 6), 222, 254–255,
        267
    historical overview, 60–70, 82–89, 118,
        130–131
    IPCC relationship with, 222
    Misc.1 papers, 64

Intergovernmental Negotiating Committee (INC) **(Cont'd)**
  Nairobi session (INC 3), 65–66, 121, 140, 209, 253, 263
  national action plans and, 328
  New York session (INC 5). *See* New York INC sessions, INC 5
  New York session (INC 7), 222–223
  NGOs and, 22, 132–133, 251–255, 264, 328
  Paris session (INC 5), 68–69, 88–89, 101–102, 164, 210–214
  PrepCom and, 154–157
  Revised Text under Negotiation, 67, 68
  role of, 93, 99
  SBSTA (IPCC) and, 327–328
  Secretariat. *See* Secretariat (INC)
  simultaneous drafting and negotiating by, 99–100
  structure of, 20
  unresolved technical issues, 327–328
  Working Groups. *See* Working Groups (INC)
  See also Climate Convention; Climate Convention negotiations; Ripert, Jean
Intergovernmental Panel on Climate Change (IPCC)
  developing countries and, 58, 100–101, 120
  establishment of, 45, 51, 52, 78–79, 98, 116, 130
  First Assessment Report. *See* First Assessment Report
  first meeting, 51–52
  future success of negotiations and, 92, 122–123, 171, 221–222, 267–268, 326
  INC relationship with, 222
  industrial countries and, 58
  NGOs urge ecological limits analysis by, 267
  politicization of, 123
  purpose of, 51, 53, 98, 130, 177
  restructuring of, 221
  role in negotiations, 60, 84, 92–93, 100–101, 171, 177, 221–222, 267–268
  SBSTA and, 20, 122–123, 171, 221–222, 327–328
  Scientific and Technical Advisory Panel (GEF) and, 123
  Secretariat (INC) and, 327
  Special Committee on the Participation of Developing Countries, 58, 100–101
  Sundsvall meeting, 57, 58–59, 80
  Working Groups, 51, 52, 54

International Energy Agency (IEA), 92, 169, 295
International Finance Corporation, 236
International Monetary Fund (IMF), 203
international negotiations
  business participation in, 229
  CFC negotiations and, 280–281
  Climate Convention negotiations and, 4–5, 16, 21–22, 31, 109, 119, 226, 329–330
  constituency groupings and, 119
  domestic political events and, 36, 183
  Law of the Sea and, 280–281
  universal participation in, 203–204, 304
  unresolved differences, 107, 324–325
  UN traditional model for, 119
  See also Climate Convention negotiations; conventions; protocols
inventories. *See* greenhouse gases, inventories
investments. *See* financing
IPCC. *See* Intergovernmental Panel on Climate Change
island states (small)
  active involvment in negotiations, 28, 105–106, 132, 209, 265
  formation of AOSIS, 28, 59, 132
  Greenpeace and, 245
  Group of 77 and China and, 138
  high risk status of, 81, 98, 171, 243, 244
  as key player. *See* key players
  strong commitments supported by, 59, 66–67, 132, 136, 138, 171, 265
  Working Group II co-chair from, 64, 106

J
Japan, 35, 179, 181–182, 191, 282
  as key player. *See* key players
  "pledge and review" and, 65, 86, 135, 136, 264
  targets and timetables favored by, 83, 137, 206–207, 268, 281, 295, 312
  targets and timetables opposed by, 56, 65, 179, 268, 295
joint implementation
  ambiguous definition of, 145
  analytic studies needed, 328
  Conference of the Parties and, 145, 167
  described, 22, 27, 87, 125, 160–161, 167, 179
  global emission increases and, 269
  NGOs critical of, 27, 256, 261, 269
  unresolved differences, 324, 325

K
key players
  Alliance of Small Island States (AOSIS).
    See key players, island states (small)
  Brazil, 35, 81, 175–177, 181–182, 191, 196
  Bureau, 29, 32, 102–103, 107, 119, 154
  business representatives, 39–40, 94–95,
    229–238
  China
    Bodansky viewpoint, 64
    Dasgupta viewpoint, 133, 138–139
    Djoghlaf viewpoint, 28–29, 104–106
    Goldemberg viewpoint, 35, 181–182
    Hyder viewpoint, 38, 209
    Nitze viewpoint, 191, 196, 198
  Communist bloc countries (former), 29,
    31, 104–105
  Cutajar, Michael Zammit, 152
  developing countries as bloc
    Bodansky viewpoint, 63, 69
    Borione and Ripert viewpoint, 27, 83,
      85
    Dasgupta viewpoint, 138
    Hyder viewpoint, 38, 39, 208, 209–210,
      211, 213–214, 215–217
    Nitze viewpoint, 36, 196–197
    Rahman and Roncerel viewpoint,
      259–260, 261–262
    See also key players, Group of 77
  developing country subgroups
    Bodansky viewpoint, 59, 66–67
    Borione and Ripert viewpoint, 27, 83
    Dasgupta viewpoint, 31–32, 132, 136,
      138–139
    Djoghlaf viewpoint, 28–29, 104–106
    Goldemberg viewpoint, 35, 182
    See also key players, island states
      (small); key players, oil-exporting
      developing countries
  Environment Ministries, 30, 115
  European Community
    Bodansky viewpoint, 63, 65, 66, 69
    Borione and Ripert viewpoint, 26–27,
      82–83
    Dasgupta viewpoint, 134, 136–137, 139
    Goldemberg viewpoint, 35, 178–179
    Kjellen viewpoint, 158–160, 163
    Nitze viewpoint, 191
  Group of 7, 180–181
  Group of 77
    Bodansky viewpoint, 26, 66–67, 68
    Borione and Ripert viewpoint, 26
    Dasgupta viewpoint, 138–139
    Djoghlaf viewpoint, 28–29, 104–106

key players (**Cont'd**)
  Group of 77 (**Cont'd**)
    Goldemberg viewpoint, 35, 177, 179
    Hyder viewpoint, 38, 39, 203, 208,
      209–210, 211–217
    Kjellen viewpoint, 33–34, 157, 158,
      162, 163, 170
  India
    Bodansky viewpoint, 69
    Dasgupta viewpoint, 133–134, 136, 143
    Goldemberg viewpoint, 35, 181–182
    Hyder viewpoint, 209
    Nitze viewpoint, 191, 196, 198
    industrial countries as bloc, 28, 106,
      135, 142, 157–163, 311–312
    See also key players, OECD
  industrial country subgroups
    Borione and Ripert viewpoint, 26–27,
      82–83, 84, 91
    Dasgupta viewpoint, 31, 134–137, 141
    Goldemberg viewpoint, 35, 178–179
    Kjellen viewpoint, 157–163
    See also key players, European Com-
      munity; key players, United States
  island states (small)
    Bodansky viewpoint, 59, 64, 66–67
    Dasgupta viewpoint, 132, 136, 138
    Djoghlaf viewpoint, 28, 105–106
    Hyder viewpoint, 209
    Kjellen viewpoint, 171
    Rahman and Roncerel viewpoint,
      243, 244, 265
  Japan, 35, 65, 83, 86, 179, 181–182, 191
  Kuala Lumpur Group, 28, 106, 182
  Kuwait, 262, 268
  NGOs
    Bodansky viewpoint, 65
    Borione and Ripert viewpoint, 27, 95
    Dasgupta viewpoint, 132–133, 136
    Djoghlaf viewpoint, 29, 106–107
    Dowdeswell and Kinley viewpoint,
      30, 119–120
    Faulkner viewpoint, 39–40, 230–232
    Goldemberg viewpoint, 36, 183–184
    Nitze viewpoint, 200
    Rahman and Roncerel viewpoint, 40–
      42, 239–266
    See also non-governmental organiza-
      tions, critical role of
  OECD
    Bodansky viewpoint, 68, 69
    Borione and Ripert viewpoint, 26–27,
      83, 92–93
    Dasgupta viewpoint, 137, 141

Goldemberg viewpoint, 178–179
Hyder viewpoint, 38, 208–209
Kjellen viewpoint, 33–34, 157–163
Nitze viewpoint, 36, 188, 191, 194, 195, 196
oil-exporting developing countries/ OPEC
Bodansky viewpoint, 59, 66
Borione and Ripert viewpoint, 82, 83
Dasgupta viewpoint, 136, 138
Djoghlaf viewpoint, 28–29, 106
Goldemberg viewpoint, 35, 178
Kjellen viewpoint, 156
Rahman and Roncerel viewpoint, 262, 268, 273
Ripert, Jean, 149, 152–154, 164
Saudi Arabia, 66, 82, 156, 262, 268
Secretariat, 104, 153
United Kingdom, 35, 64, 69, 83, 179
United States
Bodansky viewpoint, 63, 65, 68, 69
Borione and Ripert viewpoint, 26, 82, 83
Dasgupta viewpoint, 134–135, 136, 139
Goldemberg viewpoint, 35, 178, 179, 181, 183
Hyder viewpoint, 38, 208–209
Kjellen viewpoint, 33, 155, 158, 160, 162, 163
Nitze viewpoint, 36–37, 187–200
Rahman and Roncerel viewpoint, 268
Kinley, Richard, 24–25, 29–31, 113–128
Kjellen, Bo, 25, 33–34, 149–173, 214
Kohl, Helmut, 180
Koh, Tommy, 153–154, 304–305
Kuala Lumpur Group, 28, 106, 130, 182
Kuwait, 156, 262, 268

L

land use, 131, 283, 284
Langkawi Declaration, 53, 100
Latin America, 58, 262, 312
laws. *See* conventions; international negotiation; protocols
legal entry into force. *See* Climate Convention, entry into force
legal issues surrounding Climate Convention, 90–91, 329–330
Leonard, J. Amber, 3–44, 321–334
Limited Test Ban Treaty, 302–303, 316n5

M

Maastricht Treaty, 279–280

Maldives, 244
Malta, 52, 54, 79, 130, 152. *See also* Cutajar, Michael Zammit
Malta summit, 52–53
mechanisms. *See* Climate Convention negotiations, structure of; financing; Working Groups (INC), Working Group II on mechanisms
media, 84, 183
methane, 7, 10, 47, 95, 286. *See also* greenhouse gases
Mexico, 58, 262, 312
migration, 11–12, 13–14
Mintzer, Irving M., 3–44, 321–334
Misc.1 papers (INC), 64
models, 8, 9, 11, 46–47, 114
Montreal Protocol on Substances that Deplete the Ozone Layer, 16, 51, 98, 219
Climate Convention vs., 91, 229, 282–283, 284, 290–292
developing countries and, 58, 282–283, 313
dynamics of negotiations, 282–283, 284, 290–292, 309, 313–314
See also Vienna Convention

N

Nairobi INC session (INC 3), 65–66, 121, 140, 209, 253, 263
Nairobi PrepCom session, 154
national action plans
Borione and Ripert viewpoint, 90, 91–92
Dasgupta viewpoint, 144, 145
Hyder viewpoint, 211, 223
Kjellen viewpoint, 163
Mintzer and Leonard viewpoint, 18–19, 328
Nitze viewpoint, 198–199
Rahman and Roncerel viewpoint, 271–272
national strategies approach, 57
negotiation analysis, 278–279, 316n2
negotiations. *See* Climate Convention negotiations; international negotiations
"net emissions," 88, 145, 148n34, 161, 168, 178, 269
Netherlands, 52, 79, 83, 179, 197, 244
New International Economic Order (NIEO), 287, 290, 308, 310
New York INC sessions
INC 5
Bodansky viewpoint, 67–68, 69–70
Borione and Ripert viewpoint, 89
Dasgupta viewpoint, 32, 140–144
Goldemberg viewpoint, 179, 182

New York INC sessions (**Cont'd**)
  INC 5 (**Cont'd**)
    Hyder viewpoint, 207, 210, 215–216
    Kjellen viewpoint, 34, 160–165
    Rahman and Roncerel viewpoint, 254,
      256
  INC 7, 222–223
NGOs. *See* non-governmental organizations
nitrous oxide, 7, 47, 286. *See also* greenhouse gases
Nitze, William A., 25, 36–37, 187–200
Non-Aligned Movement, 53, 105, 130
non-governmental organizations (NGOs),
  239–273
  acceptance by delegates, 251
  benefit of negotiations to, 240
  business and industrial, 22, 132–133,
    229–238, 331
  Climate Action Network, 246, 250,
    254–253, 255–257
  Climate Convention criticisms, 150–151,
    240, 258, 266–267, 268–270, 293
  commitments' meaning examined by,
    268–270
  "comprehensive approach" and, 269
  consensus-building among, 243, 248,
    249–250, 251
  critical role of, 40–42, 106–107, 132–133,
    183–184, 240–242
    advisory capacity, 310
    better informed than governments,
      232, 241, 248, 265–266, 322
    bridge between delegations, 30, 39,
      41–42, 120, 248
    conscience for overall process, 40–41
    daily briefing and review process,
      265–266
    development as right espoused, 266
    economies in transition participation
      encourged, 266
    equitable participation encouraged,
      265–266
    equity issues focus, 41, 243, 247, 250–251,
      255–256, 261–262, 271, 322
    excessive consumption by North
      focus, 255, 256, 263, 270
    future negotiations and, 42, 95, 183,
      232–238, 273, 274, 296, 310, 328, 331
    momentum maintained, 257, 322
    national policy influence, 41, 231,
      248–249
    newsletters, 30, 40–41, 120, 136, 249,
      255–257, 261, 264, 265

non-governmental organizations (**Cont'd**)
  critical role of, (**Cont'd**)
    NGO representative present at all
      sessions, 265–266
    overpopulation concerns, 263
    per capita entitlement focus, 247,
      250–251, 261–262
    powerful advocacy, 133, 249
    pragmatism, 322
    public attention focused by NGOs,
      27, 36, 84, 95, 120, 161, 183, 245, 298
    scientific analyses and tracking, 41, 120,
      241, 244, 245, 247–248, 253–254, 257,
      264–265
    South's effective participation
      encouraged, 256, 265
    spurred governments to participate,
      29, 107, 266
    UN private diplomatic club undone, 107
  ecological limits issue, 252, 258, 267
  evolution of awareness of climate prob-
    lem and, 48
  evolving dynamics within, 40, 41
  financial issues and, 245, 246, 247, 253, 254,
    256–257, 259–261, 270–271, 272–273
  GEF and, 245, 246, 254, 256, 259–261,
    270–271, 272–273
  Global Forum (Flamingo Park), 183
  goals and expectations of, 239, 271–273
  historical responsibility of North and,
    240, 258, 261, 263, 270
  INC and, 22, 132–133, 251–255, 264, 328
  informal consultations, 264
  IPCC role as seen by, 267–268
  issues addressed by, 241–242, 252–255,
    257–262, 266–268, 271–273
  joint implementation and, 27, 256, 261, 269
  marginalization of, 231
  multiple viewpoints of, 41, 248
  "net emissions" and, 269
  networking among, 244–247
  newsletters, 30, 40–41, 120, 136, 249,
    255–257, 261, 264, 265
  Northern NGO viewpoints, 241, 244,
    245–247, 253–254, 258, 263, 266
  North-South relations among, 41, 240–241,
    242–243, 244–247, 249–251, 266
  "pledge and review" and, 65, 136–137,
    263–264
  post cold-war era and, 243
  roles of, 242–243, 247–249
  scientific analysis vs. political necessity
    and, 267–268
  scientific seminars, 264–265

non-governmental organizations (**Cont'd**)
  Second World Climate Conference and, 242–243, 249
  Southern NGO viewpoints, 244, 246–247, 250–251, 252, 253, 255–257, 258–261, 263, 265, 266
  United States and, 200, 254, 268
"non-papers," 121, 133
Noordwijk Ministerial Conference, 55–56, 72n24, 79–80, 100, 116, 130
  recommendations of, 55–56, 79–80, 91, 96, 110–111, 116
Nordic countries, 35, 158, 179, 208–209, 270. *See also specific countries*
"no regrets" policy, 52
North-South relations
  bargaining minimal between North and South, 141
  bargaining power of South, 216, 287–290
  business community and, 234
  Climate Convention impact on, 4–5, 16, 21–22, 31, 170–171
  dependence of North on South, 287–290
  dual rotating co-chairs and, 102–103
  equity ethic, 110
  evolution of role in negotiations, 50, 54–56, 58–60, 117–118, 133–139
  interdependence of North and South, 83–84, 110, 333–334
  joint implementation and, 22, 27, 87, 125
  linking environment and development and, 205–208
  mixed teams of experts, 309–310
  in NGO community, 41, 240–241, 242–243, 244–247, 249–251, 266
  ozone layer issues, 330
  protocol negotiation and, 307–311, 329–330
  replacement of "North-South" concept, 310–311
  technology transfer and, 170
  unresolved differences, 107, 324–325
  See also developing countries; industrial countries
Norway, 52, 79, 158, 160
  "clearing house" concept, 158, 161, 167–168, 179
  joint implementation concept and, 27, 87, 160–161, 179
  targets and timetables and, 134, 197
  See also Nordic countries

O
OASIS. *See* island states (small)

Objective of the Climate Convention, 17–18, 21, 143, 166, 172, 195, 204–205, 223, 267, 339
oceans
  carbon dioxide uptake by, 10, 46, 270, 330
  climate and, 10, 11, 15
  current alterations, 9, 57
  rise in level of
    global warming and, 9, 10–11, 80, 98
    negotiating protocol on, 267, 299
    small island states and, 132, 171, 243
  temperature at surface, 9
  See also conventions, Law of the Sea
OECD. *See* Organization for Economic Cooperation and Development
Official Development Assistance (ODA), 29, 35, 177, 205
oil-exporting developing countries
  Agenda 21 and, 156
  blocking coalitions and, 292
  energy/carbon tax and, 83, 106, 294
  "go slow" approach of, 35, 59
  as key player. *See* key players
  "net emissions" and, 178
  obstructive role of, 268, 273
  per capita entitlements and, 262
  periodic consultative meetings, 106
  position of, 35, 59, 83, 136, 138
  PrepCom sectoral approach and, 155
  targets and timetables opposed by, 83, 136, 138
  See also Kuwait; Saudi Arabia
OPEC. *See* oil-exporting developing countries
open nature of negotiations, 22, 40, 64, 82, 120, 203–204, 215, 230, 232, 265–266, 322
  future success and, 22, 203–204, 232, 326, 331
  See also universal participation
Oppenheimer, Michael, 253–254, 269–270
Organization for Economic Cooperation and Development (OECD)
  ad hoc group, 159–160, 161–163
  checklist, 159–160
  commitments of, overview of, 18–21, 165–170, 341–347
  compromise agreed on, 89, 142
  consensus with other Northern countries, 83, 137, 311–312
  coordination attempts, 157–163, 167–170, 178–179
  debates amongst members, 33–34, 83, 137, 141, 160–161, 178–179

Organization for Economic Cooperation
and Development (**Cont'd**)
developing country distrust of, 196
development in developing countries
and, 164–165
energy production in, 191
excessive consumption by member
countries, 332
Group of 77 and China conflicts, 26, 27,
33–34, 36, 157, 158, 162
historical responsiblity. *See* industrial
countries, historical responsibility
initial negotiating position, 26–27, 83
as key player. *See* key players
principles and, 209
protocol should be signed by, 166, 169
significance of, 26, 33–34, 36, 184
Special Working Group, 33–34, 159–160
targets and timetables
favored, 26–27, 68, 83, 91, 134, 137, 166,
169, 293
opposed, 137, 197, 268, 293
specifical significance of OECD, 184
U.S. conflicts, 26–27, 33–34, 36, 51, 83,
137, 158, 160, 161–162, 197–198
See also European Community; industrial
countries; North-South relations; *specific
countries*
Ottawa meeting, 53, 116
ozone layer
depletion of, 14, 48, 267, 331
negotiations, 91, 229, 282–283, 284,
290–292, 330
See also Montreal Protocol; Vienna Convention

P
Pakistan, 38, 163, 210, 219, 225–226, 261
Paris G-7 Economic Summit, 52, 54, 79,
116, 130
Paris INC session (INC 5), 68–69, 88–89,
101–102, 164, 210–214
participatory nature of negotiations. *See*
open nature of negotiations
"phased comprehensive approach," 64–
65
Philippines, 143
"pledge and review"
criticism of, 65, 66, 86–87, 126, 136–137,
183–184
described, 65, 86, 135
developing country opposition to, 86–87,
136–137

"pledge and review" (**Cont'd**)
European Community adoption of,
136–137
introduction of concept, 65, 86, 135, 136,
264
NGO "hedge and retreat" view of, 65,
136–137, 263–264
persistence of concept, 66, 137
U.S. favoring of, 181, 264
politicization
of climate change consideration, 49, 78,
116, 130–131
of climate negotiations process, 28, 57–59,
80, 117, 118, 177
of science, need to avoid, 123
population growth, 263, 284, 289, 307, 324,
331–332
poverty, 177, 217, 260, 271, 331–333
environmental degradation and, 35–36,
177, 185, 205–206, 332
See also equity
Precautionary Principle, 18, 21, 41, 110,
169, 171, 173n3, 323–324, 340
Preparatory Committee (PrepCom)
(UNCED), 101, 104, 107–108, 153–157,
177, 207
principles
general, 18, 21, 39, 64, 65, 66, 110, 212–213,
217, 339–341
Geneva INC session (INC 4) and, 209
Group of 77 and China recommendations,
209, 210, 212, 215, 216
industrial country resistance to, 164,
209, 212–213
Paris INC session (INC 5) and, 164, 212–
213
Precautionary Principle, 18, 21, 41, 110,
169, 171, 173n3, 323–324, 340
Rio Declaration principles adopted, 39,
212–213
private industry. *See* business
process orientation of Climate Convention, 21–22
"prompt start," 90, 126–127, 264
protocols
adoption of, 360
baseline, 301–303
deadline for completing, 183
European Community insistence on,
83, 91
indirect goal achievement vs., 279–280,
296–299
industrial countries should adopt, 166,
271

protocols (**Cont'd**)
   negotiating, 184, 233
      delaying tactics, 301
      postponement of, 85, 91–92, 199, 298–299
      should begin immediately, 27
      strategies for, 299–315
      time it will take, 278, 284, 293–294, 304
      Noordwijk Ministerial Conference recommendations, 91
      North-South considerations and, 307–311, 329–330
      OECD protocol, 166, 169
      "package deals," 304–307
      Second World Climate Conference recommendations, 91
      single-issue, 304–307
      small-scale, expanding, 311–314
      undesirability of, 42–43, 91–92, 145–146, 279–280, 293–294, 296–299, 303
      See also conventions; Montreal Protocol; targets and timetables
public awareness of climate issue
   Climate Convention and developing, 347–348
   cyclic nature of, 303
   expectations by public, 183, 202–203
   future success of negotiations and, 184, 280, 296–297, 298
   IPCC and, 177
   NGOs and, 27, 36, 84, 95, 120, 161, 183, 245, 298
   science and, 84
   Toronto Conference and heat wave of 1988 and, 116

R
Rahman, Atiq, 25, 40–42, 239–273
"ratchet mechanisms," 302–303
ratification. *See* Climate Convention, ratification of
Reagan, Ronald, 51, 283, 288
recession, 179, 262, 273
reciprocal commitments. *See* equity
refugees, environmental, 11, 12
regional effects, 9–10, 11, 13–14, 15, 57, 81, 329
regional groups, eclipse of, 104
Reilly, William, 193
Reinstein, Robert A., 158, 160, 162, 193, 194
reporting
   citizen and advocate mobilization and, 298

reporting (**Cont'd**)
   by developing countries
      Borione and Ripert viewpoint, 89, 90, 91–92
      Climate Convention requirements for, 18, 20–21, 354–355, 356
      Hyder viewpoint, 207–208, 211, 217, 223
      Kjellen viewpoint, 166
      Mintzer and Leonard viewpoint, 18, 19, 20–21, 328, 334
      Nitze viewpoint, 193, 198–199
      Rahman and Roncerel viewpoint, 271–272
   by industrial countries
      Borione and Ripert viewpoint, 89, 91–92
      Climate Convention requirements for, 18, 19, 20–21, 354–356
      Kjellen viewpoint, 163, 166–167
      Mintzer and Leonard viewpoint, 328
      Nitze viewpoint, 193
   citizen and advocate mobilization and, 298
   Climate Convention requirements for, 18, 19, 20–21, 354–356
   goals of, 123
   implementation process and, 122, 123
   importance of, 163, 167
   OECD ad hoc group recommendations, 163
   unresolved differences on, 324, 325
   U.S. role in commitment to, 193
   See also reviewing
resolutions. *See* United Nations General Assembly
reviewing
   Climate Convention's provisions for, 122, 123, 144, 145, 151, 172, 323, 328, 344
   Conference of the Parties and, 122, 123, 151, 172
   developing countries and, 138, 144, 211
   establishing credible capability for, 297
   goals of, 123
   importance of, 151, 163
   iterative process of, 172, 193, 323
   North's insistence on, 141–142
   OECD ad hoc group recommendations, 163
   U.S. role in commitment to, 193
   See also "pledge and review"; reporting
Revised Text under Negotiation (INC), 67, 68
Rio Declaration, 16, 17, 39, 212–213

Rio Summit. *See* United Nations Conference on Environment and Development

Ripa di Meana, Carlo, 294

Ripert, Jean, 24–25, 26–28, 77–96
 Agenda 21 and, 155
 appointed chair of INC, 61, 132
 excellence of, 149, 152–154, 164
 Extended (Expanded/Enlarged) Bureau and, 88–89, 142–143, 162
 integrated text preparation by, 68–69, 89, 102, 103, 118–119, 213–216
 NGOs included by, 230
 as Special Committee on the Participation of Developing Countries chair, 58
 Working Papers based on Revised Text, 142–143

Roncerel, Annie, 25, 40–42, 239–273

Rose, Chris, 253

Russia. *See* Soviet Union

S

Sarney, José, 175

Saudi Arabia, 155, 156–157, 252, 262, 268

targets and timetables opposed by, 66, 72n24, 82, 156, 262, 268

SBSTA. *See* Subsidiary Body for Scientific and Technological Advice

Schmidheiny, Stephen, 39–40

science
 blocking coalitions and, 290–293
 and "buying time" to study climate system, 15, 297
 as a catalyst to international action, 48
 Climate Convention provisions for research, 347
 consensus on global warming, 9, 17–18, 26, 45, 46–48, 57, 80–81, 116
 epistemic community of experts, 30
 evolution of awareness of climate problem and, 48, 78–79, 309
 as a foundation of the Climate Convention, 28, 65, 80–82, 84, 100
 future success of negotiations and, 92, 120–121, 171–172, 296–297, 309–310, 323, 326, 327–328
 importance of research, 92, 120–121, 171–172, 296–297, 309–310, 323, 326, 327–328
 NGOs and, 41, 120, 241, 244, 245, 247–248, 253–254, 257, 264–265
 oil-exporting developing country emphasis on, 83
 politicization must be avoided, 100, 120, 123

science (Cont'd)
 Precautionary Principle and, 171
 protocol negotiation and, 309–310
 public opinion alerted by, 84
 reviews of research findings, 22, 123
 risk of improved, 309–310
 sink research needed, 168–169
 success of negotiations and, 28, 100, 120–121, 322, 326
 uncertainty about greenhouse effect, 13, 14–15, 81, 114–115, 182, 189, 253–254, 329
 unresolved issues, 327–328
 U.S. emphasis on, 26, 54, 82
 See also First Assessment Report; Intergovernmental Panel on Climate Change; Scientific and Technical Advisory Panel; Subsidiary Body for Scientific and Technological Advice

Scientific and Technical Advisory Panel (STAP) (GEF), 123, 184, 221, 237

seas. *See* oceans

Sebenius, James K., 25, 42–43, 277–320

Second World Climate Conference, 28, 100, 105, 242–243, 249
 as backdrop for framework convention, 57–59, 80, 117
 recommendations of, 57–58, 91, 98–99, 130

Secretariat (Climate Convention), 20, 351

Secretariat (GEF), 220, 224

Secretariat (INC), 20
 Agenda 21 and, 104, 222
 business viewpoints included by, 40
 described, 29
 diminished role and importance of, 29, 103–104, 223
 establishment of, 60, 152
 GEF and, 327
 interim role of, 222–223
 IPCC and, 327
 leading role of, 89, 104, 153
 logistics should be coordinated by, 327
 review of national action plans and, 328
 "streamlining" by, 140–141
 structuring for COP, 221
 Subsidiary Body for Implementation and, 123–124
 UN Commission on Sustainable Development and, 327
 See also Cutajar, Michael Zammit

signing. *See* Climate Convention, signing of

sinks. *See* greenhouse gases, sinks of

small island states. *See* island states (small)

social costs, 10–13, 92

Soviet Union, 56, 161, 168, 191, 268, 284, 310. *See also* Communist bloc (former); economies in transition
Special Working Group (OECD), 33–34
STAP. *See* Scientific and Technical Advisory Panel
Stockholm Environment Institute, 241, 267
"streamlining," 32, 140–141
Subsidiary Body for Implementation (SBI), 20, 123–124, 353
Subsidiary Body for Scientific and Technological Advice (SBSTA), 20, 100, 122–123, 171, 221–222, 327–328, 352
Sundsvall IPCC meeting, 57, 58–59, 80
Sununu, John, 37, 54, 68, 189, 192, 193–194, 279, 288
sustainable development
    Brazil's funding argument for, 35
    Brundtland Commission and, 183
    Climate Convention and, 4–5, 16, 17, 18, 21, 22, 81–82, 93, 113
    Commission on Sustainable Development, 93, 221, 222, 310
    defined, 114, 311
    defining to suit own agenda, 38, 205
    developing countries and, 58, 59, 256
    economic costs of, 115
    linking of environment and development, 16, 17, 21, 38, 81–82, 177, 205–208, 266
    link with other issue, 331–333
    necessity of, 114, 204–205
    as new ideological template, 311
    UN General Assembly Resolution 44/29 on, 205–207, 211
    World Bank and, 259
    See also Business Council for Sustainable Development; development
Sweden, 160, 161, 163, 168, 170–171
    position of, 135, 158, 163, 268
    See also Nordic countries
Switzerland, 182, 268

T
targets and timetables
    "Action Plan" vs., 297
    baseline, 301
    blocking coalitions and, 293–294
    business and, 184, 237
    compromise position on, 69, 302–303
    Conference of the Parties and, 145–146
    deadline for completing, 183
    difficult to implement, 91, 197, 294–295

targets and timetables (**Cont'd**)
    within European Community, 294
    favored by developing countries, 66–67, 132, 136, 171, 211
    favored by industrial countries
        Australia, 137, 268, 281, 312
        at Bergen Conference, 56–57
        Canada, 137, 268, 281, 312
        European Community. *See* European Community, targets and timetables
        France, 83, 134
        Germany, 83, 137, 179
        Group of 7, 79
        Japan, 83, 135, 137, 206–207, 268, 281, 295, 312
        Netherlands, 83, 179
        at Noordwijk Conference, 56, 79–80, 91
        Nordic countries, 83, 134, 179, 206–207, 268
        OECD, 26–27, 68, 83, 91, 134, 137, 166, 169, 293
        United Kingdom, 83, 135, 294
        United States, 63, 134, 160, 198, 199, 281
    as goal of negotiations, 293
    importance of, 166, 184
    indirect achievement of goals vs., 279–280, 296–299
    industrial countries should adopt, 166, 271
    middle-ground positions, 35, 66, 135, 158
    negotiating, 184, 233
        delaying tactics, 301
        postponement of, 85, 91–92, 199, 298–299
        should begin immediately, 271
        strategies for, 299–315
        time it will take, 278, 284, 293–294, 304
    NGOs and, 240, 271–272, 293
    opposed by developing countries, 66, 83, 136, 138, 184
        Saudi Arabia, 66, 72n24, 82, 156, 262, 268
    opposed by industrial countries
        European Community, 83, 197, 258, 268
        France, 65
        Japan, 56, 65, 179, 268, 295
        OECD, 137, 197, 268, 293
        Soviet Union, 56
        Turkey, 293, 294
        United Kingdom, 65, 69, 179, 258, 294
        United States. *See* United States, targets and timetables opposed by

targets and timetables (**Cont'd**)
   "pledge and review" and, 65, 126, 136–137,
      264
   reduction of alarm about, 87
   sea level rise, 267
   Second World Climate Conference rec-
      ommendations, 58, 91
   setting without means of achieving,
      237
   undesirability of, 42–43, 91–92, 145–146,
      279–280, 293–294, 296–299, 303
   vagueness of commitments, 18–19, 90,
      126, 143, 144, 145, 188, 202, 268–270
   See also protocols
taxes (carbon/energy), 92, 236–237, 294
   European Community and, 27, 83, 84,
      96, 163, 294
   negotiating protocols on, 299, 300, 302,
      312–313
   opposition to, 29, 83, 84, 106, 294
technologies, co-development of, 22
technology transfer
   Brazil's funding argument for, 35
   business sector recommendations, 236
   Climate Convention requirements for,
      18–19, 20, 144, 345
   developing country requirement for, 84,
      135, 139, 164, 170, 215, 217
   horizontal or reverse, 22
   importance of, 164, 170, 184
   INC structure for addressing, 63
   industrial country commitments, 20,
      62, 217, 223, 345–346
   industrial country reticence about, 107,
      109, 135, 194
   intellectual property rights and, 135,
      139, 195, 288
   joint implementation and. *See* joint
      implementation
   U.S. position on, 194
temperature increase. *See* greenhouse effect, as problem
timetables. *See* targets and timetables
Tokyo declaration, 100
Tolba, Mostafa, 53, 237, 302
Toronto Conference, 49–50, 53, 56, 78, 96,
   100, 116, 130
tradeable emission permits, 161, 167–168,
   179, 236–237, 299, 300
transfer of technology. *See* technology
   transfer
transparency. *See* open nature of negotiations

transportation, 172, 284, 285, 286, 292, 299,
   300, 329–330
Turkey, 293, 294

U
UN. *See entries beginning with* United Na-
   tions
UNCED. *See* United Nations Conference
   on Environment and Development
UNDP. *See* United Nations Development
   Programme
UNEP. *See* United Nations Environment
   Programme
United Kingdom
   Chantilly meeting and, 62
   compromise position of, 35, 135, 179
   domestic opposition to emissions
      reductions, 294
   as early coordinator of negotiations, 54
   as key player. *See* key players
   Met Office model, 8
   "phased comprehensive approach"
      and, 64–65
   pivotal role of, 83, 258
   "pledge and review" and, 65
   targets and timetables and, 65, 69, 83,
      135, 179, 258, 294
   UN Economic and Social Council
      proposal of, 111
United Nations
   business participation in, 229–230
   Clinton administration and, 226
   universal participation pros and cons,
      203–204
United Nations Commission on Sustain-
   able Development, 327
United Nations Conference on Environ-
   ment and Development (UNCED), 15–
   16, 34–35, 117, 175–176, 232. *See also*
   Climate Convention, signing of; Prepa-
   ratory Committee
United Nations Conference on the Human
   Environment, 204
United Nations Development Programme
   (UNDP), GEF and, 197, 218, 219, 260
United Nations Environment Programme
   (UNEP)
   establishment of, 204
   GEF and, 197, 218, 219, 260
   IPCC establishment by, 51, 78–79, 116, 130
   "law of the atmosphere" model and,
      53
   role in negotiations, 54, 59–60, 81

United Nations General Assembly
  as conductor of climate negotiations,
    60, 81–82
  Maltese declaration on climate (1988),
    52, 79, 130
  Resolution 44/28, 205–207, 211
  Resolution 45/212, 60–61, 84–85, 99,
    101, 105, 130, 203
  resolution on climate change (1988),
    53–54, 58
  Resolution INC/1992/1, 203
United States
  Climate Convention reflects position of,
    36–37, 165–166, 188
  "comprehensive approach" and, 55,
    168, 188
  compromise agreed on, 89, 142
  consensus with other Northern countries,
    311–312
  deference to, 83, 188, 258
  developing country distrust of, 196
  development of position of, 36–37,
    188–194, 288
  disappointment in, 279, 281
  domestic policy of, 37, 50–51, 183, 191–192,
    199
  economic cost emphasis of, 51, 82, 188
  economic cost of reducing/stabilizing
    emissions, 189, 192, 198, 257, 292
  European Community conflicts, 26–27,
    35, 51, 83, 136, 137, 141
  example should be set by, 199–200
  financial assistance commitments and,
    180, 196, 198–199
  financial assistance position, 135, 141, 194
  foreign policy issues, 190–192, 199
  forest convention sought by, 178, 181
  future negotiations and, 199–200, 288–289
  GEF and, 181, 195–196
  goals of, 199
  Group of 77 and China conflicts, 26, 27,
    33–34, 36
  historical responsibility rhetoric
    opposed by, 134, 181
  intial negotiating position, 26, 82, 134
  as key player. *See* key players
  lack of progress in negotiations and,
    187, 279
  Law of the Sea and, 280, 282, 287, 288,
    291, 305, 307
  leadership of
    Bush administration failure, 37, 179,
      187–200, 279
    Clinton administration, 198, 199,
      226, 279

United States (**Cont'd**)
    leadership role of, 153, 191–192, 197–198,
      199
  missed opportunities of, 189, 192, 195–198
  Montreal Protocol and, 282, 283, 290–292
  National Action Plan on Climate
    Change, 198
  national strategies approach and, 57, 199
  NGOs and, 200, 254, 268
  OECD conflicts, 26–27, 33–34, 36, 51, 83,
    137, 158, 160, 161–162, 197–198
  ozone negotiations and, 283, 290–292,
    302
  "pledge and review" and, 181, 264
  PrepCom sectoral approach and, 155
  principles and, 164, 212–213
  process-oriented approach of, 211
  protocol negotiation and, 308
  recession in, 179, 262, 273
  role in climate negotiations, 187
  scientific study emphasis of, 26, 54, 82
  stabilization commitment of, 195, 197, 200
  "stabilization" wording opposed by,
    162
  targets and timetables favored by, 63,
    134, 160, 198, 199, 281, 293
  targets and timetables opposed by
    general, 34, 36–37, 281, 293, 294
    Geneva (INC 4), 158
    Goldemberg meeting in Washington,
      181
    initial negotiating position, 82,
      134–135
    New York (INC 5), 34, 68, 69, 141
    NGO viewpoint on, 268
    Noordwijk Conference, 56, 80
    other countries may join over time, 293
    other countries not entirely unhappy
      about, 195, 197
    political costs of, 37, 195–198, 200
    predisposition toward, 26, 82, 134,
      141, 188–189
    reasons for, 179, 189, 190, 193, 198
    Resolution 45/212 and, 91
  technology transfer position of, 194
  See also Bush, Georg; Clinton, Bill;
    Reagan, Ronald
universal participation
  in creating and solving global environ-
    mental problems, 333–334
  pros and cons of, 203–204, 304
  See also open nature of negotiations
universal right to development, 16, 206,
  212, 216, 266

V

Van Lierop, Robert, 64, 154
Vanuatu, 85, 105–106
Vienna Convention for the Protection of the Ozone Layer, 54, 81, 82, 98, 284, 292–293, 304. *See also* Montreal Protocol
Villach Conferences, 47–48, 51, 116, 130
voting, 70, 132, 360–361

W

war, rapid climate change and, 11–12
Washington INC session. *See* Chantilly INC session
weather, extreme conditions
    Climate Convention and, 4
    effects of, 10–14
    multiple disaster scenarios, 241–242
    rapid climate change and, 11, 12–13
    recent examples of, 3, 12, 15, 48
winning coalitions. *See* blocking coalitions
withdrawal from Climate Convention, 363–364
Working Groups (INC)
    additional proposed, 73n35
    confusing anomalies of, 214
    Consolidated Working Document, 67
    dual rotating co-chairs, 29, 30–31, 64, 102–103
    establishment of, 61, 62–63, 85, 132, 209
    replacement by three groups, 69, 103, 142–144, 164, 214–215
    "streamlining" and, 32, 140–141
    Working Group I on commitments, 31, 102–103, 209
        Chantilly session, 62, 63, 85
        deadlocking of, 122
        Geneva session, 64, 138–140
        Nairobi session, 65–66
        New York session, 140–141
        replacement of, 69, 103, 142–144
    Working Group II on mechanisms, 30–31, 102–103, 209
        Chantilly session, 62–63, 85
        Geneva session, 64, 65, 139
        incrementalism strategy of, 121
        innovation in, 124–126
        method of working, 154
        Nairobi session, 65, 121
        "non-papers" and, 121, 133
        significance of, 121–122
        Vanuatu as chair of, 106
Working Groups (IPCC), 51, 52, 54

Working Groups (PrepCom), 154
workshops, 183, 235, 309–310, 328
World Bank, 109–110, 203, 216, 256, 259
    GEF and, 196–197, 218, 219–220, 259, 260, 306
World Meteorological Organization (WMO), 48, 51, 59–60, 78–79, 81, 116, 130
World Wide Fund for Nature, 245, 246

Y

Yeutter, Clayton, 193–194

Z

Zoellick, Robert, 194